# HARVARD ECONOMIC STUDIES

I. The English Patents of Monopoly. By William H. Price. 8vo.

II. The Lodging House Problem in Boston. By Albert B. Wolfe. 8vo.

III. The Stannaries: A Study of the English Tin Miner. By George R. Lewis. 8vo.

IV. Railroad Reorganization. By Stuart Daggett. 8vo.

V. Wool-Growing and the Tariff. By Chester W. Wright. 8vo.

VI. Public Ownership of Telephones on the Continent of Europe. By Arthur N. Holcombe. 8vo.

VII. The History of the British Post Office. By J. C. Hemmeon. 8vo.

VIII. The Cotton Manufacturing Industry of the United States. By M. T. Copeland. 8vo.

IX. The History of the Grain Trade in France. By Abbott Payson Usher. 8vo.

X. Corporate Promotions and Reorganizations. By A. S. Dewing. 8vo.

XI. The Anthracite Coal Combination in the United States. By Eliot Jones. 8vo.

XII. Some Aspects of the Tariff Question. By F. W. Taussig. 8vo.

XIII. The Evolution of the English Corn Market from the Twelfth to the Eighteenth Century. By N. S. B. Gras. 8vo.

XIV. Social Adaptation: A Study in the Development of the Doctrine of Adaptation as a Theory of Social Progress. By L. M. Bristol. 8vo.

XV. The Financial History of Boston, from May 1, 1822, to January 31, 1909. By C. P. Huse. 8vo.

XVI. Essays in the Earlier History of American Corporations. By J. S. Davis. 8vo. 2 volumes.

XVII. The State Tax Commission. By H. L. Lutz. 8vo.

XVIII. The Early English Customs System. By N. S. B. Gras. 8vo.

XIX. Trade and Navigation between Spain and the Indies in the time of the Hapsburgs. By C. H. Haring. 8vo.

XX. The Italian Emigration of Our Times. By R. F. Foerster. 8vo.

XXI. The Mesta: A Study in Spanish Economic History, 1273–1836. By Julius Klein. 8vo.

XXII. Argentine International Trade under Inconvertible Paper Money: 1880–1900. By J. H. Williams. 8vo.

XXIII. The Organization of the Boot and Shoe Industry in Massachusetts before 1875. By Blanche E. Hazard. 8vo.

XXIV. Economic Motives. By Zenas C. Dickinson. 8vo.

XXV. Monetary Theory before Adam Smith. By Arthur E. Monroe. 8vo.

XXVI. Canada's Balance of International Indebtedness 1900–1913. By Jacob Viner. 8vo.

HARVARD UNIVERSITY PRESS
CAMBRIDGE, MASS., U.S.A.

HARVARD ECONOMIC STUDIES

PUBLISHED UNDER THE DIRECTION OF
THE DEPARTMENT OF ECONOMICS

VOL. III

LONDON: HUMPHREY MILFORD

OXFORD UNIVERSITY PRESS

# THE STANNARIES

## A STUDY OF THE ENGLISH TIN MINER

BY

GEORGE RANDALL LEWIS, Ph.D.

AWARDED THE DAVID A. WELLS PRIZE FOR THE YEAR 1906-07, AND PUBLISHED FROM THE INCOME OF THE DAVID A. WELLS FUND

CAMBRIDGE
HARVARD UNIVERSITY PRESS
1924

PRINTED AT THE HARVARD UNIVERSITY PRESS
CAMBRIDGE, MASS., U. S. A.

# PREFACE

THE following monograph, the outcome of a thesis for an undergraduate course at Harvard University, is the result of three years' investigation, one in this country and two in England, — for the most part in London, where nearly all the documentary material relating to the subject is to be found. For facilitating with ready courtesy my access to this material I am greatly indebted to the officials of the British Museum, the Public Record Office, and the Duchy of Cornwall Office. I desire also to acknowledge gratefully the assistance of Dr. G. W. Prothero, Mr. Hubert Hall, and Mr. George Unwin. My thanks are especially due to Professor Edwin F. Gay of Harvard University, under whose supervision my work has been done.

HOUGHTON, MICHIGAN, November, 1907.

# CONTENTS

## INTRODUCTION



## CHAPTER I



## CHAPTER II

# INTRODUCTION

THE main object of this essay is to show from the history of the Cornish and Devon tin mines, accompanied by references to conditions in other mining districts, the organization of an important part of the English mining industry, — in other words, to bridge the gap in the economic history of England, which, in that of Germany, has attracted numerous investigators.[1]

Two reasons have suggested this choice of subject. In the first place, apart from any connection with mining itself, the commodity tin is a subject which merits historical treatment. The fact that it formed, with lead and wool, the great bulk of England's exports during the Middle Ages, with an importance in international trade hardly less than that of the spices of the East, makes it all the more strange that up to the present day the subject has not been more thoroughly studied. If one excepts the controversial pamphlets in which antiquarians have threshed out the question of the Phœnicians' alleged visitations to Cornwall, comparatively little has been written concerning the history of tin.

A second reason for giving the tinners a central place in the history of the free mines of England is the wealth of historical material for the stannaries, in sharp contrast with the paucity of evidence from other mining districts. The documents relating to the stannaries form a broken but valuable series from the year 1156 down to the present day. They comprise not only the laws of the stannaries, as ratified and extended by some half-dozen local convocations, but also the tinners' charters, the rolls of the mine courts, and the Pipe Rolls, the accounts of the receivers and the auditors of the Duchy of Cornwall, and of the coinage officers. These accounts give, in addition to other statistics, the figures for the annual production of tin, from the twelfth or thirteenth century down to modern times,

[1] See among others, Zycha, *Das böhmische Bergrecht des Mittelalters;* Achenbach, *Das gemeine deutsche Bergrecht;* and Schmoller's masterly summary, *Die Geschichtliche Entwickelung des Unternehmungs*, Jb. für Gesetzgebung, xv, 661–710, 963–1028.

which will be found collected in the appendices of this work. Mention should be made also of the Rolls of Parliament, which contain the petitions and counter petition of miners and landlords, and the Statutes of the Realm, the Patent, Close, and Charter Rolls, the Domestic State Papers, and the Treasury Papers, all of which throw considerable light upon the relations between the miners and the Crown. Other series which here and there yield stray bits of information are the so-called Ancient Petitions, the Petitions in Chancery, the Lay Subsidy Rolls, and the reports of the Royal Historical Commission. In the British Museum the manuscript collections of Harley, Lansdowne, Cotton, Hargrave, and Stowe, as well as the series of the Additional Manuscripts, contain useful material concerning the stannaries, although chiefly in connection with the Elizabethan and Stuart tin monopolies.

The order of treatment for the various topics under the main heading of this investigation, I have, after some hesitation, arranged as follows. In the first chapter an attempt will be made to set forth in brief the history of the technical side of tin mining and smelting, including an account of mediæval mining operations, the change from alluvial to lode tinning, and the improvements in mining methods which revolutionized the production of tin in the seventeenth century. The second chapter treats of what may be called the external history of the mines, beginning with the period immediately preceding the tinners' first charter, and including, among other topics, the freeing of the miners from villeinage, the growth of the tin industry, and an account of the trade in tin, in England as exemplified by the history of the pewterers, and abroad as shown by the direction of the exports. In the third chapter I have endeavored, by a brief study of the origins of mineral law in England and elsewhere, to coördinate the history of the British stannaries with that of other mining districts in England, France, Germany, and the Low Countries.

The second half of the essay is concerned more particularly with the miners. Chapter IV takes up the judicial relations between the tinners and the Crown and Duchy authorities, including a study of the mineral courts, their conflicts with outside, and especially with manorial jurisdictions, and finally a treatment of the so-called "parliaments" of the tinners. The following chapter, upon the fiscal connections of stannaries and Crown, reviews the forms assumed by the stannary taxes, the manner of collection, and the amount yielded. In the sixth chapter we proceed to a consideration of the privileged status of the tinners and of the trade rules under which they worked, with special reference to the character of these rules as contrasted with that of the mediæval gild law. Finally, in the last and in subject the most important chapters, I have attempted to piece together what little information exists with regard to the forms of industrial organization in the stannaries, viewed in the light of similar developments in the mines of Germany, the economic and social status of the mining classes, and their relations with the middlemen, especially the tin merchants and smelters.

Certain limitations of this work should be here indicated. Concerning many fundamental points in mining history it is impossible to obtain enough evidence to warrant definite conclusions. Such a case, for example, arises in connection with the origin of free mining in England, a subject of importance not merely in its bearing on mining history but on that of early property rights in land and on other matters of wide interest. A similar haze surrounds such matters of stannary constitutional history as the administration of the stannaries previous to 1198, and the origin of the tinners' parliaments. But I have endeavored carefully to distinguish between conjecture, inference, and fact, and I trust that this study, despite its manifest imperfections, may throw some new light on the history of a large and little known element in English industrial life, the free miners of England.

# TABLE OF ABBREVIATIONS

| | |
|---|---|
| Add. MS | Additional Manuscripts, British Museum. |
| Anc. Pet. | Ancient Petitions, Record Office. |
| Bodl. MSS. Add. | Bodleian Manuscripts Additional. |
| Bdle. | Bundle. |
| Bk. | Book. |
| Cal. | Calendar. |
| Camb. Univ. MS. | Cambridge University Manuscripts. |
| Chanc. R. | Chancery Enrolments. |
| Chart. R. | Charter Rolls. |
| Close | Close Rolls. |
| Coll. Proc. | Collection of Proclamations. |
| Coll. Sign Man. | Collection of Sign Manual Warrants. |
| Compl. Min. Laws Derb. | Compleat Mineral Laws of Derbyshire. |
| Convoc. | Convocation (or parliament of tinners). |
| Cornw. | Cornwall. |
| Ct. R. | Court Roll. |
| Curs. Rec. | Cursitor's Records (Durham). |
| Cust. Accts. K. R. (or Cust. K. R.) | Customs Accounts, Exchequer, King's Remembrancer. |
| D. B. | Domesday Book. |
| D. O. | Duchy of Cornwall Office. |
| Duchy Accts. Excheq. Aug. Pt. | Duchy of Cornwall Accounts, Exchequer, Augmentation Department, portfolio. |
| Duchy of Lanc. Misc. | Duchy of Lancaster, Miscellanea. |
| Eng. Min. Almanack | English's Mining Almanack. |
| Escheat R. | Escheators' Accounts, Enrolments of. |
| Hales MS. | Hales Manuscripts, Lincoln's Inn Library. |
| Hargrave | Hargrave Manuscripts, British Museum. |
| Harl. | Harleian Manuscripts, British Museum. |
| Hist. MSS. Com. | Historical Manuscripts Commission. |
| H. O. Let. Bk. | Home Office, Letter Book. |
| Hund. R. | Hundred Rolls. |
| Jb. | Jahrbuch (für Gesetzgebung, Verwaltung, und Volkswirthschaft im deutschen Reich). |
| K. R. | King's Remembrancer, Exchequer. |
| L. T. R. | Lord Treasurer's Remembrancer, Exchequer. |

| | |
|---|---|
| L. & P. Hen. VIII. | Letters and Papers, Foreign and Domestic, Henry VIII. |
| Lans. | Lansdowne Manuscripts, British Museum. |
| Lay Subs. R. | Subsidy Rolls, Lay. |
| Min. Accts. | Ministers' Accounts. |
| Orig. R. | Originalia Rolls. |
| P. C. | Privy Council. |
| Parl. R. | Rolls of Parliament. |
| Pat. | Patent Rolls. |
| Pet. | Petitions. |
| Proc. | Proceedings. |
| Publ. | Publications. |
| Receiver | Receivers' Rolls. |
| Rec. Com. | Record Commission. |
| Reg. | Register. |
| Rep. | Report. |
| Rep. Stan. Act Amend. Bill | Report on Stannary Act Amendment Bill. |
| S. P. Dom., Col., Ven | State Papers; Domestic, Colonial, or Venetian Series. |
| Soc. | Society. |
| Stat. | Statute. |
| Trans. | Transactions. |
| Treas. Pap | Treasury Papers. |

NOTE. Abbreviated titles of periodicals may be readily identified in the list of periodicals included in the Bibliography.

# THE STANNARIES

# THE STANNARIES

## CHAPTER I

### TECHNICAL CONDITIONS IN THE TIN INDUSTRY, PAST AND PRESENT

CORNWALL, the farthest west of all England, a wedge of land jutting some eighty miles into the Atlantic, has for a backbone a central ridge of rock running longitudinally through the country from east to west and throwing out ramifications which meet the sea on either side in the rugged outlines that render the country so attractive to the tourist and the artist. This ridge gives rise to numerous streams, flowing for the most part from north to south, and running through small valleys which broaden in places into moorlands of considerable extent. Here it was that tin mining had its origin.

The ore occurs either in veins in rocks, or in the form of gravel or sand in alluvium. These detrital tin deposits are easy to explain. The lodes have disintegrated and their contents have been washed out. The specific gravity of tin is so high[1] that, as the carrying force of the water moderated, it sank to the bottom of the streams in beds.[2] This is the origin of the "stream tin" deposits in the valleys of Cornwall, especially those to the south of the watershed, and to a lesser extent in the valleys of the Dartmoor rivers in Devon. Attention was probably first directed to deposits of stream tin by an agency similar to that leading to their formation. Streams and rivers swollen by rains would cut deeper gutters through the alluvium of their valleys and expose layers of tin stone, pebbles, and gravel. What was thus shown to occur in several valleys would be anticipated and sought in similar situations elsewhere, although the surface indications might not precisely correspond, and by degrees discovery would become an art. Nor could stream works be long in operation without showing some evidences of their connection with the lodes in adjacent hills. The early miners might

[1] 6.8.

[2] Journ. Plymouth Inst., v. 126; Worth, 5, 8–10; Pryce, 66.

not recognize the fact that the quantity of tin stone washed down into the valleys and moors was a measure of the denudations of the more elevated regions of the county, but they could not fail, as they worked upward, to discover some traces of the veins from which stream tin had been derived. Hence, unquestionably, arose the practice of "shoding."[1]

Pryce[2] in 1778 described the ores of tin, as "shode, stream, and mine.[3] The shode is adjunct to, and scattered to some distance from, its parent lode, and consists of pebbly and smoothly angular stones of various sizes from a half-ounce to some pounds in weight. Stream tin is the same as shode but smaller in size and arenaceous, and in that state is formed of small pyramids of various planes, broad at the base and tapering to a point at the top. Stream tin ore is the smaller, looser particles of mineral detached from the bryle or backs of sundry lodes situated on hilly ground, and carried downward into the vales by the retiring waters of the floods. In the solid rock of the valley there is no tin ore, but immediately upon it is deposited a layer of stream tin of various thicknesses, perhaps over that a layer of earth, clay, or gravel, and upon that another stratum of tin ore, and so on continuously, stratum on stratum, according to their gravity and the different periods of their coming. Mine ore," he goes on to say, "is the original lode, buried usually in rocky substances in the hills or the cliffs."

We cannot end this description of the tin beds, so essential to a proper understanding of the history of Cornish mining, better than with an account of an old stream work discovered a century ago and mentioned by the historian Polwhele. "They (the Porth stream works) were situated near the shore of Trewardreth Bay; the ore was of the purest kind and contained two thirds metal. The pebbles from which the metal was extracted varied in size from sand-like grains to that of a small egg and were included in a bluish marl mixed with sand and containing various marine excuviæ. The depth of the principal bed was nearly twenty feet and its breadth six or seven. This appears to have been worked at a very remote period, and before iron tools were employed, as large pick-axes of oak, holm, and box have been found there. . . . In the St. Blazey, St. Austell, St. Stephens in Brounell, and St. Ewe

[1] Worth, 5. [2] Pryce, 66. [3] Lode.

are many old stream works which were commonly attributed to Jews. . . . The most considerable stream of tin in Cornwall is that of St. Austell Moor, which is a narrow valley about a furlong wide (in some places somewhat wider), running near three miles from the old town of St. Austell southward to the sea. On each side and at the head, above St. Austell, are many hills, betwixt which are little valleys which all discharge their waters and whatever else they receive from the higher grounds into St. Austell Moor, whence it happens that the ground of this moor is adventitious for about three fathoms deep, the shodes and streams from the hills on each side being here collected and caught into floors according to their weight, and the successive dates of their coming thither. The uppermost mat consists of thin layers of earth, clay, and pebbly gravel, about five feet deep, more stony, the stones pebbly-formed and with a gravelly sand intermixed. These two coverings being removed, they find great numbers of tin stones, from the bigness of a goose egg and larger down to the size of the finest sand. The tin is inserted in a stratum of loose, smooth stones, from a foot diameter down to the smallest pebbles. From the present surface of the ground down to the solid rock or 'karn' is eighteen feet deep at a medium. In the solid rock there is no tin. This stream tin is of the purest kind, and a great part of it, without any other management than being washed on the spot, brings thirteen parts for twenty at the melting-house."[1]

From the shallowness of the stream tin deposits and the comparative ease with which the ore could be shovelled out, as contrasted with the difficulties of approaching the native lode, it goes without saying that, of the two methods of mining, the former was first to be adopted.[2] All discoveries of ancient tin mines have been made in alluvial ground,[3] and it may be stated with some

[1] Polwhele, bk. ii, 10.

[2] This is a fact not peculiar to Cornwall. In all young mining districts, even at the present day, the first operation is the digging of the alluvial deposits, and not until after these are exhausted is the parent lode likely to be tapped. See the description given of gold mining in Scotland in the early seventeenth century by Atkinson (*Discoverie and Historie of Gold Mines in Scotland*, 15). The German stannaries passed quickly from the alluvial state to that of the lode (Reyer, 32, 35, 36, 79, 237).

[3] Trans. Roy. Geol. Soc. Cornw., iv, 47–56.

degree of certainty that in the early and Middle Ages stream tinning prevailed to the exclusion of lode mining, save possibly when the latter was carried on upon remarkably rich lodes and in shallow depths.[1] The tools of the ancient and mediæval miners, which, as we have seen, were of wood and unfitted for piercing rock, the fact also that whenever mention is made in the old documents of the specific nature of a tin work it is invariably described as a moor or stream work but never as a mine work,[2] the mining customs of Cornwall and Devon, which in their total lack of provision for the occurrence of veins of mixed metals are evidently adapted only to stream tin works, all point to this as the proper solution of the question.

A similar conclusion may be drawn from the continual complaints of the landlords of Cornwall and western Devon concerning the destructiveness of the stannary works to their standing crops. To take a single instance, in 1361 John de Treeures complains to the Prince and his Council "that whereas the tinners have warrant of the Prince to dig and raise tin where they can find it and have dug and collected it for a long time on the moor waste of the said John and his ancestors in the vill of Treeures, who received from the tinners a third part of the tin for toll according to ancient ordinance for the damage done to the lord of the place, but now of late more than sixty tinners have entered on his demesne and soil bearing corn, barley, oats, beans, and peas, as fine as any in Cornwall, and have conducted water to the vill of Treeures over his demesne and soil, so that by reason of the great quantity of water they deluge the land there where they work upon the moor and

[1] Cf. Polwhele, i, suppl., p. 64.

[2] Cf. Smirke, 26. April 10, 33 Edward III. Writ from the Prince to the seneschal of Cornwall reciting the complaint of Henry Nanfan and his associates that they were disturbed in working for tin on the moor of Lamorna.

July 5, 33 Edward III. Grant to certain tinners to dig and work for tin in the moor of Nansmora for twelve years.

So we read of the suit of John Thomas, a Cornish laborer, against Wyse and Clark for having disseised him of his tin work in Crukbargis Moor (Proc. Chanc. Eliz., i, p. xiii).

Again in the early sixteenth century the statutes of Henry VIII against the choking of rivers with silt from the mines refer expressly to stream workers as the offending parties. (Stat. 23 Hen. viii, c. 8; 27 Hen. viii, c. 23; see also Ct. R., bdle. 156, no. 26).

nothing remains of the good land there after they have worked but stones and gravel, so that corn will not grow there; that the tinners refuse to give more toll for waste done to the demesnes than for damage on the waste moor; wherefore the said John prays for the love of Christ that you may be pleased to ordain a remedy, that is, to increase the toll in the demesne beyond the toll in the waste, in proportion to its greater value."[1] Complaints of this sort, so numerous during the Middle Ages,[2] could not have been occasioned by the driving of a shaft in a rocky stanniferous ledge, but by the wholesale trenching and excavating for alluvial deposits in the soil of the Devon and Cornish valleys.

This is borne out by a comparison of data as to the yield of ore. Stream tin, as we know from the testimony of Thomas Beare,[3] himself a veteran tinner and a stannary official of the sixteenth century, was considered far superior in quality to mine tin. Three "foot-fates" of the former, about eight quarts ordinary measure,[4] sufficed for 105 pounds of refined metal. If we turn to the sole surviving Pipe Roll of Edmund of Cornwall, we read, in the account of the latter's preëmption of "black tin" in 1297,[5] that having purchased his ore at 1s. 6d. per foot-fate, to produce a thousandweight of "white tin" he used twenty-eight and one-half feet of black; figures which tally almost exactly with the account given by Beare three centuries later. The inference is that Edmund's tin — and he seems to have preëmpted the entire output — was obtained from stream works.

Detailed accounts of former methods of prospecting date back only to the seventeenth century, although in all probability the operations then in vogue had for the most part been in use for centuries. Omitting from consideration the use of the divining rod,[6] dreams,[7] and other popular superstitions as guides to tin deposits, we learn that the tinners' first aim was to discover shodes, or tin stones.[8] "Where we suspect any mine to be," writes an anonymous

[1] Smirke, 25, citing the White Book of Cornwall.
[2] Cf. Parl. R., i, 297, 312, 382; ii, 190.
[3] Harl. 6380.
[4] *Ibid.*, fol. 35.
[5] Excheq. K. R., Bailiff's Accts. of Edm. of Cornw., 24–25 Edw. I.
[6] Phil. Trans., 1671, p. 2101.
[7] Childrey, 6.
[8] Carew, ed. 1811, pp. 8, 9; Phil. Trans., 1671, pp. 2097, 2098.

Cornishman in 1671, "we diligently search that hill and country, its situation, the earth, or grewt, its color and nature, and what sort of stones it yields; the reason thereof being only this, that we may the better know the grewt and stones, when we meet with them at a distance in the neighbouring valley, for mineral stones may be found 2, 3, 4, 5 miles from the hills or lodes they belong unto.

"After any great land flood (in which it is supposed there are some new frets made in the sides of the banks), we go and diligently observe such frets (which usually after such floods are very clean), to see if happily we can discover any metallic stones in the sides or bottoms thereof, together with the cast of the country (*i. e.* any earth of a different colour from the rest of the bank), which is a great help to direct us which side or hill to search into. Neither will it be much amiss in this place to subjoin the few but sure characters of mineral stones, by which we know the kind of metal, and how much it yields. The first is by its ponderousness, which easily informs us whether it be metal or no. The second is its porosity, for most tin stones are porous and not unlike great bones, almost thorough calcined; yet tin sometimes lies in the firmest stones. The third is by water, which we term vanning, and that is performed by pulverising the stone or clay or what else may be suspected to contain any mineral body, and placing it on a vanning shovel; the gravel remains in the hinder part and the metal at the point of the shovel, whereby the kind, nature, and quantity of the ore is guessed at; and, indeed, most commonly without any great deception, especially if the vanner have any judgment at all." He goes on to describe other methods of prospecting, by turning a stream of water through trenches on a hillside or by sinking an ascending series of essay "hatches" or pits.[1] By one of these means the tin at last was found and mining operations were fairly launched.

The first form assumed by the ancient mines was that of pits or quarries open to the sky, the mineral at this stage appearing at the surface and requiring only to be shovelled out like gravel or hewn as in a quarry.[2] Thus Pliny, in a reference to either tin

[1] Phil. Trans., vi, 2097-2100.

[2] This seems to have been the case in Derbyshire as well (Farey, i, 368; Eng. Min. Almanack, 1850, p. 121). Cf. also the form assumed by the early coal pits (Galloway, 19, 191).

or lead, states that "in Britain it runs in great abundance in the uppermost crust of the ground, so that, by an express act among the islanders, it is not lawful to dig and gather ore above such a proportion set down by stint." In similar fashion Polwhele describes the remains of some ancient tin works in the Scilly Isles.[1] "On the downs in the Isle of Trescaw, we saw a large opening made in the ground and dug about the depth of a common stone quarry, and in the same shape. There are several such in the parish of St. Just in Cornwall . . . and they show that the more ancient way of mining was to search for metal in the same way that we at present raise stones out of a quarry, which, as the metals bear no proportion to the strata of stone in which they lie, must have been tedious and expensive." [2] This method has been followed where suitable almost continuously since the day of its adoption, examples of the present day being at hand in Carclase, near St. Austell, and the Gwennap Pit.[3]

Another form of "daylight mining" is that of following the course of lodes by opening trenches known as "coffins," a good instance of the survival of which is still to be found in the Goonbarrow lode, a little to the north of Rock Hill near St. Austell.[4] "Costeaning" was a method of mining adopted by the early miners much as it was used centuries later by the tinners of Banca in the East Indies.[5] A succession of small pits was sunk from six to twelve feet deep, and drifts carried from one to the other across the direction of the veins or tin layers.[6]

Probably subsequent to the introduction of these methods came that of the "shammel," seemingly a mode of transition from open workings to mining proper, which was carried on both in open pits and in stream works or underground lodes. It is perhaps best described by the anonymous sixteenth century writer previously quoted. The lode found, "we sink down about a fathom and then leave a little long square place called a shamble, and so continue

[1] Polwhele, bk. ii, 10, n.; i, 175. Numbers of these ancient works are still to be found in the Forest of Dean (Journ. Roy. Arch. Inst., xxvii, 227).

[2] Hunt, 418. [3] Worth, 10. [4] Hunt, 418. [5] Foster, 57.

[6] Cf. Worth, 7; Borlase, *Nat. Hist. of Cornw.*, 166; Pryce, 124, 126; Polwhele, i, supplement, p. 63; Schmoller, Jb., xv, 664; Inama-Sternegg, iii, bk. iv, p. 153. The so-called "bell pits," common in the early history of coal mining, are only another form of the costeaning method (Galloway, 32–34).

sinking from cast to cast (*i. e.* as high as a man can conveniently throw up the ore with a shovel), till we find the lode grow too small or degenerate into some kind of weed. . . . Then we begin to drive either west or east as the goodness of the lode or convenience of the hill invite; which we term a drift, three foot over and seven foot high, so as a man may stand upright and work; but in case the lode be not broad enough of itself, as some are scarce one-half foot, then we usually break down the deads, first on the north side of the lode, and then we begin to rip up the lode itself."[1] The shaft was thus divided into a series of step-like stages, each as high as a man could conveniently heave stuff with a shovel from one to the next above.

All of these processes proving useless for the discovery and raising of any tin beyond a certain shallow depth, it became necessary to contrive some other way of following the tin stones. Thereupon they sunk shafts down upon the lode to cut it at some depth and then drove and stoped east and west along its course. Thus by a process of gradual transition there crept in the system of mining such as now exists in Cornwall to the exclusion of almost every other method, namely, the drilling of deep shafts with ramifications to tap the ore in the bed rock.[2]

Shaft mining in Cornwall is probably of great antiquity, although Pryce did not think it to have been introduced earlier than the year 1450.[3] But although we may, perhaps, admit the existence at an early date of examples of mining in the modern sense, the tin was for the most part still obtained from alluvial deposits, and the shafts were no deeper than was necessary to reach the stanniferous gravel. The transitional period, during which the approaching exhaustion of the stream tin rendered necessary the tapping of the lode itself, occurred probably in the sixteenth and seventeenth centuries,[4]

[1] Phil. Trans., vi, 2102. Cf. Pryce, 141; Farey, i, 359. For examples of old shammel works see Polwhele, ii, 10 n.; i, 175.

[2] Cf. Pryce, 141.

[3] It seems to have been employed somewhere in England as early as 1366, for Bartholomæus Anglicus, who wrote in that year, has described it in terms which show that shaft mining had already passed its infancy (*De Proprietatibus Rerum*, ed. 1582, p. 212).

[4] The Bailiff of Blackmore (Harl. 6380) speaks of stream tinning only. Carew (1602) refers to both methods, and Merrest and the anonymous writer in the Philo-

at about which period we find unmistakable signs that mining was being pursued at depths which taxed to their utmost the rude machines for drainage then in vogue.[1]

The stream works were all of limited depth,[2] it being merely a question of digging down to the bed rock through the substratum, a distance which would vary according to the locality but which could not very well be greater than fifty or sixty feet. Thirty-six feet is the depth to which the old miners had driven a tin stream work exhumed about half a century ago,[3] and from what has already been said we know that the tin gravel might be found at even slighter depths, — in fact immediately beneath the surface. With the advent of shaft mining in the solid rock, however, pits were sunk forty, fifty, and sometimes sixty fathoms,[4] and at once the question of drainage assumed the overwhelming importance that has clung to it ever since.[5]

In the earlier stream works wooden bowls seem to have been used for bailing purposes,[6] or the "level," a deep trench running from

sophical Transactions refer to lode mining only (Phil. Trans., vi, 2107; xiii, 949). On the great development of mining engineering in the coal fields in the seventeenth century, see Galloway, 74, 75, 174, 176, 177, 180, 181, 202, 203.

[1] The sixteenth and seventeenth centuries were distinctly the "wet period" for miners not only in Cornwall but elsewhere in England. The colliery owners were engaged at this time in one continuous fight against the incoming waters (Galloway, 74, 157), and the exigencies of the situation are illustrated by the vast number of applications taken out for drainage devices (cf. Law Quart. Rev., xvi, 47, no. xxxv, 46, no. xxix; S. P. Dom. Eliz., xlvi, 69; xxxvi, 73; Price, *English Patents of Monopoly*, 63). Mr. W. H. Price has called my attention to no less than seventy-five of these would-be mine drainers.

[2] Journ. Plymouth Inst., v, 131–134. [3] Trans. Roy. Geol. Soc. Cornw., iv, 47–56.

[4] Phil. Trans., 1678, p. 949; Childrey, 8; Worth, 19. Shafts were frequently sunk thirty fathoms deep in Scotland for the mining of gold (Harl. 251, fol. 109b), and the royal silver mines at Combe Martin in Devon had been driven to a depth of thirty-two fathoms (Atkinson, 52), while the same or nearly the same depth was reached in the coal pits (Galloway, 120, 121, 176, 219), and in the Derbyshire lead mines (Duchy of Lanc. Misc., bdle. 17, no. 6). Bermannus (Schmoller, Jb., xv, 975), speaking for Germany, states that in the sixteenth century shafts of 500 fathoms existed in the Kuttenberg silver mines. Agricola, however, gives the customary depth of mines as fourteen fathoms.

[5] The increased price of materials added to the expense of drainage to bring about a period of great depression throughout the stannaries (S. P. Dom. Chas. I, cccxxii, 1).

[6] Worth, 28.

the stream work to the river, served to clear it of water.[1] After that came the windlass, turned at first by human power and bringing up the water in leathern bags or buckets.[2] There followed the use of small hand or force pumps [3] and at the same time in the larger works the adit, similar to the level but in the form of a drainage tunnel, driven through the hillside to meet the shaft at its bottom.[4] The importance of the adit, not only in a technical sense, but as an incentive to permanent investment of capital in mining, cannot be exaggerated,[5] and its introduction was encouraged by mineral law.[6] But it was too expensive an improvement to be within reach of all, and even when it was employed its usefulness was limited, since when the shaft was driven deeper than the level of free drainage, pumps and windlass had to be employed to bring the water to the adit head.[7]

In the meantime the windlass took various developments as regards the application of its power, the best known being the horse-whimsey, or whim, in which the rope from the shaft passed around the barrel of a huge upright drum turned by a team of horses.[8] In other mines recourse was had to a rag-and-chain pump,[9] consisting of an endless chain broadened out at intervals by leathern

[1] Gent. Mag., xiii, 696.

[2] Trans. Roy. Geol. Soc. Cornw., iii, 48. Cf. Arch. Journ., xxvi, 314; Galloway, 57, 102; Farey, i, 360. For its use in later centuries see Sinclair, 298; Add. MS. 6682, fol. 141; Schmoller, Jb., xv, 975.

[3] Worth, 30; Eng. Quart. Min. Rev. iii, 303–305; Law Quart. Rev. xvi, 47, no. xxxv.

[4] The adit as well as many other devices for mine drainage was known to the ancients (Pulsifer, 36; Del Mar, 63, 72), then went out of use with the general decline of the arts and sciences, and was rediscovered in the thirteenth century by the Germans (Schmoller, Jb., xv, 667–669) and by the English perhaps somewhat later (cf. Arch. Journ., xxvii, 133; Pat., 19 Rich. II, pt. i, m. 4 d; Cal. of Pat., 1301, p. 623; Galloway, 56–71), as well as by other nations. For later mention of the adit in connection with the history of mining see Bushell, *Tracts on Mines*, c. 2; J. Houghton, *Collection for the Improvement of Husbandry and Trade*, April 21, 1693; Farey, i, 360; Galloway, 56, 74, 124, 161; Duchy of Lanc. Misc., bdle. 17, no. 6.

[5] See p. 178.

[6] Houghton, 17, art. xxxiv; Smirke, 86, 89, 113.

[7] Galloway, 71; Farey, i, 360.

[8] Worth, 30. Cf. Sinclair, 298; Galloway, 175. Agricola, in his treatise on the German mines in the sixteenth century, describes elaborate variations of the windlass, including tread-mills, both for men and for horses (*De Re Metallica*, 131).

[9] Worth, 30.

binding, to fit snugly into a long pipe of from twelve to twenty-two feet in length. It was worked by a windlass from the surface[1] and catching up as it did a series of short columns of water served very well to clear small mines of water, its chief drawback being the severity of the labor which it entailed upon the men working it. To drain a mine of any depth a series of these pumps was necessary,[2] and a four-inch pump drawing twenty feet employed from twenty to twenty-four men working five or six at a time in six-hour spells.[3] For the introduction of hydraulic drainage engines[4] it is impossible to fix a date. They usually took the form of overshot water-wheels twelve or fifteen feet in diameter,[5] turning in shallow shafts and operating rag-and-chain pumps or their improvements, the plate-and-chain and the bucket-and-chain.[6] In deep mines half a dozen of these wheels, one above another, might be called into service.[7]

As far back as we have definite information most of the above drainage devices seem to have been in contemporaneous use, and to a certain extent they still are to-day. The successive dates of their introduction can be gauged only approximately. The level, practicable only in the most shallow works, had probably been familiar from prehistoric times.[8] The introduction of the adit into the tin mines cannot be traced back beyond the beginning of the seventeenth century,[9] although Carew refers to it in terms which seem to imply that it was then no innovation.[10] Rag-and-chain pumps appear first at a somewhat later period.[11] The typical mine described in the Philosophical Transactions of 1671 seems to have been drained when on a hillside by an adit, to the head of which

[1] Cf. Agricola, 131 *et seq.;* Galloway, 157, 158. The power, of course, might be applied by horses (Galloway, 167).

[2] Sinclair, 299; Galloway, 159.

[3] Pryce, 150.

[4] A possible instance, the earliest which I can find, belongs to the year 1480 (Accts. Excheq. K. R., bdle. 266, no. 25). Water wheels came into use in the German mines in the fourteenth century if not earlier (Schmoller, Jb., xv, 666).

[5] Worth, 29.

[6] Cf. Galloway, 157–159, 167; Sinclair, 298, 299.

[7] Sinclair, 299; Pulsifer, 36; Galloway, 158.

[8] Polwhele (ii, 10 n.) has found the remains of one at the end of a prehistoric tin work in the Scilly Isles.

[9] Convoc. Cornw., 12 Chas. I, c. 28, 31.

[10] Carew, 12.

[11] Convoc. Cornw., 2 Jas. II, c. 5. Cf. Pryce, 141. Polwhele (iv, 136) makes it a century earlier.

the water was lifted by windlass and buckets.[1] At about the close of the eighteenth century the famous Wherry Mine at Penzance was drained by a rag-and-chain pump worked by thirty-six men, a mode of drainage still at times resorted to in shallow pits.[2]

Aside from the forms assumed by drainage, certain other features of the early tin mine deserve mention. For raising the ore and rubbish from the works, buckets or "kibbles" were used in Carew's time[3] and have been employed ever since. In the older mines a simple hand-windlass[4] lifted and lowered them; later, perhaps, came the horse whim[5] and the water-wheel, just as was the case with drainage. In some mines the same means served for the descent and ascent of the workmen. In shammel workings the shammels or terraces themselves furnished a means of getting up or down. Lode works, however, required the adoption of special facilities. Carew tells us that the workmen were let up and down in a stirrup operated by two men who wound the rope at the top.[6] This system, for a long time the only one in use outside of ladders, was practicable only in perpendicular shafts. Ladders, in the early small, single shaft concerns, would have taken up too much space, but when levels and winzes became properly developed they grew to be indispensable and in time universal.[7] Among the chief advantages which resulted from their use was the economy of lifting power and the avoidance of the mechanical difficulties of stopping cages or buckets at the entrances of different levels; but as the shafts deepened the use of ladders brought with it a great increase in the miners' toil, although it was not until after the eighteenth century that this drawback became very apparent.

The ventilation of the tin works was probably not a pressing question until the sixteenth or seventeenth century, when long galleries began to be driven and shafts extended in depth. The

[1] Phil. Trans., vi, 2107. Cf. Sinclair, 298; J. Houghton, April 21, 1693; Galloway, 71–74, 157–167

[2] Rep. Stan. Act Amend. Bill, 1887, Q. 291.

[3] Worth, 27; Carew, ed. 1811, p. 11; Phil. Trans., vi, 2104.

[4] Cf. Galloway, 54; Schmoller, Jb., xv, 976.

[5] Cf. Galloway, 74, 168, 178, 185; Pettus, *Fleta Minor*, 307; Phil. Trans., iii, 370.

[6] Carew, ed. 1811, p. 10. This was the method of descent and ascent employed in some coal mines as late as 1765 (Galloway, 282, 283).

[7] Worth, 26. Cf. Galloway, 185; Phil. Trans., iii, 770.

old lode workers were much troubled by foul air,[1] and went as far from the shaft only as the air would allow them to breathe. When it failed they sank another shaft and left Nature to do the rest.[2] With a few trifling exceptions, such as perhaps the use of ventilating bellows at St. Agnes in 1696 [3] after the manner in which Bushell had purified his Cardigan mines fifty years before,[4] these few words sum up the subject of mine ventilation in the stannaries until a comparatively modern date.[5]

The primitive nature of Cornish mining until modern times can be seen by a glance at the tools of the ancient stream tinner,[6] which consisted simply of a pick and shovel with perhaps a bowl for bailing.[7] Discoveries in exhumed stream works[8] show that as late as the sixteenth century wooden implements were not uncommon, although by Carew's time the pick was usually of iron and the shovel iron-shod.[9] In the lode works, before the utilization of gunpowder, the only additional tools were gads and wedges to split the rocks, the pick being flat at one end for driving them. A few

[1] Carew (ed. 1811, p. 11) speaks of "unsavourie damps which here and there distemper their heads."

[2] As late as the seventeenth century the English coal miners scarcely knew anything different (Galloway, 160), but they gradually evolved a system of natural ventilation by the use of partitions and ventilating doors, until a fair state of perfection was reached (Galloway, 179). Occasionally we hear of attempts made to purify a mine by the lighting of fires (Galloway, 194).

[3] With their superior technical skill and initiative the German miners had early grappled with the problem of ventilation. Agricola (p. 159) in the sixteenth century gives pictures of pits connected with shafts, galleries, or chimneys, into which the air was forced by huge bellows worked by horses or by water power. This practice was in the early seventeenth century imitated by Thomas Bushell in his mines in Wales (*Tracts on Mines*, D2; Fuller, 4).

[4] Worth, 33.

[5] Cf. Childrey, 8. Beating the air with a cloth, one of the most primitive methods of dispelling noxious gases (Pliny, *Nat. Hist.*, xxxi, 28), seems to have been unknown in the stannaries although practiced in the coal mines (Galloway, 521).

[6] For tools used in other mediæval mines, see Arch. Journ., xxvii, 314–322. Cf. Galloway, 53, 74; Eng. Min. Almanac, 1850, p. 221; Schmoller, Jb., xv, 665.

[7] Journ. Plymouth Inst., v, 127.

[8] Cf. Harl. 6380, fol. 1; Carew, ed. 1811, pp. 8, 11.

[9] Journ. Plymouth Inst., v, 121. These shovels were rude but elaborate. The handle was stuck slanting-wise into a hole in the face or, as in the case of another specimen in the Truro Museum, the entire shovel was of one piece, shaped like a huge wooden spoon.

stone hammers have been found in Cornwall and Devon. In most other mining districts they abound, but in tin streaming they were not needed and in lode mining the poll pick answered all purposes until the introduction of blasting. The ore was broken out by the use of wedges. Into holes bored in the same way as at present, except that the bit ended in a quadrangular point instead of a single edge, were put two semi-cylindrical rods of iron or steel called "feathers," just of a length with the hole itself. A steel wedge was then driven between them and the rock broken off bit by bit. Sometimes also wooden wedges were driven into clefts and then soaked with water to cause the wood to swell, and when the ground was unusually hard the miners wore away its face in the same manner as that by which masons cut stone for building.[1]

The process spoken of broadly as smelting comprises two operations, the preparation of the ore and its conversion into white tin. In the infancy of mining only the more massive and productive pieces were dealt with before melting. The richest tin stones were smelted in the block and the metal disengaged by the direct action of heat. The poorer were subsequently pounded and washed. The first improvement upon this method was the use of something like a mortar and pestle. Next came the use of mills to reduce the ore to a still finer state of pulverization. In Loe Pool Valley are still to be seen boulders of hard elvan with surfaces indented into deep hollows, where the tin stone was rudely battered preparatory to its reduction in the furnace;[2] and Polwhele has left an account of a mill discovered in the islands of Scilly,[3] as well as the relics of an old "buddle" or washing place.[4] Twenty years ago there existed at Retallack Farm in Cornwall interesting remains of such an ancient "crazing mill."[5]

[1] Trans. Roy. Geol. Soc. Cornw., iv, 85; Tilloch's Phil. Mag., v, 357. The use of fire for this purpose was quite common in most mines although no mention of it occurs as customary in Cornwall. In the German and the Derbyshire mines its use was restricted by law to certain hours of the day, so as not to discomfort one's neighbors (Schmoller, Jb., xv, 665; Houghton, 20, art. xl). On this general subject see also Galloway, 187; Phil. Trans., iii, 769.

[2] Worth, 35. [3] Polwhele, i, supplement, p. 64. [4] *Ibid.*, 65.

[5] The walls of the house measured twenty feet by thirteen, and in the gable wall was an opening, two feet square, where the axle of the water wheel passed. Within and without were several granite millstones three or four feet in diameter, grooved

The first detailed account of the preparatory process known as tin dressing is given in 1602 by Carew.[1] The ore was broken small with hammers and then carried in carts or on horses to a stamp mill of three or sometimes of six iron-shod heads, driven by a water-wheel. The previous practice had been to stamp the tin while dry, but wet stamps had by this time come into use with the result that only the roughest part of the ore had now to go from the stamp to the crazing mill, whereas in the dry method all the ore must go through the double process.

The next operation no longer survives in Cornwall. The water, after it had left the crazing mill or stamp mill, was made to descend a series of stages at each of which it fell upon "greene turfe, three or four feet square and one foot thick." Here the sandy ore was laid and gently tossed to and fro so that the lighter particles of waste might wash away and the tin remain entangled in the fibres. Finally it was washed "in a wooden dish, flat and round, being two feet over, and having two handles fastened at the sides, by which they softly shogge the same to and fro in the water between their legs as they sit over it, until whatever of the earthy substance that was yet left be flitted away." "Some, of later times," adds Carew, evidently referring to the present process of buddling, "with a slighter invention and lighter labor, do cause certain boys to stir it up and down with their feet, which worketh the same effect."[2]

The conversion of the prepared ore into white tin, or smelting properly speaking, was in the earlier period carried on by the miners

on the face in a circular direction. In the vicinity were also found stones with basin-shaped hollows similar to many found in different parts of Cornwall and Devon, probably used for pounding the ore, and also one stone, a rough granite block four feet in length by fourteen inches in breadth and depth, which showed by the regularity of the hollows worn in it that the pounders were probably worked by machinery, like the modern stamp. Other stones were found, apparently used for pulverizing the ore by hand, and also a rough stone buddle, or washing trough, about two feet in diameter (Journ. Roy. Inst. Cornw., vii, 213–214).

[1] Carew, ed. 1811, pp. 39–40. For tin dressing in 1586 see Harl. 6380, fol. 106.

[2] The process here described as buddling was adopted in the Derbyshire lead mines only in the late seventeenth century, or possibly not until the eighteenth. Before that the ore was simply washed in a trough (Add. MS. 6682, fol. 141; Farey, i, 377). In the Mendip Hills the methods were equally primitive (Phil. Trans., ii, 527). In Cornwall buddles had already been mentioned in the Act of Henry VIII to restrain tinners from filling up rivers with their silt (Stat. 23 Hen. VIII, c. 8).

themselves. A small pit was dug and a fire kindled close to the spot where the ore was found. Upon this the stones were thrown and the metal afterward gathered from among the ashes and sand.[1] Several antiquarian discoveries in Cornwall have led Pryce to the conclusion that this was the form of operation prevailing at the time when the Phœnicians supposedly visited Britain.[2] When Diodorus Siculus wrote, however, an advance had been made. The "astragalus" block which figures so prominently in his account of the Cornish tin trade[3] must have been the product of a furnace from which the flow of metal could be directed. Of such there are many remains, varying much in character but passing under the common name of "Jews' houses." Some were built up into the shape of inverted cones of hard clay[4] about three feet broad at the top and three feet deep. A blast of air conveyed by common bellows to the lower part of the furnace served to create an intense heat, and the molten tin was discharged from a small opening at the foot. Others were of granite and dome-shaped,[5] and with these have usually been associated the granite moulds of which many have been discovered on Dartmoor[6] and other ancient stream-tin fields.[7] Another fairly advanced but probably exceptional smelting furnace has been discovered in the Land's End district, consisting of a bronze caldron resting upon a layer of charcoal.[8]

The first written record dealing with the course of smelting operations is the De Wrotham letter of 1198,[9] which speaks of two smeltings. The first was probably a rough process taking place near the

[1] Trans. Roy. Geol. Soc. Cornw., vi, 43. See Pryce, 281; Louis, 6.

[2] Pryce, 281.

[3] Diod. Sic., v, 21, 22.

[4] Trans. Roy. Geol. Soc. Cornw., vi, 44, 45.

[5] Cf. the early iron furnace, as described by Fairbairn (p. 4), and the "boles," or primitive lead furnaces used in Derbyshire, which did not disappear until the seventeenth century (Phillips and Darlington, 16; Percy, *Metallurgy of Lead*, 213–216; Childrey, 112; Rep. Hist. MSS. Com., Hatfield House MSS., pt. ii, 523).

[6] Rep. Roy. Cornw. Polytech. Soc., 1872, p. 149.

[7] Gent. Mag., lxi, 34. Some time ago, in East Cornwall, there was unearthed an entire mining village belonging to this period, containing three granite furnaces in various stages of preservation, while scattered about were pieces of slag and occasionally of metallic tin (Journ. Roy. Inst. Cornw., iv, 75, 76).

[8] Penzance Nat. Hist. and Ant. Soc., i–ii, 347–351. This has been attributed to the Phœnicians.

[9] App. A.

mine, while the second, for reasons of stannary taxation, was permissible only at certain towns designated by the warden. With the advent of improved methods of smelting, it became no longer necessary to fuse the tin twice in order to obtain a proper fineness, and from the first and second smeltings instanced by De Wrotham arose the single "blowing-house" process known to Beare and to Carew.[1] To set even an approximate date for the introduction of the blowing house is impossible, but it seems to have been in common use by the middle of the fourteenth century, as we find the Black Prince sharing in the profits of several at Lostwithiel in 1359.[2] In 1426 occurs the case of John Aunger of Cornwall, "husbandman and blower,"[3] and in 1495 the new ordinances of Prince Arthur dealt, among other objects, with the entering of blowing-house marks and the swearing in of blowers.[4]

The blowing-house at which the smelting of the ore finally took place was a rude structure, probably of rock and turf, with a thatched roof, the whole being so inexpensive that every few years it was burned down in order to save the particles of tin which the blast had driven up into the thatch.[5] Here the prepared ore was made into parcels according to its quality,[6] and then smelted on the hearth[7] of the granite furnace by a charcoal fire fed by a blast from a large pair of bellows worked by a water-wheel. The molten metal was cast into slabs and blocks of from two hundred to three hundred pounds each in weight.[8] Abundant evidence exists that the white tin produced in this crude fashion was as pure in quality as that produced by the smelters of to-day.[9]

[1] A lead blowing-house succeeded in Derbyshire the primitive bole (Farey, i, 383–384).

[2] White Bk. of Cornw., 32 Edw. II, c. 89d. The ironmaster's furnace was quite similar (Plot, 161–164; Eng. Hist. Rev., xiv, 513 *et seq.*). A seventeenth century account of iron smelting is given by Powle (Phil. Trans., vi, 931–936). The blast furnace for iron smelting is said to have originated in the Rhine Provinces at about the beginning of the fourteenth century (Swank, 80–83; Scrivenor, 80–82). Prior to the adoption of the blast furnace the English ironmasters used the Catalan forge, or the high bloomery. The iron was smelted direct from the ore and when finished was wrought, or bar iron (cf. Ellis, i, 137 n.; Swank, 45–47; Ray, 134–137; Dudley, 37–38).

[3] Cal. of Pat., 1426, p. 308.

[4] Add. MS. 6713, fol. 101–104.

[5] Fuller's Worthies, 195.

[6] Harl. 6380, fol. 107.

[7] *Ibid.*, 108.

[8] Accts. Excheq. K. R., Coinage Rolls.

[9] *Cornish Mining*, 13.

With this account of tinning, corroborated a few years later by those of Norden[1] and Childrey,[2] we reach the end of the distinctively mediæval period of stannary operations, a period marked on the whole by very sluggish progress in mining and in smelting. A certain acceleration of pace is, however, observable as the period draws to a close. Especially during the first half of the sixteenth century some improvements were introduced into the Cornish mines. These came from the continent. The English kings at an early period had been greatly impressed with the superior skill of the Germans in mining and metallurgy, and repeated instances are to be found in mediæval and Elizabethan state documents of mineral concessions made to foreign workmen to induce their immigration.[3] It may have been the royal patronage given these foreigners and the reports of their great skill which induced Sir Francis Godolphin, a large tin producer, to send for the person mentioned by Carew as Burchard Craneigh, by whose aid were effected the improvements which he notes in the management of the great Godolphin tin works.[4] These were probably the use of the hydraulic stamp, already considerably employed in the German mines,[5] and the various improvements in the dressing of tin ore previously described. There should perhaps be added the use of charcoal for smelting instead of the peat mentioned in all stannary grants of privilege[6] as the usual fuel.

The changes of the sixteenth century, important though they were as signs of industrial awakening, were not fundamental or far-reaching. The great period of technical advance dates from the latter half of the seventeenth and the early eighteenth century, when a series of improvements and inventions initiated a revolution in the tin districts and sent the annual output by successive

[1] Norden, 13.
[2] Childrey, 10.
[3] See p. 41, n. 6; Journ. Roy. Inst. Cornw., xiii, 430–434; Atkinson, 18–20, 33; Watson, 58; Calvert, 52, 87, 98, 103, 109, 130, 131, 139, 144; S. P. Dom. Eliz., xxxvii, 62; xl, 73, 81; clxvii, 24; Addenda, xiii, 32; Reyer, 126; Hunt, 123, 166 n.; Pettus, *Fodinae Regales*, 20, 29, 30; Scrivenor, 34 n.; Fourth Rep. Hist. MSS. Com., pt. i, 517; Price, 49, 55.
[4] Carew, ed. 1811, p. 13. It may have been the elder Hochstetter (Bushell, *Tracts on Mines*).
[5] Louis, 14; Reyer, 81; Schmoller, Jb., xv, 666.
[6] Chart. R., 36 Hen. III, m. 18; 33 Edw. I, nos. 40, 41.

leaps to double its previous figures. From five hundred to seven hundred tons a year had been the usual production during the first half of the seventeenth century. In the second half the average yield ran from a thousand to fifteen hundred tons. The following half-century sees the average carried well up to the two thousand mark, and during the fifty years ending with the year 1800 the output varied between two thousand and thirty-five hundred tons.[1]

That the new movement links on to the technical advances of the preceding period, illustrating anew the continuity of economic progress, is apparent from the direction of its first efforts in perfecting the processes for dressing and smelting the ores. According to the well-informed anonymous writer of 1671 already cited, the ore-dressing was done chiefly by boys. In his day,[2] after the stones had been broken to a convenient size, they were carried to a stamping mill whose mechanism by this time had become so improved as to work for a couple of hours without attention. One John Tomes, when a boy thirty years before, patented a device by means of which when there was not enough ore in the coffer the water was turned off, whereas before this a bell had been used to give warning when the coffer was empty, by which time the mill was often damaged. One wheel could now work the stamps in three or four coffers. On leaving the stamp heads the crushed tin ore was subjected to a series of operations the object of which was to grade and concentrate it by the application of running water.

From the stamps the ore was washed out through a grate into a "launder," or shallow trench, where it was divided into "forehead," "middle," and "tails," according to its specific gravity. After having been "trambled" or buddled, it was "sezed," "dilleughed," "crazed," or "framed," as required. The buddle is described as a long square "tye" of boards or slate, about one foot in depth, six feet long and three feet broad, wherein stood a man barefooted, who with a "trambling-shovel" cast up ore upon the buddle head as high as his middle. The stuff was worked both with the shovel and with the feet, and as the buddle was traversed by a gentle flow of water the effect of the operation was to separate the ore into several qualities, the heavier stuff remaining at the head and the lightest being deposited at the foot. For "retrambling"

[1] App. J.

[2] Phil. Trans., vi, 2108–12.

there were also used "drawing-buddles" which had no tye but a plain sloping board. "Sezing" consisted in the use of a hair sieve instead of the drawing buddle to grade the tin. "Dilleughing" was performed by putting the forehead of the doubly trambled tin stuff into a canvas sieve and shaking it in a large tub of water. The tails or leavings from the buddle were thrown into "strakes" or tyes, of which there were commonly three or four in succession, where the "slimes," or finer ore, were separated from the coarser "roughs." The latter were crazed and retrambled; the former were framed, the frame being a rack six feet long by three and one half broad, suspended on two pivots like a cradle.

In this account of 1671 mention is made for the first time of calcining, designed to burn away the impurities of the ore. It was done in a square kiln heated by furze, the ore being spread above the fire on a flat granite slab over which the flame played. Having been continually stirred on the slab with a rake, the ore was finally pushed into the fire, the fireplace, when choked up, emptied, and the mixture of ore and ashes retrambled. Rude as many of these operations doubtless were, they show a great advance in mining skill since the time of Carew. Dr. de Merrest, a few years later, describes the tin stuff as being dried in a furnace on an iron plate[1] before being crazed, which seems to be an inexact reference to calcining. He says also that tin stuff not worth working was thrown into heaps which in the course of some six or seven years would be fit for reworking. The germ of truth in this statement, undoubtedly made in all honesty, seems to be that just at this date the advances in dressing were unusually rapid, and every few years with the advent of more economical methods it became profitable to use inferior ores previously rejected.

Considerable improvements had also been made in smelting. The slovenly habit, described by Fuller in 1662 as customary, of burning down the blowing-house to catch the tin in the thatch, was made unnecessary by the construction of chambers in the chimneys for the deposit of the metallic dust.[2] And although pit coal was still unknown in the melting-house, a difference had been

[1] Phil. Trans., xiii, 952.

[2] Worth, 50. This invention was probably introduced from Germany, where it had been known for more than a century (Agricola, 320, 322).

made in the fuels used for varying grades of ore. Moor, or stream tin, was fused by charked peat, lode tin by charcoal and peat mixed, and slag was remelted by charcoal alone.[1]

Following closely upon this seventeenth century advance in ore dressing came the invention of improved devices for mine drainage. As the tin fields of Cornwall became more and more developed, tinning took on more of the character of lode mining. The stream works were still largely in evidence in 1765,[2] but in 1778 Pryce gives us to understand that they were distinctly of minor importance. No great changes had taken place in their operation since the days of Carew. The adventurer sank a hatch three or five fathoms to the shelf of clay on which the tin stones were deposited. When the rough washing on the point of a shovel had indicated that it was "paying" tin, the work was drained by a level and worked by the tinner with the aid of a few helpers until the spot was exhausted.[3] But in lode mines the accumulation of water called for more advanced methods. At the beginning of the eighteenth century John Coster of Bristol had taught the Cornishmen the use of one large water-wheel forty feet in diameter instead of the half-dozen smaller ones then used for a single mine,[4] but his invention, important as it doubtless was, was quickly overshadowed by that of the steam engine. It is not clear at precisely what date or at what place the steam engine first appeared in Cornwall.[5] It was still believed that water could be raised only thirty-two feet, and at first the new invention took the form of a series of steam suction pumps[6] which at great depths were so multiplied that the expenditure in initial outlay and subsequent cost was enormous.

The scale on which the Cornish mines were operated and the increasing amount of work thrown upon the engine soon rendered some forcing arrangement imperative. Morland had patented

[1] Phil. Trans., 1671, p. 2113.
[2] Jars, iii, 187.
[3] Pryce, 132–133.
[4] *Ibid.*, 307.
[5] Pryce dates its introduction at about the year 1702 (p. 153). Carne declares that one was at Wheal Vor from 1710 to 1714 (Trans. Roy. Geol. Soc. Cornw., iii, 50), and Redding says that the earliest was erected in 1725, at Wheal Rose (*Yesterday and To-day*, i, 128). The project of a steam engine had been seriously considered long before these dates (Galloway, 236–238; Cal. S. P. Dom., 1629–1631, pp. 382, 483).
[6] Cf. Farey, i, 325; Eng. Quart. Min. Rev., iii, 303, 304.

the plunger in 1675,[1] but its development was slow, and the first note we have of its being adopted in any mine is in 1796, in the United Mines at Gwennap.[2] In the meantime, Savery's engine of 1696[3] had been superseded in 1705[4] by that of Newcomen. Yet so conservative were the tinners that in 1742 only one steam engine was to be found in all Cornwall.[5] Then came a rapid advance, and in the next thirty-six years more than sixty were erected, and more than half had been rebuilt and enlarged.[6]

Newcomen's engine, effective as it was in comparison with previous methods of drainage, was completely displaced, late in the eighteenth century, by that of Boulton and Watt.[7] Their first engine in Cornwall was erected in 1777[8] at Chacewater. In five years twenty-one had been set up and only one of Newcomen's remained, which disappeared in 1790.[9] Further improvements at the hands of Trevithick, Hornblower, and Woolf, brought the Cornish mine engine to a high state of efficiency in the early decades of the nineteenth century,[10] while the practice of draining the surface of the mines and the greater attention given to the tightness of the adits and pit work lessened materially the work required of the engines.[11]

The immediate result of these improvements was an increase in the depths at which tin could be mined. From 1720 to 1778 ninety fathoms seems to have represented the maximum depth attained by the aid of the Newcomen engine,[12] but the advent of Watts's improvements was marked by an increase of the maximum to about two hundred fathoms in the years approximately from 1778 to

[1] A similar kind of pump was known to the ancients but had lacked the most important part of Morland's invention, the stuffing box.

[2] Rep. Roy. Cornw. Polytech. Soc., 1874, p. 135.

[3] On the subject of Savery's invention see Galloway, 196, 197, 238.

[4] Galloway, 238–240.

[5] Worth, 22.

[6] Pryce, xiv. For the spread of the steam engine in the collieries, see Galloway, 242–244, 251–253.

[7] Cf. Galloway, 285–289, 309, 310.

[8] Pryce, 313.

[9] Worth, 22.

[10] Trans. Roy. Geol. Soc. Cornw., iii, 52, 53, 56; Worth, 23.

[11] Trans. Roy. Geol. Soc. Cornw., iii, 66.

[12] The English collieries had in many cases attained far greater depths. Whitehaven in 1755 contained mines 780 feet deep (Galloway, 350). In the Forest of Dean the pits were still primitive affairs of 75 feet or thereabouts (*ibid.*, 340; Nicholls, 239).

1812. During the following quarter-century two hundred and ninety fathoms was reached,[1] and the progress during the remainder of the century was correspondingly rapid. The Dolcoath Mine in 1900, the deepest in Cornwall, had attained a depth of four hundred and seventy fathoms below adit, and several other Cornish tin works were almost as deep.[2]

Almost contemporaneous with the great advances in mine drainage and ore dressing appears a change in the apparatus of mining itself. A description of an ordinary tin miner's tools is given in the Philosophical Transactions for 1671, showing that, with the sole exception of tamping-iron and borer, they were practically the same as to-day. A "bele," or Cornish "tubber," was used, with double points, eight or ten pounds in weight, sharpened at both ends and well steeled. A sledge weighing ten to twenty pounds and with proper care lasting seven years, steel-pointed gads, or wedges, of two pounds weight, and the ubiquitous shovel and barrow, constituted the tinner's kit.[3] But the most important addition to the miner's equipment was gunpowder, which rendered obsolete the tedious drilling and splitting away of the lode. Blasting seems first to have been used in Hungary or Germany about the year 1620,[4] but England did not take it up until 1670, when we find it introduced into the copper mines at Ecton[5] in Staffordshire by Prince Rupert's German miners. From there it spread into Somerset in 1684[6] and soon afterward entered Cornwall, where it seems to have been first employed at St. Agnes at the beginning of the eighteenth century.[7] For more than a century blasting was carried on in Cornwall in a very dangerous way.[8] After the powder had been introduced and tamped, an iron rod called the "needle" was

[1] Worth, 24.

[2] Records of London and West County Chamber of Mines, i, pt. i, pp. 16, 18.

[3] Phil. Trans., vi, 2104.

[4] Cf. Galloway, 226. Dr. Edward Brown writes in 1669 of blasting as carried on in the copper mines of Hungary (Add. MS. 5233, fol. 54).

[5] Tilloch's Phil. Mag., v, 357; Worth, 17; Plot, 165; Trans. Roy. Geol. Soc. Cornw., iv, 84; iii, 35–85.

[6] Tilloch's Phil. Mag., v, 357.

[7] Worth, 17; Trans. Roy. Geol. Soc. Cornw., iv, 86. It was not used in the mining of coal until 1719 (Galloway, 227, 285, 321, 348, 485).

[8] Trans. Roy. Geol. Soc. Cornw., i, 78, *et seq.*

driven through the tamping, and in the aperture thus made a hollow rush was placed, filled with powder, to act as a fuse. Sometimes quills were used for fuses, in which case the tamping was put around them and the needle not inserted. The iron needle and the tamping-bar were the cause of many casualties among the workmen, but it was not until within a century that the safety fuse and safety tamping-bars shod with copper were suggested, and even then it required some time for them to enter into general use.[1] In 1850 the present safety fuse was invented, and the use of dynamite or gun-cotton has within the last few decades driven out that of gunpowder in all open workings or in wide levels. Rock-borers, worked by compressed air,[2] have likewise made considerable headway in the larger mines, but the more primitive methods of the pick and drill are still far from being completely displaced.

Contemporaneous with the introduction of the steam engine, but much more rapid in its development, came the reduction of the tin ore by the use of pit coal as fuel. This problem — and the growing scarcity of wood fuel made it a serious one not only for tin but for other metallic industries [3] — had long been a favorite hobby of inventors.[4] As early as 1632 Dr. Jorden had asserted its

[1] Worth, 18. [2] Hunt, 507–528, 537–539, 561; Worth, 19, 20.

[3] Cf. Stat. 35 Hen. VIII, c. 17; 23 Eliz., c. 15; 27 Eliz., c. 10; Rep. Hist. MSS. Com., x, App., pt. iv, p. 7; Sturtevant, 5; Dudley, 36; Malden, 275.

[4] Naturally enough, lead, as the least refractory of metals in the melting-pot, was the first to be smelted by the pit coal process. As early as 1526–1528 we hear of an experiment of this sort under the direction of Cardinal Wolsey (Galloway, 124), although whether or not successful we do not know. In 1589 an unsuccessful effort was made by Thomas Proctor and William Peterson "to make iron, steel, or lead, by using of earth coal, sea coal, turf, peat, or some of them" (Lans. MS. 59, fol. 196). Bushell's attempt at smelting with peat some years later also failed (*Tracts on Mines*, A3), but inventors were not lacking in the seventeenth century, and although Grandison and Hodge's attempts in 1678 and 1690 probably failed, Thomas Wright's reverberatory pit coal furnace was successfully smelting Flintshire lead by 1698 (Percy, *Met. of Lead*, 216–218. See also Pilkington, ii, 118–123; Phil. Trans., 1729, no. 647, p. 31).

More interest, however, attaches to the use of pit coal in iron smelting. The first patent for this purpose seems to have been granted to Robert Chantrell in 1607 (Cal. S. P. Dom., 1603–1610, p. 346), and a second to Simon Sturtevant in 1611, but there is no evidence of their ever having been put into practice (Sturtevant, 22; Percy, *Met. of Iron and Steel*, 882). Sturtevant's idea was probably that of coking the coal (Sturtevant, 9). His privileges in 1613 were transferred to John Rolvenson,

practicability and had tried unsuccessfully to solve it;[1] and at about the same period an attempt to the same end was made by Sir Bevil Grenville,[2] but this also failing of accomplishment the matter seems to have dropped from public notice. In the mean time, with the decadence of the stream works, came an added incentive to the supersession of the still primitive charcoal blast. The alluvial ore, occurring as it did in rounded masses and grains in a high state of purity, was especially adapted to this method. The charcoal ashes formed the only necessary flux, while the fuel contained no elements capable of injuring the tin. But for lode tin this method was less satisfactory because of certain chemical changes in the ore, and this fact, added to that of the scarcity of charcoal, resulted in a series of further attempts to use the cheaper fuel.

The invention of the process for smelting tin with pit coal has usually been ascribed to John Joachim Becher, a German chemist residing in Cornwall in the latter years of the seventeenth century,[3] but little if any use was made of it for some years, and Becher's claim to the invention, depending as it does merely on his own assertion in the preface to one of his works,[4] is possibly open to question.

but in his case also we have no evidence of any satisfactory performance (Rolvenson, 6, 10, 11), although Rolvenson seems to have gone a step beyond Sturtevant and to have conceived the idea of a reverberatory furnace (Percy, 882, 883).

After some desultory experiments in the Forest of Dean (Galloway, 253), the next and one of the greatest names connected with this work is that of Dud Dudley. In his book, *Metallum Martis* (2–24), he published an account of his labors, but carefully refrained from disclosing his method. Whatever this was, however, it evidently was for a time strikingly successful (cf. Malynes, 269, 270), but he suffered so from the attacks of rival ironmasters (Dudley, 12, 13) and dishonest partners that he finally abandoned the enterprise in 1651 and never resumed it.

We are told by Plot that after Dudley the smelting of iron with raw coal in reverberatory furnaces was again attempted in Staffordshire and Lancashire (Plot, 128; Galloway, 189), but without success. The intense heat necessary to smelt iron made the problem much more difficult than in the case of the softer metals, tin and lead, and until the year 1735 all English iron was smelted by charcoal. Abraham Darby finally solved the problem (Percy, 885–889; Galloway, 259; see also Phil. Trans., 1747, no. 482, p. 370), and this with subsequent improvements by Cort and Roebuck (Scrivenor, 83) led to the extinction of the old charcoal furnaces and to the revolution of the entire industry. The effect of the cheapening of iron upon the mining industries is incalculable (Galloway, 256–260).

[1] Jorden, 50; Fuller's *Worthies*, Devon; Galloway, 215.

[2] Pryce, 282.

[3] For an account of his life, see Trans. Roy. Geol. Soc. Cornw., iv, 87–91.

[4] *Alphabetum Minerale* (1689).

It was not until 1705, a date coincident with Newcomen's invention of the steam engine, that a Mr. Liddell obtained from Queen Anne a patent for smelting black tin with fossil coal in iron furnaces.[1] Close upon the heels of this patent came the invention of the reverberatory furnace of masonry in which the finely divided ore could be smelted easily and at the same time direct contact with the fuel be avoided. This is the origin of the so-called Cornish method of tin smelting. The ore, mixed with finely crushed anthracite or culm, was charged upon the bed of the reverberatory furnace and heated until reduction was complete. The less fusible and pasty slag was then drawn from the furnace, while the completely melted portion, or "glass," was tapped out with the liquid metal. The former was then removed and the metal purified.[2] Pryce's description of this method in 1778 shows that in all essentials it was the same as now, except that his furnace charge was but five or six hundred-weight of ore, while nowadays a two or three ton charge is preferred.[3]

The first reverberatory furnace is said by tradition to have been set up at Treloweth, although according to another statement it was established in 1706 at Newham by Monatt, a German chemist, and after a few years transferred to Calenick.[4] Many years were still to elapse before blowing-houses were superseded. Charcoal tin, being of a slightly superior grade, brought a higher price,[5] and as long as sufficient stream tin could be found to charge them blast furnaces still persisted. In 1765 Jars found both methods used side by side;[6] Pryce's testimony in 1778 is to the same effect;[7] and it was not until about sixty years ago that the last blowing-house — at St. Austell — was finally discontinued.[8]

These great seventeenth century improvements in ore dressing, mine drainage, and smelting must not blind us to smaller advances in mine economy, which contributed to the development of the industry. First of all we have to note an almost unbroken series

[1] Pryce, 282.
[2] Louis, 8.
[3] *Ibid.*, 9; *Cornish Mining*, 13, 14.
[4] Worth, 50.
[5] Borlase, *Nat. Hist. Cornw.*, 182.
[6] Jars, iii, 212, 213.
[7] He tells us that the blowing-house smelted from eight to twelve hundred-weight of tin in twelve hours, by the use of from eighteen to twenty-four sixty gallon packs of charcoal (Pryce, 136).
[8] Worth, 51; *Cornish Mining*, 13–15.

of improvements in ore dressing, which have continued from the first spurt above noted down to the present day. Hardly had the eighteenth century begun when the crazing mill fell into disuse,[1] owing to improvements in stamping and dressing which rendered it unnecessary. Between 1671 and 1778, the year in which Pryce wrote, the former process had undergone great changes, the stamps becoming probably six times as effective as those they supplanted. The lifters were then of ash and their iron heads weighed 140 pounds.[2] Pryce's description also seems to show that all the heads in a coffer operated upon the stuff in succession, the blow of the first forcing it on to the second, and that of the second to the third, after which it was permitted to emerge.[3]

We find buddling, sezing, dilleughing, and framing[4] practiced as before, but with greater delicacy of manipulation. Trunking also had been introduced[5] for the stamped tin stuff which ran from the coffer to the two farther pits. At the semicircular head of the trunk, — a pit like the buddle, — a boy stirred these slimes with a small shovel, so that the water which ran in might wash away both the filth and the tin over a cross stick or board about ten inches deep, from which it passed into the body of the trunk. What remained at the head was framed, and the residue trunked again and then framed also.[6] The calciners, formerly of moorstone, were now built of brick,[7] and the burnt leavings which, until 1735, had been thrown away as good for nothing were after that date reduced to metal.[8] The modes of dressing the leavings were so various that Pryce could not detail them without danger of prolixity.[9]

In mining itself more scientific methods of prospecting came into vogue during the eighteenth century. Costeaning and shoding,[10] although now abandoned,[11] were still practiced in the days of Pryce, but already they had been supplemented by the practice of boring[12] and by a better knowledge of geology. Our ancestors were satisfied to pursue a single vein, without suspecting that others might

[1] Jour. Roy. Inst. Cornw., i, 179.
[2] Worth, 39.
[3] Pryce, 221.
[4] Add. MS. 6682, fol. 294–296.
[5] Pryce, 133–135; Ray, 132, 133.
[6] Add. MS. 6682, fol. 295.
[7] Pryce, 224.
[8] *Ibid.*, 230.
[9] Worth, 39.
[10] Ray, 131.
[11] Trans. Roy. Geol. Soc. Cornw., iii, 74, 75.
[12] Add. MS. 6682, fol. 281.

exist near at hand, or, if aware of their existence, they were apt, from want of capital or from disinclination to invest it or, perhaps, from want of a greater spirit of initiative, to leave them unexplored. At present the lodes are more speedily and fully searched by the practice of driving across the country from north to south, and *vice versa*, as well as by other methods too technical to be here described. Ventilation, in the eighteenth century, was extended somewhat by the provision of boarded channels in the bottom of adits, by which currents of pure air were carried into mines.[1] Another method was that of a stream of water passing into one of the shafts, the accompanying air being carried by a pipe close to the discharge of the water to the extreme end of the level where required.[2] This process, still used in 1860, was sometimes assisted by small fans worked by boys.[3] Other appliances have been suggested from time to time,[4] but none have proved very effective, and the ventilation of the tin mines, even to-day, is largely natural, the air finding its way in by certain channels and out by others.[5] Save where a drift is very long the ventilation is good.

One effect of these improvements was an increase of health to the laborer and an added efficiency to his work. In 1586 and in 1602 four hours out of the twenty-four was the longest time during which a tinner could remain underground,[6] but by Pryce's time it had become possible to extend this period to six hours,[7] and some fifty years later to eight.[8] Another result of better ventilation was that instead of being forced to sink air shafts at a distance of about thirty fathoms from one another, the miners, by the end of the first quarter of the nineteenth century, could proceed one hundred fathoms from a shaft without feeling discomfort.[9]

The old-fashioned method of descent into the mines by means

[1] Pryce, 146, 147.

[2] Trans. Roy. Geol. Soc. Cornw., iii, 64.

[3] Worth, 32.

[4] *Ibid.*, 33. Ventilation by fire was apparently not known, although it was introduced at about this time with success in some of the British coal mines (Galloway, 253, 254, 282, 326, 327).

[5] Rep. Stan. Act Amend. Bill, 1887, Q. 366. Cf. the ventilation by "coursing the air" as practiced in the collieries (Galloway, 279).

[6] Childrey, 8; Carew, ed. 1811, p. 11; Worth, 58.

[7] Pryce, 178.

[8] Trans. Roy. Geol. Soc. Cornw., iii, 64; Worth, 58.

[9] Trans. Roy. Geol. Soc. Cornw., iii, 63, 64.

of long ladders, which had been so injurious to the health of the men, has within the last century been superseded in the larger mines by the use of the man-engine, first introduced in 1842, but on account of the expense not yet universally adopted. In other mines is used the wire rope-and-chain method of descent so well known in the collieries.[1]

At the same time as the improvements in drainage came the use of steam engines for drawing ore and rubbish from the mines, a work previously done by the application of horse power.[2] A saving in expense of fifty per cent hastened its adoption, especially in view of the fact that modern mines of any depth could not employ horses enough to raise their rubbish.[3] Kibbles, or heavy iron buckets, are still clung to in many of the works.[4] In others this primitive and clumsy apparatus has been supplanted by the use of "skips," which travel between guides much after the fashion of ordinary freight lifts.[5]

The transportation of ore from the mines had been effected by means of pack horses,[6] but as mines became deeper and more extensive this primitive method grew not only costly but entirely inadequate. So much ore was raised in 1750 at Polberrow, St. Agnes, that carts had to be pressed into service, while the Fowey Consols, one of the larger mines, maintained in its service mules by the hundred.[7] Tramways superseded the pack horse, beginning in Cornwall in 1818,[8] but in the course of a few decades we find

[1] Worth, 26; Rep. Stan. Act Amend. Bill, 1887, Q. 710, 711.

[2] This was first done away with in the collieries at Whitehaven in the late eighteenth century (Galloway, 354, cf. Galloway, 253, 274–277, 296–302, 314).

[3] Trans. Roy. Geol. Soc. Cornw., iii, 61.

[4] The use of corves, or baskets, was given up in the collieries in 1840 or thereabouts (Galloway, 314).

[5] Hunt, 596, 597. Cf. Galloway, 314, 323, 483, 484.

[6] This was a method employed in the collieries (Galloway, 57, 102) and especially in the Forest of Dean, where the use of carriages for ore was forbidden by the mine court (Nicholls, 45).

[7] Worth, 48.

[8] There is no mention of any use in Cornwall of self-acting planes and other devices which preceded the tram in the northern coal fields (Galloway, 318, 329–331, 370). The Cornish miners were very tardy in the introduction of the tram. It had been employed in the Tyne coal fields as early as the middle of the seventeenth century (Galloway, 154–156, also 243, 248, 257, 283, 304, 361, 318, 169).

them pushed aside by the steam railway with its branch lines reaching to the shafts' mouths.

With the deepening and better drainage of the tin mines came various improvements in their general engineering. Originally the lodes were followed from the shafts at points where they seemed promising, without any attention to order or regularity. The workmen, throwing the deads behind them into the worked-out places as they proceeded, were led on by a bunch of ore, and when that failed their work was done. Obviously this system was not one for making discoveries. Where lifting power was limited it was undoubtedly the most economic mode of procedure, but it remained at best a hand-to-mouth sort of arrangement, destined inevitably to give way to other and better methods.

The difficulty of pursuing this method where the water was "quick" led probably to that of stoping downward from the shaft, in which the lode was hewn away in steps of six or eight feet in height, one man following another. Working in this way downward, or under foot, is called "stoping the bottom," upward, or overhead, is called "stoping the back." On this system, as soon as the shaft is sunk six or eight feet under the adit, if the lode is productive, the first stope is commenced, a second follows it, and a third as soon as the shaft is sufficiently deep.[1] The facilities for exploring the lode and making new discoveries were scarcely greater under this system than by the former, and a further improvement soon followed, namely that of driving levels or horizontal galleries on the lode from the shafts and stoping the lode downward from one level to another. On this plan, although the mine was explored by the levels, the ore was taken away almost as fast as the shaft was sunk, and if the unexpected happened, — if, for instance, the lode should even for a short space become unproductive, — the mine had no resources in itself to furnish the means of paying its ordinary expenses. With hand-to-mouth mining this was often fatal.[2] Independently of the risk this method was expensive, for, in the first place, even if all the lode were ore, a mass obviously can be taken from above at much less cost than from below. Of more importance, furthermore, was the fact that in stoping downward, the whole lode, good or bad,

[1] Trans. Roy. Geol. Soc. Cornw., iii, 67.

[2] *Ibid.*, 69.

had to be removed, as it was impossible to get at the ore without removing the dead ground also, all of which work had to be done before the lode was properly drained. The mixture of ore with rubbish meant extra expense in dressing and caused considerable waste, since with the additional washing the finer parts of the ore were liable to be carried off by the water.[1]

The downward stope began to be abandoned toward the end of the eighteenth century[2] for the system which prevails at present, namely that of driving levels and stoping upward. As soon as a shaft is sunk to sufficient depths beneath the adit, a level is commenced upon the lode and carried both east and west. If the latter is rich at the commencement of the level, as soon as the workman goes forward another is employed to stope or dig down the ore above the level, and as he makes progress a third follows him in another stope, and so they proceed until the intermediate part of the lode — or as much of it as is productive of ore — is wholly removed.[3] Meanwhile the shaft becomes deep enough for several other levels long before the ore above the first is exhausted. If the lode is poor in the first level, nothing more is done. If it becomes productive in some parts at a distance from the shaft, there the miners begin to stope.

The advantages of this system are several. In the first place it is easy to find what part of the lode is rich and what part is barren, and the miners have it in their power to take away the valuable parts and leave those which are worthless. Even the latter are useful, as they serve the purpose of timber in keeping the mine open. Every part of the works is much better drained. The ore, taken from the lode where comparatively dry, is more easily kept separate from the worthless ground, and is, therefore, subject to little waste. The riches of the lode may be extracted with greater speed and the product is far less fluctuating. It is possible, also, to make greater efforts for the discovery of new bunches of ore in other parts of the lode, and, as a general consequence of the system, the mine is not only more profitable but more permanent.[4]

The great work of the nineteenth century has been the provision

[1] Trans. Roy. Geol. Soc. Cornw., iii, 69, 70.
[2] Worth, 14.
[3] Trans. Roy. Geol. Soc. Cornw., ii, 170.
[4] *Ibid.*, iii, 70, 71; Hunt, 602–603.

of more precise and efficient arrangements for dressing, chiefly by the substitution of automatic mechanism for mechanical labor, the motive power in almost every instance being derived from the steam engine. The stamps, for example, which are heavier and more numerous, running in many cases to forty-eight in a set, are worked almost entirely by steam.[1] Among other improvements introduced have been the crushing-mill, the stone breaker, the "sizing-trommel," the classifier, the continuous jigger,[2] the round buddle, the automatic frame, and the self-acting calciner. The crushing-mill was introduced shortly after 1806 by John Taylor and from that time to this has formed the chief apparatus for reducing ore for the jigger, buddle, and other concentrating apparatus.[3] Trunking by machinery was introduced at St. Ives. Framing has been so far improved that one hundred frames can now be managed by a girl and a boy.[4] The buddle, formerly a shallow oblong trench flooded with water, is now a circular concave or convex frame revolving slowly beneath a jet of water, centrifugal force shifting the ores by their weight.

Yet in spite of the progress which has taken place and the proverbial skill and cleverness of the Cornish miner, the wastage at the smelting-works is still enormous, and the tin mining industry as a whole lags far behind metallic mining elsewhere in England.[5]

[1] Worth, 40; Trans. Roy. Geol. Soc. Cornw., iii, 61, 62. The first steam stamp was made by Woolf, in 1812, at Wheal Vor (Hunt, 735).

[2] The first jigging machine erected in Cornwall was introduced by Richard Taylor at the Consolidated Mines, Gwennap, in 1831, and the first continuous jigger was patented in 1843 (Hunt, 694, 695. Cf. Proc. Min. Inst. Cornw., i, no. 3, 33, 34).

[3] Hunt, 693.

[4] Worth, 40, 41.

[5] *Cornish Mining.*

# CHAPTER II

## EXTERNAL HISTORY OF THE STANNARIES AND THE TIN TRADE

If we omit the extensive collection of literature upon the question as to the relation of the Phœnicians to the early British tin trade, — a subject which has long excited the ardor of antiquarians, — there remains little evidence, either monumental or documentary, which can enlighten us upon the history of tin or of any other form of mining in England before the Norman Conquest. That the stannaries — here meaning, of course, the mines themselves rather than the political organism which arose about them — must date to a remote period in antiquity, at least to the Bronze Age, is practically undisputed.[1] As to whether they came under Roman administration, opinions differ.[2] One would naturally be led to believe that the Romans would not have failed to seize properties of such value, yet scarcely a trace of their presence has been found in the older mines, or even in Cornwall itself, so that this as well as the Phœnician question must remain unsettled.

Passing from the Roman period to that of the Saxon régime in England, we find distinct evidence, in the discovery of Anglo-Saxon bracelets and ornaments in old workings,[3] that tin mining was not suffered to lapse,[4] either before or after Athelstan's conquest of Cornwall, in 937. Another indication, pointing in the same direction, is found in a seventh century life of St. John, patriarch of

[1] Journ. Plymouth Inst., v, 130–140; Trans. Roy. Geol. Soc. Cornw., iii, 86–112.

[2] Worth, 12; Polwhele, i, 175; Journ. Roy. Inst. Cornw., xiii, 430–434; xv, 18; Trans. Roy. Geol. Soc. Cornw., vi, 45; iv, 72, 73; Borlase, *Antiquities of Cornw.*, 279; Gent. Mag., xiii, 696. For the Romans in the lead mines, see Pulsifer, 27–33; Hunt, 27 *et seq.*; Camden (Gough's ed.), 556; Stevens, 45; Sopwith, 19; Del Mar, 114. On the Romans in the copper mines see Del Mar, 14; Postlethwaite, 50; Trans. Roy. Geol. Soc. Cornw., iii, 36; Eng. Min. Almanack, 1849, p. 162. On the Romans in the iron mines see Scrivenor, 28–30; Wilkins, 6, 7.

[3] Journ. Roy. Inst. Cornw., ii, 292.

[4] *Ibid.*, xiii, 430, 431. For the Saxons in the lead mines see Pulsifer, 48; Derb. Arch. and Nat. Hist. Soc. Journ., vii, 65, 66; Pilkington, i, 98, 99.

Alexandria, which mentions tin as a commodity obtained by a voyage to Britain.[1] There is, indeed, no valid ground, despite the lack of direct evidence, to doubt the continued operation of the tin mines throughout the Saxon period. Yet the Domesday book, in which occurs mention of both iron and lead mines much less valuable than those of tin, not once refers to the stannaries.[2] For this a possible reason may be found in the fact that the great survey was designed for ascertaining the value of the estates of the country for purposes of taxation, while tin, being considered a royal property, was not noted. It may have been also that the continual incursions of the Danes upon a remote shire so exposed to attack by sea, and the ravages of Godwin and Edmund, the sons of Harold, in 1068,[3] reduced the mines to temporary inactivity. At all events it is not until 1156 that the documentary history of the tin mines may be said to have begun.

The mines receive their earliest authentic mention in brief entries on the Pipe Rolls of taxes gathered from the owners, but under date of 1198 we find a long letter from the Warden of the Stannaries to the Justiciar, Hubert Walter,[4] by means of which it is possible to deduce the previous status both of mines and miners. It would seem that in 1156 the production of tin was small and for the most part confined to western Devon.[5] From 1156 to 1160 the tax on output, 30d. per thousand-weight in Devon and 5s. in Cornwall, was farmed by the sheriff of Devon for the annual sum of £16 13s. 4d.,[6] showing a production of about 133 thousand-weight[7] of tin. During succeeding decades the farm was raised from time to time to keep pace with the increasing yield of the mines. Using the same basis of estimate, the yield rose to 183 thousand-weight in 1163,[8] 533 in 1169,[9] and 640 in 1171.[10] The miners them-

[1] Leontius, *Vita Sancti Joannis Eleemosynarii*, in Migne's Patrologia (Greek Series), xciii, 1625.

[2] Ellis, i, 136–138.

[3] Cf. Lyson, iii, pp. xi, xii.

[4] App. A.

[5] Pipe R., 2 Hen. II, Devon, and for following years. It is probable that these entries, apparently relating only to the Devon mines, comprise the then less important mines of Cornwall.

[6] *Ibid.*, 2–6 Hen. II, Devon. Occasionally we find some mines farmed separately to private individuals or to the sheriff and one or two others (Pipe R., 16–21 Hen. II, Devon).

[7] The thousand-weight of 1200 lbs.

[8] Pipe R., 9 Hen. II, Devon; App. J.

[9] *Ibid.*, 15 Hen. II, Devon; App. J.

[10] *Ibid.*, 17 Hen. II, Devon; App. J.

selves were as yet not far removed in political and social status from the villein class. They were subject to the same customary payments and services, owed suit to the manor and hundred courts,[1] and probably varied their underground pursuit with that of agriculture. Around the industry, however, had already grown a certain customary law, not only for miners but for smelters and dealers,[2] and of this the provision which perhaps more than any other tended to raise the tinner above the ordinary agricultural laborer was the so-called right of "bounding," [3] or of freely searching for tin wherever it might be suspected, regardless of the rights of the landlord. Had the mines remained attached to the ownership of the soil, perhaps nothing could have saved the stannaries from a régime of capitalism such as that which in all likelihood prevailed in the collieries.[4] Where bounding prevailed, however, it was open to the poorest villein to become his own master simply by laying out a claim and registering its boundaries in the proper court.

Until the year 1198 we see nothing like the stannary administration which in later years regulated mining in the two counties. The sole connection between the mines and the Crown seems to have been the collection of the annual tax or farm of the stannaries and an occasional use of the right of preëmption.[5] In 1198, however, we find the mines placed under the supervision of a warden appointed by the King. De Wrotham, the first incumbent, seems to have combined in himself not only executive but legislative powers, since he issued ordinances at pleasure for the regulation of mines and miners. Among other things he introduced a series of innovations in stannary procedure, which initiated the more important changes soon to follow. As yet these had to do almost entirely with taxation. He summoned to his assistance juries drawn from the miners of the two shire moots,[6] and with their aid, at the collection of the old tax, rectified the weights used for the

[1] No doubt they answered to the royal courts for breaches of mining custom. Cf. the "Assize of Mines" mentioned in Pipe R., 1 Rich. I, Cornw.

[2] App. A. See p. 96.

[3] For later rules of bounding see Pearce, *passim;* App. B. D.; Laws of Stannary of Devon, *passim.*

[4] See p. 216.

[5] See p. 133.

[6] This shows the non-existence, as yet, of stannary courts.

official measurement of the tin blocks. He imposed a further tax of a mark per thousand-weight of tin, appointed collectors, controllers, and treasurers, and promulgated a stringent code calculated to bring all tin under the view of the royal officials. The annual output had now advanced to nine hundred thousand-weight[1] which, under the new system, yielded Richard far more than all the rest of his revenue from Cornwall.[2]

But the supply was subject to extreme fluctuations. In the year 1200 it had fallen to eight hundred thousand-weight,[3] and it may have been to protect an industry which brought him so much profit that John in 1201 issued the first charter of the stannaries.[4] The provisions of this important document were brief. It confirmed the ancient privileges of "digging tin and turfs for smelting it at all times, freely and peaceably and without hindrance from any man, everywhere in moors and in the fees of bishops, abbots, and counts, . . . and of buying faggots to smelt the tin without waste of forest, and of diverting streams for their works, and in the stannaries, just as by ancient usage they have been wont to do." More important was the clause removing the tinners from pleas of villeins.[5] Over them no magistrate or coroner had jurisdiction save their warden. "We have granted likewise," it reads, "that the chief warden of the stannaries and his bailiffs through him have over the aforesaid tinners plenary power to do them justice, and to hold them to the law, and that they be received by them in our prisons, if it shall happen that any of the aforesaid tinners ought to be seized or imprisoned for the law; and if it shall happen that one of them be a fugitive or an outlaw, then let his chattels be delivered to us through the hands of the warden of our stannaries, for the tinners are of our farm and always in our demesne."

The issue of this charter was followed somewhat tardily by a renewed interest in mining, and the output of tin, which from 1201 to 1209 had fallen to 600 thousand-weight per annum,[6] now touched

[1] Chanc. R., 1 John, Cornw.; App. J.
[2] See Pipe R., Cornw.
[3] *Ibid.*, 2 John, Cornw.; App. J.
[4] Chart. R., 36 Hen. III, m. 18; App. B.
[5] This protection from pleas of serfs was enjoyed by all villeins on the King's demesnes. The writ *de nativo habendo* directed the sheriff to seize the fugitives "nisi sunt in dominico nostro" (Smirke, 8 n.).
[6] Pipe R., 8 and 11 John, Cornw.; App. J.

800 in 1211,[1] 1000 in 1212,[2] and two years later made the record yield of 1200 thousand-weight,[3] or about 600 long tons. But the disastrous effects of the new charter upon the manorial lords, offering as it did complete freedom to any villein who would turn miner or make a pretence of mining,[4] brought its practical revocation at the instance of the barons. Devon had been disforested in 1204,[5] and eleven years later, shortly before Magna Charta, the King restored to the men of Cornwall the liberties which they had enjoyed under Henry and promised that "they should not lose by reason of the stannaries aught of the services or customs which they are accustomed to have from their men and serfs; but their men, whoever will, may go for tin, saving their services and customs which they have of others, their serfs, who do not go for tin." [6] In the following reign, however, the tinners' charter was solemnly confirmed in all its provisions,[7] and inasmuch as even before that we find the tinners of Devon in possession of a court of their own,[8] it is a question whether the provisions of the charter were ever in practice abrogated.[9]

The thirteenth century has left little evidence as to the administration of the stannaries, for the reason that, beginning with 1215, the King resumed the practice of farming his tin revenues,[10] thus

[1] Pipe R., 13 John, Cornw.; App. J.

[2] *Ibid.*, 14 John, Cornw.; App. J.

[3] *Ibid.*, 16 John, Cornw.; App. J.

[4] The great diffusion of tin ore in Cornwall and Devon as shown by the mediæval subsidy rolls (cf. Lay Subs. R., bdle 95, no. 12) must have made this nuisance widespread.

[5] Chart. R. (Rec. Com.), 28 May, 5 John.

[6] App. C. The process of disforestation had been partially carried out in Cornwall in 1204 (Chart. R., 5 John, m. 9).

[7] App. B.

[8] Pipe R., 27 Hen. III, Devon.

[9] Apparently, however, the tinners of Dartmoor at least were not fully emancipated by 1250. Cf. Lyson, vi, p. cclxxx, citing Pat., 35 Hen. III. In 1222 (Close, 6 Hen. III, m. 6), the sheriff of Devon is ordered to put the stannary of Devon into the condition in which it was before John's wars with the barons, and in a previous roll the same directions are given the wardens of the Devon stannary (Close, 5 Hen. III, m. 3).

[10] Close, 17 John, m. 12, 16; 1 Hen. III, m. 23; 4 Hen. III, m. 1, 16; 5 Hen. III, m. 6, 8; 8 Hen. III, m. 14; 9 Hen. III, m. 4; 10 Hen. III, m. 27; Pat., 1 Hen. III, m. 5; Hen. III, pt. i, m. 1; 5 Hen. III, m. 4, 6, 8; 8 Hen. III, m. 11; 19 Hen. III, m. 16; 37 Hen. III, m. 18; Fine R., 5 Hen. III, m. 7; 6 Hen. III, m. 2; Cal. of Orig. R., 38 Hen. III, ro. 3; 6 Edw. I, ro. 4; 32 Edw. I, ro. 7.

removing them from the subjects embraced by the Pipe Rolls and other state documents. From 1225 to 1300 Cornwall and the Cornish [1] and Devon [2] stannaries were under the earls Richard and Edmund. Their mining ordinances and charters, if any were granted, have wholly disappeared. It is necessary to emphasize this point in view of the fact that most of the older histories of Cornwall contain the statement that the two earls favored the tinners with charters of privilege,[3] which in 1305 were merely confirmed by the King. Not only is there no evidence to verify this assertion but there is good reason to question its truth. Possessing as we do the earlier stannary documents, it is surprising that we should find no trace at all of a charter or charters which, if issued, must have been of great importance, while the charters of 1305 are explained by the petition of the Cornish tinners in 1304 requesting their charter of liberties, not conjointly with the men of Devon, *juxta confirmationis Regis Henrici.*[4] This reference can be only to the confirmation in 1252 of John's instrument,[5] and the fact that it is here referred to as the tinners' palladium is strong evidence that no grants, certainly none more favorable, were made in the intervening period.

The administration of the stannaries probably varied little from that of 1198. A few minor changes took place in the fiscal officials, and the appointment of a warden was sometimes accompanied by that of clerk wardens,[6] one for Devon and one or sometimes two for Cornwall, who in all likelihood performed the actual work, while the real warden, often as well the farmer of the stannaries, remained in London.[7] The institution of stannary courts called for the appointment of court stewards and of head bailiffs, one

[1] Close, 9 Hen. III, m. 7, 12; 2 Edw. III, m. 24; Pat., 9 Hen. III, m. 9; Orig. R., 12 Edw. II, ro. 11; 19 Edw. II, ro. 7; 1 Ed. III, ro. 7.

[2] Close, 9 Hen. III, m. 7, 9; Pat., 37 Hen. III, m. 18; 19 Hen. III, m. 16; Chart. R., 19 Hen. III, m. 4; Orig. R., 6 Edw. I, ro. 4. The Devon mines were for the most part farmed as above stated, but occasionally the lease expired before a fresh one was issued and the stannary receipts then recorded in the Pipe Rolls afford us some idea of the way in which the properties were managed.

[3] Carew, ed. 1811, p. 17; Add. MS. 28079; 6682, fol. 501. See also Delabeche, 526.

[4] Parl. R., i, 164.

[5] App. B.

[6] Chanc. R., 3 John; Pipe R., 11 John. They were paid a mere pittance of £6 1s. 8d. each.

[7] Pat., 9 John, m. 16. Cf. also Pipe R., 13 and 14 John.

for each of the eight judicial divisions of the stannaries, who acted for them as the sheriffs for the counties.[1]

Statistics for the production of tin in the late thirteenth century are not plentiful, and those that exist refer mostly to the output of Devon, which by this time had fallen far behind her sister county. So far as the figures go they point to a distinct decrease in the annual product, as compared with that in the reign of John. This may have been the effect of several causes. Heavy taxation under Richard of Cornwall,[2] concerning whose actions and measures, however, we are entirely in the dark, may have discouraged the miners, or the banishment of the Jews from England in 1290 may have seriously upset the machinery by which the mines were promoted.

Whatever may be the true causes of the depression in the mining industry, it is probable that, in addition to the desire of the Cornish tinners to be separated in administration from those of Devon, an economic motive, the desire to see the stannaries extended and his own revenues thereby increased, induced Edward I to issue the charters of 1305.[3] These mark an important step in stannary government, not merely because they contained new features making for the further differentiation of the tinners from the rest of the community, but because, with one amendment, they remained from that year down to within less than a century the real constitution of the stannaries. To paraphrase them briefly, they confirmed the customary rights of bounding, freed the tinners from ordinary taxation, confirmed the already existing tin coinage, or stampage, and preëmption, and attempted (subsequent chapters will show with what success) to give precision to the jurisdiction of the warden and his lieutenants.

As was the case a century earlier, the issue of the two charters of 1305, one to Cornwall and one to Devon, was followed by a renewed outburst of mining activity which lasted until the Black Death. The tin production of Cornwall in 1337, namely 1328 thousand-weight,[4] was up to that time the greatest on record.

[1] Cf. Pat., 11 Edw. III, pt. i, m. 39; 9 Rich. II, pt. i, m. 7; 1 Hen. VI, pt. iii, m. 10; White Bk. of Cornw., 32 Edw. III, c. 93 d; i, c. 24, 78; Accts. Excheq. K. R., bdle. 264, no. 2.

[2] He is said to have drawn enormous revenues from the stannaries.

[3] App. D. Each county received a charter.

[4] App. J.

The plague, however, almost ruined the stannaries. Thorold Rogers does not believe that the Black Death extended into the extreme west.[1] He might have been of another opinion had he seen the stannary tax rolls for the years immediately before and after 1350. A single instance will suffice. Tribulage, a poll-tax laid on two districts of the Cornish stannaries, produced in 1349 £1 10s. 3d., in 1350, 15s., and in 1357 only 9s. 4d.[2] So great was the demoralization among the tinners that the Black Prince in 1351, then as the Duke of Cornwall[3] governing the stannaries, was obliged to give orders that on pain of forfeiting their works tinners must expend on them the same cost and labor as they had done in times past.[4] The actual amount of tin mined in 1355 in Cornwall was but 496 thousand-weight, a significant contrast to the figure for 1337.[5]

Not until the closing years of the fourteenth century did the stannaries regain their former prosperity, and then not for long, as another period of depression set in which lasted through the fifteenth century, — the annual yield, which in 1400 had been almost sixteen hundred thousand-weight, falling to eight hundred thousand-weight in 1455 and not rising much above ten hundred until forty years later. No explanation entirely satisfactory can be given of these irregular ebbs and flows in the tide of mining prosperity. The fact that the same phenomenon may be observed in the German tin districts throughout the Middle Ages,[6] suggests the hypothesis that a cause common to both was the capriciousness of nature in revealing the ore, especially the stream tin, which might be found in abundant "pockets" for a limited number of years and then disappear almost entirely. In the case of the English stannaries in the fifteenth century an attendant factor may

[1] Rogers, i, 601, 602.
[2] App. O.
[3] Until 1338 Cornwall had been an earldom, this title carrying with it no control over the mines, which were reserved for the Crown unless otherwise stated. The earldom of Cornwall, held now by one branch of the royal family, now by another, was finally created a duchy and made over to the Prince of Wales and his heirs (Chart. R., 11 Edw. III, no. 60, m. 28). This grant carried with it, among other things, whatever control the Crown had over the stannaries of Cornwall and Devon.
[4] White Bk. of Cornw., i, Feb., 25 Edw. III.
[5] App. J.
[6] Cf. statistics for the production of tin in the Zinnwald district, Germany (Reyer, 35), and in Graupen (*ibid.*, 32, 33).

have been the Wars of the Roses, the concomitant anarchy, and the continual inroads of the French.[1]

The opening years of the sixteenth century brought the stannaries a renewal of their charters, with the grant of an additional privilege. In consequence of alleged violations of certain rules promulgated by Prince Arthur[2] for the administration of the mines,[3] the Cornish tinners had seen their charters confiscated by Henry VII. In 1508, however, in consideration of a fine of £1000 the royal miser was induced to restore them with the additional right of vetoing, through the stannary convocation, any statute or proclamation in which the tinners were concerned.[4]

The granting of this concession and the frequent meetings of the convocations in the following decades are almost the only bright spots in the history of the stannaries during that and the following century. Economic difficulties followed thick and fast, and the output of tin, which in the first half of the sixteenth century had averaged over sixteen hundred thousand-weight, dropped to small proportions in the reigns of Elizabeth and the early Stuarts.[5] For this several reasons may be assigned. The general European rise of prices which marks the period was for the English tinners much more pronounced in the materials necessary for mining than in the products of the mines themselves; the price of tin rose absolutely, but was stationary and even declining if considered relatively to that of timber, rope, iron, and corn. By a fatal coincidence, the mines at about the same period began to attain a depth at which water flowed in faster than the clumsy drainage devices then in use could get rid of it. The Elizabethan outburst of mining activity under royal patronage, and the careers of the Mines Royal Company, the Mineral and Battery Works, and the new Mines Royal Company which succeeded both, present for this narrative little interest.[6] They were typical of the ordinary monopoly

[1] Cf. Polwhele, ii, bk. iv, c. 39.

[2] Eldest son of Henry VII, and Duke of Cornwall.

[3] Add. MS. 6713, fol. 101–104.

[4] Pat., 23 Hen. VII, pt. ii, m. 29–31.

[5] App. J.

[6] The Society of the Mineral and Battery Works was founded by royal charter in 1564 to Humphrey and Schutz, a German mining expert. Shortly afterward a second corporation, the Mines Royal Company, in which another German, Hochstetter, was a leading figure, was established, and the two companies between them

of the time, possessed no rights whatever in the stannaries or the other free mining districts of England as against other Englishmen, save as regards gold and silver, and concern the present subject only in so far as their general effect was to stimulate mining enterprise and improve mining technique by means of the importation of skilled workmen. What influence the latter may have had upon the mining of tin has been mentioned in a previous chapter, but neither this nor the great advance in mining skill in the later seventeenth century much availed the stannaries to meet the economic depression.

There is little worthy of note in the general history either of the administration or of the output of the tin mines after the seven-

seem to have monopolized English mining as regards gold and silver, calamine, copper, and other minor metals, and the manufacture of iron wire. The two companies for a time carried on mining operations on a considerable scale, especially noteworthy being the Mines Royal Company's works at Keswick, which are said to have employed several thousand men down to the Civil War (Stringer, 247). The Society of the Mines Royal, as the name implies, concerned itself also with the mines of silver and gold, of which it was the Crown's lessee. In 1668 the two corporations were combined into the Society of the Mines Royal. They seem not to have attempted mining operations on their own account as a general rule, but instead granted licenses to capitalists and corporations to dig for ore in various places, thus being the licensers among others of such famous mining characters as Sir Hugh Middleton, Thomas Bushell, Sir Carberry Price, and Sir Humphrey Mackworth. The influence of the company in the seventeenth century grew to be pernicious, tending as it did to check all private mining enterprise and leading landowners to conceal traces of ore upon their grounds rather than to exploit the property. The Act of William and Mary (see p. 76, n. 2), by narrowing the definition of mines royal, weakened the already decrepit company, which, however, did not disappear until 1852.

Numerous references to these two companies and to their lessees are to be met with in works on mining history. The following, among others, may be mentioned: Grant-Francis, 38–41, 44, 57, 60, 65, 67, 68, 81, 82, 84, 104; Calvert, 41, 64, 65, 69, 70, 77, 78, 99, 113, 140; Hunt, 90, 91, 151, 152, 153, 841; Atkinson, 35–40; Journ. Roy. Inst. Cornw., iv, 155–159; 160; Stringer, esp. ii, iii, ix, 101, 145, 217 247; Heton, 21–24146, 168; Waller, 8, 31; Pettus (*Fleta Minor*), *Essay on Metallick Words;* Watson; Bushell, *Tracts on Mines;* Price, 49–61. Among the documentary evidence may be cited: S. P. Dom. Eliz., xviii, 18; xx, 103; xxiii, 6; xxxiv, 58–60; xxxv, 3; xxxvi, 59, 62, 91, 95; xxxvii, 34, 44, 69; xxxix, 57; xl, 14, 79, 87; xlii, 27, 68; xliii, 56; xliv, 15, 16, 39, 45; xlvi, 8; xlvii, 14, 52; xlviii, 1; lxxiv, 21, 47; lxxvii, 29; lxxxv, 63; cxliv, 32, 33; ccliii, 103; cclxxi, 40; cclxiv, 30; cclxxv, 145; Addenda, xiii, 38; Jas. I, clxxiv, 14, 15; Chas. I, cccxxvi, 68; cccclxxv, 48; Harl. MS. 1507, fol. 40; Lans. 22, fol. 5; 26, fol. 12, 13; 81, fol. 2, 4, 9; Pat., 7 Eliz., pt. ix; Treas. Papers, ccxliii, 30.

teenth century. In the stannary constitution whatever developments took place were the result not of specific enactment but of changes inevitable with the general economic advance. As the mines slowly developed into great capitalistic enterprises, the administrative duties of the stannary authorities changed accordingly. The court leet, with its analogies to the leet of the manor, fell into disuse. In the stewards' and the vice-warden's courts old principles were altered to meet new emergencies arising from the more complex nature of the mines themselves. The stannary convocations were abandoned. Of the various forms of taxation most of the petty dues were tacitly dropped, but the entire machinery of collectors, controllers, receivers, and assay-masters drifted on until the reform acts of the early nineteenth century set matters upon a business basis.

Quite another point of view is that furnished by the story of the trade in tin. Here the subject falls naturally into two divisions, the domestic trade, a large part of which must be studied in connection with the history of the pewter industry, and the commerce with foreign parts. A factor common to both was the initial concentration of the tin market in the coinage towns, their location and volume of trade being naturally dependent upon the situation of the mines themselves.

During the Middle Ages the scene of operations had steadily shifted from east to west. In the twelfth century the rich alluvial deposits of southwest Devon had produced nearly all the tin of Europe and whatever there was of Cornish mining lay near the Devon boundary. In 1198 De Wrotham had held inquests at Exeter and at Launceston.[1] In Devon he had twenty-six witnesses and in Cornwall only eighteen, a proportion which indicates the relative importance of these tin districts. The richer stream tin deposits of Devon, however, were soon exhausted, and in the thirteenth century Cornwall came to the forefront.[2] Devon, which in 1189 yielded over six hundred thousand-weight, produced only

[1] App. A.

[2] In 1220 the Devon stannaries were farmed for but 200 marks, and the Cornish for 1000 (Fine R., 5 Hen. III, m. 7; Pat., 5 Hen. III, m. 4, 8; Close, 5 Hen. III, m. 8, 9; 9 Hen. III, m. 4, 9; 10 Hen. III, m. 27). In 1235 the stannaries of Devon were leased to the bishop of Exeter for 90 marks, plus £10 per annum (Pat., 19 Hen. III, m. 16).

seventy-four in 1243.[1] Although in later centuries it sometimes exceeded this amount and throughout the Middle Ages could muster over a thousand tinners all told,[2] nevertheless Cornwall never failed to maintain its easy predominance. Within Cornwall itself the centres of activity moved ever westward. Out of a total yield for Cornwall of 1850 thousand-weight in 1305, the tin coined at Lostwithiel and Bodmin, the two eastern markets, amounted to 616 thousand-weight, while the western parts, represented by Helston and Truro, produced only 134.[3] During the forty or fifty years of the accounts taken in the reigns of Elizabeth and James I, the average annual product of the two eastern stannaries, as represented by the coinages at Lostwithiel and Liskeard, was but 135 thousand-weight, as compared with 807 for the west.[4] Penzance, in the Land's End district, was first made a coinage town in 1663,[5] and in 1778, according to Pryce, coined more tin each quarter than all the towns of Liskeard, Lostwithiel, and Helston for a whole year.[6] The concentration of mining in the west even more recently is shown by the fact that in 1892, when the output for West Cornwall was 7752 tons, that of East Cornwall was but 628 tons, while Devon produced only 96 tons.[7]

These facts receive an added significance from the system of tin coinage or stampage, by virtue of which the tin, directly after smelting, must be taken to one or another of a number of county

[1] App. J.

[2] Lay Subs. R., bdle. 95, no. 33 (47, 48 Edw. III), where the stannary population of Devon was divided as follows among the four districts: Tavistock, 100; Plympton, 209; Chagford, 338; Ashburton, 403. See also Lay Subs. R., bdle. 95, nos. 29, 30; Trans. Devon Assoc., viii, 311–322; Journ. Roy. Inst. Cornw., iii, p. xxvii. A large percentage of the Devon tin mines were probably on Dartmoor (Westcote, 65; Trans. Devon Assoc., viii, 223, 227; Gent. Mag., xiii, 697; Polwhele, i, 175; Tilloch's Phil. Mag., v, 359; Rep. Roy. Cornw. Polytech. Soc., 1872, p. 149; Lyson, vi, p. cclxxvi). Cf. Pat., 16 Rich. II, pt. i, m. 19 (Grant of the offices of forester and "tyn bailiff" in the forest of Dartmoor). Only about twenty tinners, however, all of the poorer sort, actually resided on the moor (Lay Subs. R., bdle. 95, nos. 29, 33).

[3] Accts. Excheq. K. R., bdle. 261, nos. 1, 3.

[4] D. O. Receiver's Rolls. An examination of the tribulage accounts (App. O) shows, especially in the fifteenth century, an enormous increase in tinring in the Penwith and Kerrier (Land's End) district and a falling off in Blackmore, in the east.

[5] Cf. Pearce, 73.

[6] Journ. Roy. Inst. Cornw., xiii, 432.

[7] *Ibid.*, iv, 189; xiii, 432.

towns, there to be stamped and taxed before any sale was permitted. The mediæval coinage towns comprised Chagford, Tavistock, Plympton, and Ashburton, for Devon, and Bodmin, Liskeard, Lostwithiel, Helston, and Truro, for Cornwall, the tin going to one or another according to the proximity of the mine. With the decay of the Devon stannaries the Cornish coinage towns early assumed greatest importance. Lostwithiel seems to have been the centre of the trade as long as mining was confined to East Cornwall, but with the shifting of the industry toward Land's End the western towns came in for the greatest share of the coinage, a condition which lasted until the final appointment of Penzance by Charles II and the subsequent concentration of tin mining in the surrounding districts.[1]

From these market towns the block tin was carried the length and breadth of England. Some found its way to local consumption in the shape of small bars for use as solder; [2] some was melted down for the casting of bells; but by far the greater part of the domestic consumption was in the manufacture of pewter. In this branch a London gild came to play a predominant part from the fact not only that it was the largest association of pewterers in the kingdom but also that it possessed the entire supervision over the manufacture elsewhere in England. The history of English pewter is largely the history of the London pewterers.[3]

The origin of the gild in question is unknown, but there seems little doubt but that it was in existence prior to 1348,[4] when as the "Craft of Pewterers" it issued an elaborate code of ordinances, designed in true mediæval spirit for the enforcement of a high standard of purity of material and skill in workmanship. Further evidence appears from time to time, testifying to the growth of the company in wealth and influence. In 1438 we find its court of

[1] The geographical distribution of the tin can be seen in the coinage rolls and in the receiver's accounts. Cf. Journ. Roy. Inst. Cornw., iv, 187–190; xiii, 432, 433; iii, p. xxvii; S. P. Dom. Chas. II, lxxvi, 68, 89; Entry Bk., xxv, fol. 169. In 1837, the year when the coinage of tin was discontinued, the towns were Calstock, St. Austell, Truro, Helston, Hayle, and Penzance.

[2] Rogers, i, 600.

[3] For the earliest mention of English pewter, apart from connection with the London company, cf. Pat., 10 Edw. III, pt. i, m. 20 d; 17 Edw. III, pt. i, m. 34 d.

[4] Welch, introd.

aldermen approving an ordinance for the regulation of the tin trade. Six years later an order of the Common Council gave the company the right of purchasing one fourth of all the tin brought up for sale in London. In 1451, with the issue of its audit books, we find it arrogating to itself the supervision of all matters connected with the trade, the concerns of its members, and the control or prohibition of foreign imports.[1] In 1473 came the first charter of the pewterers, which, besides conferring the usual privileges of incorporation, recognized its right to search for false ware anywhere in England and pledged for this purpose the assistance of royal officials. As an immediate result of these favors, the aggrandizement of the London company at the expense of all others was assured, and it was further accelerated by the addition to its membership of many of the country masters.[2]

The Londoners, then, at the close of the fifteenth century were practically the arbiters of the pewter industry in every part of the kingdom, and their strength was enhanced in the first half of the sixteenth by a series of statutes designed to screen them from home and foreign competition. In 1504, for instance, an Act of Parliament forbade all sale of pewter save at fairs or in the house of the manufacturer,[3] a blow aimed directly at the growing trade of chapmen and itinerant traders, who purchased their tin at the coinage towns, worked it into pewter and hawked it about the country. This Act was made perpetual in 1512,[4] and a further provision added which appointed as searchers and examiners for tin and pewter vessels the mayor of each city and the master and wardens of the London gild. Confirmations in 1534[5] and 1542[6] placed the statute in the final form in which it lasted for three centuries as the company's chief instrument of authority.

Independently of these artificial aids, the London company maintained a firm control of the pewter industry by its practical monopoly of the exports.[7] The trade, which seems to have been made

[1] Welch, introd.

[2] *Ibid.*; Unwin, 61. In 1474 the receiver of the stannaries was a member (Welch, i, 39).

[3] 19 Hen. VII, c. 6.

[4] 4 Hen. VIII, c. 7.

[5] 25 Hen. VIII, c. 9.

[6] 33 Hen. VIII, c. 4.

[7] In the customs accounts the pewter exports seem to have come almost entirely from London.

up of tin in vessel, small bars, and an assortment of pewter mugs, plates, and candelabra, had reached a considerable height in 1402,[1] and continued to flourish during the Elizabethan period, which for the London company was one of great prosperity and even splendor.

There are evidences, however, that its prosperity rested upon an unstable basis. The export trade owed much to the scrupulous purity which, thanks to the care of the Londoners, the English wares usually maintained. Still more was owing to the fact that for centuries the pewterers of the continent were inferior in technique to those of England, besides being handicapped by a higher price of tin. Early in the sixteenth century, however, the art was improved in France and Flanders by runaway English apprentices;[2] the German stannaries attained a considerable output;[3] the monopoly of Cornwall was broken; and the growing competition from the continental pewterers was felt so keenly in England, that the statute passed in 1534 not only prohibited the purchase of foreign pewter and authorized its seizure wherever found, but forbade the taking of alien apprentices, forbade any alien to become a pewterer in England, and considered as aliens all English pewterers who travelled or remained abroad.

The crux of the whole difficulty is probably to be found in the clauses denouncing the foreign pewter as poorly mixed and adulterated. The charge has a familiar ring; we have already seen it invoked against the chapmen and peddlers in 1504.[4] To appreciate the situation one must take into consideration the cost of the ordinary English pewter, manufactured under the company's supervision. The price of tin throughout this period may safely be stated as never less than 20s. per hundred-weight at the coinage towns. The London pewterers bought it at, say, 25s.,[5] the country pewter-

[1] Cust. K. R., bdle. 72, no. 4; bdle. 79, no. 12.

[2] Stat. 25 Hen. VIII, c. 9, preamble. On the tin and pewter trade of France, Spain, Germany and Italy see Bapst, 211–240.

[3] The Saxon and Bohemian tin mines reached their most flourishing period in the sixteenth century (Louis, 5; Reyer, 35, 237), but were ruined by the Thirty Years' War. The quality of the German tin is open to dispute (Malynes, 268; Charleton, 40). For the history of the German stannaries cf. Reyer, 32, 33, 36, 42, 43, 48, 35, 54, 55, 59, 60, 64, 69, 79–81, 236, 238, 256; Inama-Sternegg, iii, bk. iv, 142.

[4] 19 Hen. VII, c. 6.

[5] Welch, i, 77.

ers at 30s.[1] the hundred-weight, and having alloyed it slightly for pewter sold at an advance of twenty per cent.[2] The high price of pewter utensils and the necessity for their constant renewal, — for the metal was soft and soon wore out,[3] — must have constituted a serious charge upon the alehouse-keeper or landlord, and he doubtless welcomed the opportunity of buying cheaper pewter, even if it were not so pure. In the long run the body of consumers is the best judge of what is good for it, and the London pewterers, by refusing to meet the demand, doomed their trade to eventual decay.

At the opening of the seventeenth century a new turn of affairs took place, with the formation under royal patent of the tin monopoly, which with occasional interruptions continued down to the Civil War. Unless the pewterers could obtain the suspension of the patents, their sole alternatives were either to get the monopoly into their own hands, or else establish their right as manufacturers to a portion of the material of their trade upon special terms. As a matter of fact they eventually tried all three methods.

The short-lived patent of Brigham and Wemmes in 1601 was marked by a bitter warfare between the patentees and a section of the pewterers' company, headed by four of the wealthier members, Richard and Roger Glover, Thomas Smith, and Nicholas Collier, who are said to have held the entire foreign trade in pewter.[4] It is not clear whether these men were really backed by the company as a whole, or whether, as their enemies claimed, they were working for their own interests and against those of a majority of their fellow-craftsmen. We find the entire company contributing for the suit for the revocation of the monopoly; yet on the other hand it is difficult not to give some credence to the charges which were laid in the Star Chamber against the Glovers and their associates at the trial in 1606.[5]

"First, upon Her Late Majesty's taking the preëmption into her hands, they combined with other merchants-adventurers and bought up the tin, and so raised the price from £23 10s. to £25, and then to £27, and finally to £31 10s. per thousand-weight. After that they

[1] Rogers, iii, 375. In 1503 it was sold at Oxford at 3d. per pound, and in 1504 at York for 4½d. per pound. App. U.

[2] In the declarations of value in the customs accounts of London, pewter is usually set down at £12 per thousand-weight and tin at £10.

[3] Welch, introd. [4] S. P. Dom. Jas. I, vi, 79. [5] *Ibid.*, xxiii, 57.

devised to deliver great sums to different tinners to receive tin of them on account as prices go, at the next coinage, which price by their such forestalling they raised and lowered at pleasure, and thus when Her Majesty took the preëmption into her own hands the tin was so high that she could get no profit." They then endeavored to arrange matters so that the Queen and her patentees could not meddle with tin save at a great loss, and accordingly depressed the market once more; when tin had fallen to £24, they again engrossed and forestalled. This state of affairs lasted for three years, "much to the distress of the poor tinners and the loss of the Crown and the people," until at last the Queen granted the sole right of emption to Brigham and Wemmes. "The Glovers then gave out in speeches that this was very harsh and would make tin scarce, and so procured the common people to cry out for liberty in the pewterers' hall, in the pewterers' courts there held, and in other places in London and elsewhere. They also bought up all the tin in London and in Cornwall and Devon, and raised the price so that the patentees could not buy without danger of loss." Upon James's accession the whole company, apparently at the Glovers' instigation, petitioned that the patent of preëmption be withdrawn,[1] to which the King acceded, but only to exercise the right in his own behalf. Thereupon, — so the charge runs, — the Glovers resorted to their old practices of forestalling and engrossing, and when they had brought matters to such a pitch that to relieve the poorer members of the company the King was obliged to supply them with his own tin at cost, they attempted a miniature *coup d'état* in the pewterers' administration, were haled before the Star Chamber, and fined and imprisoned.[2]

This episode taught the pewterers that in an age of monopolies they must swim with the tide or else drown. Thereafter, although petitioning against the issue of patents for the preëmption of tin, they used their periodic recital of grievances chiefly as a lever by means of which to obtain little monopolies for themselves. This

[1] S. P. Dom. Jas. I, ii, 4. The petition charges the farmers with selling tin of imperfect quality and with exercising a meddlesome oversight on the pewterers' business, together with other grievances. For the rejoinder of the patentees see S. P. Dom. Jas. I, ix, 75.

[2] *Ibid.*, xxiii, 57.

was particularly true with regard to their petitions for the sole privilege of casting tin into bars. The running of the heavy blocks into small pieces for household and other uses had in all probability once been done mainly by the poorer members of the London gild,[1] but perhaps because of the rise of the French and Flemish pewterers tin came to be shipped wholly in block for use on an extensive scale, or else was run into bars at the point of disembarkation. As might have been expected, the London company before long demanded that all exports of tin leave England *via* London, there to be run into bars by the petitioners.[2] This cool request in 1593 was not immediately granted,[3] but by the Brigham and Wemmes patent of monopoly and by many of those which followed, the pewterers obtained that and other privileges. Under the first-named patent, for example, they were to make into wares fifty thousand-weight of tin each year, which the patentees must purchase at an advance of 14s. per hundred-weight over and above the price of the tin itself.[4] Later, during the Eldred-Moore-Freeman patent of 1608, the pewterers obtained the sole right of casting all tin for export,[5] and the same privilege remained theirs in the patents of 1623[6] and 1628,[7] and continued until the expiration of the latter contract.

[1] Add. MS. 36767, fol. 219; Welch, ii, 36, 152. "Shoten" (molten) tin formed no small part of the tin and pewter exports of London (Customs Accts., bdle. 113, no. 7; bdle. 73, no. 12; and numerous other references of the same sort throughout the fifteenth century).

[2] Welch, ii, 14.

[3] Strype's Stow, ii, 207. Their patent was made out, but for some reason was pigeon-holed (Lans. 86, fol. 68). The pewterers were still suing for it in 1598.

[4] S. P. Dom. Jas. I, xxiii, 57.

[5] Add. MS. 36767, fol. 219.

[6] S. P. Dom. Jas. I, lxxi, 24. If a merchant wished to transport tin in block, he bought it direct from the farmers. If he wished it in bars, he first notified the farmers of the amount he required. The farmers then delivered the same to the London gild, where it was distributed among the members of the company elected for the purpose to melt it into bars at a fixed price (in 1676 not less than 2s. 6d. per hundred-weight. Welch, ii, 152). These men, under bonds to the king against adulteration of the product, were furthermore bound by agreement to pay a certain sum per hundred-weight (4d. in 1603, 6d. in 1608, Welch, ii, 36; Add. MS. 36767, fol. 219) to the master and wardens of the company, who distributed it among the poorer members. In 1603, the melting was done by four of the wealthier pewterers, including the Glovers and Nicholas Collier. When the privilege of casting these bars was not enjoyed by the company, the king occasionally granted £200 per annum from the Treasury to be distributed among the poorer members (Treas. Papers, cxxii, 17; Welch, ii, 131).

[7] S. P. Dom. Chas. I, ccccxiii, 31; Treas. Papers, cxxii, 17.

Another line of attack along which the pewterers advanced was to make demand, as manufacturers, for a supply of tin at rates lower than those at which it was sold to the public. This was usually allowed them. In 1608 they were given five hundred thousand-weight,[1] but the amount was later reduced to three hundred for home use at £4 per hundred-weight, with a further clause that a greater quantity might be allotted at the pleasure of the Lord Treasurer.[2] In 1612, in reply to one of their numerous petitions on the subject of their grievances, they were given all the tin they could use, at a price of £4 5s., provided they exported none of it unwrought.[3]

In 1615 the company obtained for itself the preëmption of five hundred thousand-weight of tin per annum, or about half the output. Nominally the patentees were some half-dozen of the more wealthy members;[4] actually the necessary capital was raised partly by these, partly by small susbcribers, and partly by the corporation.[5] This lasted until the Catcher-Bland-Cockayne patent of 1621,[6] by which the pewterers, once more relegated to a secondary position, received three hundred thousand-weight a year for home use at £4 7s. and one hundred thousand-weight at £5 for export.[7] Upon their attempting to misuse their privilege by exporting tin unwrought, they were in the following year made to pay £5 for their entire supply.[8] Some years later they tried to oust the monopolists by offering to take half the tin output themselves.[9] The

[1] Add. MS. 36767, fol. 219; S. P. Dom. Jas. I, xxxviii, 50. This arrangement was an afterthought and contrary to the farmers' patent, which they resigned in consequence.

[2] They petitioned constantly for better terms (Welch, ii, 58, 65).

[3] *Ibid.*, ii, 58. In 1624 their privileges were removed for exporting tin unwrought (Privy Council Reg., Jas. I, v, 433).

[4] S. P. Dom. Jas. I, Grant Book, p. 135; Welch, ii, 67–71; Lans. 1215, fol. 226–230; S. P. Dom. Jas. I, lxxxi, 24.

[5] Welch, ii, 68. £800 came from the corporation, £1000 from small subscribers, and £6000 from three or four wealthy members.

[6] The pewterers promoted a bill in Parliament in 1621 which should give them the preëmption and the sole right of casting tin into bars and of preventing its export. The bill was thrown out (Rep. Hist. MSS. Com., iv, 121).

[7] Privy Council Reg., Jas. I, vi, 238, 239; S. P. Dom. Chas. I, ccccxx, 38; Jas. I, cxviii, 83; Welch, ii, 81.

[8] Privy Council Reg. Jas. I, vi, 238, 239. Cf. vi, 317.

[9] Hargrave, 321, fol. 689–693 (date uncertain, but probably 1626).

offer was rejected, but it led to another bargain with the monopolists by which the pewterers received one hundred thousand-weight of their quota at a reduction of 2s. per hundred-weight.[1] The Harbys, the next farmers, in their first patent allowed the pewterers the same terms, but by a secret renewal of the patent raised the price to £5 12s. on all but the hundred thousand-weight,[2] at which point it remained[3] until the Civil War and the Commonwealth did away with the tin monopoly altogether.

For the London pewterers the Restoration marked a turn for the worse. Not only was the monopoly renewed in the face of their protests, but they received no allowance from the first patentees[4] and a bare two hundred thousand-weight a year from the second.[5] Their influence was waning. The country pewterers revolted against their supervision, with the result that most of the control of the Londoners over the trade vanished, never to return.[6] Nor did they continue to receive consideration from the state. The change is shown by the contract of preëmption drawn up in 1688 but never put into operation, in which no mention is made of the London company, while the handful of pewterers in Cornwall received an allowance of fifty tons a year at reduced rates.[7] In the first contract made by Anne in 1703 small allowances were made to the Cornish contingent, but none to the Londoners. The latter made one final complaint and the matter was referred to the officers of the Mint, who advised them to ask for their quota as a favor rather than demand it as a right.[8]

The company met with no greater success in its effort to revise the tariff. The relative export duties on tin and pewter had been the subject of much debate ever since the publication of the Book of Rates. In 1610 they had stood at 7s. 4d. and 5s. 2d. per hun-

[1] Welch, ii, 86.
[2] *Ibid.*, ii, 91; S. P. Dom. Chas. I, ccccxx, 38.
[3] Welch, ii, 94.
[4] *Ibid.*, ii, 128, 129.
[5] Twice a year one hundred thousand-weight at 18 per cent below the general highest price (Welch, ii, 132). This patentee, Richard Ford, the pewterers helped bring to grief by purchasing and remelting hundreds of tons of old pewter after the Great Fire (S. P. Dom. Chas. II, ccxxx, 75).
[6] Welch, ii, 103.
[7] Treas. Papers, ii, 41. Another account makes them receive 30 tons at £4 6s. per hundred-weight (Hargrave, 321, fol. 41).
[8] Treas. Papers, ccviii, 30; Add. MS. 6713, fol. 437–442. Thirty tons per annum at £4 6s. per hundred-weight were allowed each time.

dred-weight respectively; in 1645 at 7s. 4d. and 5s.; in 1657, 20s. and 5s.; and in 1660, 7s. 4d. and 5s. once more.[1] Naturally the pewterers desired a heavy export duty on tin and a light one, or none at all, upon their own commodity.[2] The government, however, had evidently become of the opinion that it was better policy to extend the tin trade than to bolster up that of pewter, and in 1697 the duty upon the latter commodity was reduced to 2s., while that on tin fell to 3s.,[3] an act which called forth a chorus of remonstrance from the unfortunate craftsmen.[4]

It is not difficult to see that their position was not that of former years. With their well sustained standards of technical skill and purity of metal they might cater to the more fastidious of foreign consumers, but much of their trade was probably ousted by the cheap pewter of France and the Low Countries.[5] We find the Londoners complaining that their wares were either entirely excluded from continental markets by prohibitive tariffs, or, if admitted, bought up by the foreign pewterers, who with more shrewdness than honesty stamped them with their own trade-marks and put a British stamp upon inferior metal in order to ruin their rivals' reputation.[6] All these causes, taken in connection with the cheap tin which the Dutch were then importing from Siam,[7] coöperated to

[1] See Treas. Papers, lxvi, 2.

[2] For some of their petitions on this subject, see Welch, ii, 29, 65, 115, 118, 121.

[3] Treas. Papers, lxvi, 2; Stat. 8 and 9 Will. III, c. 34. Cf. also Stat. 27 Geo. III, c. 62, sec. 1. Duties were made to cease by Stat. 27 Geo. III, c. 13, sec. 1; 43 Geo. III, c. 98, sec. 1.

[4] Welch, ii, 195; Treas. Papers, lxvi, 2; *The Case of the Pewterers Stated.*

[5] This was said to be 25 per cent lead. See report of the mint officers, Treas. Papers, cxxii, 17; *The Case of the Pewterers Stated.*

[6] Treas. Papers, lxvi, 22; Cotton, Titus, B, v, fol. 389.

[7] The East Indies had always produced tin, but during the Middle Ages had dropped out of sight as far as the European trade was concerned. The discovery of the Cape route to India promised to change this condition of affairs, and as early as 1513 we find the Portuguese beginning to import small amounts of tin from the Indies (Cal. Ven. S. P., ii, 336). In 1605, when Holland first appeared as a factor to be reckoned with in oriental trade, a large part of the eastern tin was mined in Malacca. Most of it found a ready market in China, but about one hundred tons annually reached Europe in Dutch bottoms and contributed to some extent to break the British monopoly (Reyer, 157 *et seq.*; Beer, ed. 1862, ii, 191; Collins, 15, 16, 37). Finally in 1692 the Dutch secured a treaty with the King of Siam, who ran the Siamese tin mines as a royal preserve, by virtue of which they got the entire output on terms which enabled them to bring it into Holland at a price of 44s. per

reduce the once famous company to one of comparatively minor importance.

The passing of the London pewterers as the predominant factor in the domestic tin trade modified the thinly disguised hostility which had long existed between them and the tinners. This friction was as inevitable as the opposition in interests of consumer and producer, but intensified by the practices resorted to by the miners to avoid the laws of inspection and the consequent officious meddling on the part of the pewterers. The latter usually obtained their tin by individual purchase from dealers, but there appears in the sixteenth century along with this another method, probably used only by the richer members, namely the purchase of tin, not through dealers, but direct from the tinners themselves, the pewterers despatching their own factors to Cornwall to make the necessary arrangements with the mine owners.[1] In addition, the company as such habitually used its available cash balance for the purchase and subsequent resale of tin to its members at a slight profit. Thus in 1483 we find in its books an entry of tin to the amount of thirty-one hundred-weight, bought at 24s. and sold to various gildsmen at from 25s. to 27s., while six hundred-weight was bought at 22s. and sold for 23s.,[2] the difference in the prices charged for the tin probably varying with its quality. That the members continued as a rule to supply themselves is evident from a regulation of 1555 forbidding any to buy Cornish tin without demanding a certain allowance for "scrap." [3] In 1560–1561, however, it was agreed that "four honest men of the company shall have the buying of all such bargains of tin as hereafter shall be by any manner of means come to any of the company by brokership or

hundred-weight, the English price at that time being 63s. (Reyer, 158; *Aggravii Venetiani*). In 1710 Banca and Biliton developed fertile mines (Reyer, 167; Eng. Min. Almanack, 1850, p. 204; Flower, 20), and from that time onward the Cornish output, although the chief source of European supply, no longer was the sole factor in determining prices (cf. Trans. Min. Assoc. and Inst. of Cornw., i, 144, 145; Trans. Roy. Geol. Soc. Cornw., iii, 248–254; vi, 69–75; Louis, 5). A very slight supply of tin has been found in France (Journ. Roy. Inst. Cornw., ii, 343–345), and in Spain (Borlase, *Tin Mining in Spain*), but nothing of commercial importance.

[1] S. P. Dom. Eliz., ccliii, 46.

[2] Welch, introd. This practice seems to have lasted until the later Stuart period (*ibid.*, iii, 80, 90, 91).

[3] *Ibid.*, i, 194.

any other shift and the party shall send the broker or other party to one of the said four and they by their good advice shall make bargains in the name of the company and the tin shall be kept in the hall to be sold."[1] Yet even this arrangement for collective bargaining apparently related only to exceptional opportunities. Most of the tin used in the trade was still obtained by individual dealings with the London merchants, among whom the haberdashers appear to have been most important.[2]

Whether dealing with the tinners indirectly or directly, however, the pewterers not only insisted upon the strict enforcement of the laws against impure metal,[3] but maintained their own assayer[4] in London and endeavored to try each block before it was made into pewter. The faulty pieces were returned to Cornwall and the wrong-doers prosecuted in the stannary courts at the company's expense.[5] Occasionally the pewterers succeeded in obtaining the appointment of one of their number as assay-master or deputy assay-master of the stannaries.[6] Thus Henry Cowes, a London pewterer appointed deputy assay-master by Charles I, went down into Cornwall "to reform sundry abuses in that office"[7] and to search for bad tin on behalf of his fellow craftsmen. In view of the fact that the tinners, entrenched behind their own judiciary, were as completely a unit in opposing the enforcement of the inspection laws as the pewterers were in demanding it, one can hardly be surprised at the bad blood between the two.

Great as was the domestic consumption of tin, its export to meet the foreign demand was even larger. Unfortunately, however, this phase of the history presents almost unsurmountable difficulties not merely from the fact that much of the tin trade was illicit and therefore impossible to trace, but also from the paucity of accounts in the few old customs rolls that still survive. How great was the export of tin at any one time, it is difficult to say. From the fact that the production of tin was, roughly speaking, no greater under James I than under John we may be right in assuming that in the earlier centuries, when the infant English industries were

[1] Welch, i, 217.
[2] *Ibid.*, i, 268; ii, 10.
[3] Cf. *ibid.*, ii, 175; Camb. Univ. MS., Ff, vi, 26.
[4] Welch, ii, 116.
[5] *Ibid.*, ii, 9, 10, 124, 125, 141, 169, 177, 178.
[6] *Ibid.*, ii, 100, 128, 129.
[7] *Ibid.*, ii, 92.

inferior to those of the continental and especially of the Mediterranean states, foreign trade in raw tin played a greater rôle than in after years. That it was at all times an important feature of England's commerce is attested by papers of the sixteenth and seventeenth centuries, of doubtful value perhaps as regards exact data, but fairly trustworthy if used for general statements. Two of these documents, written in 1595 by different individuals, are in tolerable harmony. Lord Buckhurst, in his petition to the Crown for the lease of the preëmption of tin, asserted that the entire Cornish product was shipped abroad, while England was supplied by that of Devon only.[1] Some months later, Oxford, in a counter petition to Burleigh, declared that out of a total of 1500 thousandweight all but 200 was exported.[2] The development of English industries in the century which followed probably reduced this proportion, as Tresilian, the author of "Aggravii Venetiani," writing in 1697, places the export of unwrought tin at only two thirds of the total output.

If the amount of tin exported remains a matter of such general estimate rather than of accurate statistics, the trade routes it took, especially in earlier centuries, must be described in a manner hardly more precise. If one is willing to take the word of the antiquarians, it would appear that the Phœnicians carried their tin from Cornwall to Cadiz,[3] which from say 1000 to 200 B. C. remained the entrepôt for most if not for all of the Cornish product, receiving the tin from Phœnician galleys and probably dealing it out to general traders, who supplied the shores of the Mediterranean and carried it possibly as far as India.[4] In the Second Punic War Cadiz entered into a close alliance with the Romans, and with this protection escaped the fate which later befell the mother city, Carthage. Already, however, she had ceased to monopolize the trade in tin with Britain, for a land route had been opened, as we learn from Diodorus,[5] who seems to have drawn his information from Polybius or Posidonius. By the Gallic merchants who conducted this new venture the tin was landed on the continental side of the

[1] S. P. Dom. Eliz., ccli, 71; Cotton, Titus, V, v, 402.
[2] S. P. Dom. Eliz., cclii, 49. Cf. Egerton Papers (Camden Soc.), 284.
[3] Reyer, 119, 234; Trans. Roy. Geol. Soc. Cornw., iii, 118.
[4] *Ibid.*; Pliny, *Nat. Hist.*, xxxiv, c. 17.
[5] Diod. Sic., v, c. 21, 22, 38.

channel and thence taken overland to Marseilles, which acted as a distributing centre.[1]

How long this intercourse between Cornwall and the southern provinces of the Roman Empire existed, we have no reliable means of ascertaining. If it began as early as Diodorus alleges, it perhaps lasted for eight centuries. We are assured, at any rate, of the long-continued commercial importance of Marseilles after the fall of the Roman Empire, and we may infer a European demand for tin in the early centuries of the Christian era due in part to the increasing manufacture of church bells [2] for use in western Europe.[3] From this point on, the history of the tin trade, although unfortunately no less obscure in details, becomes fairly authentic. Once established, the trade was probably continuous, as there seems to have been at all times a steady demand for the commodity throughout the civilized world. In addition to its use in bell-metal and an increasing manufacture of pewter-ware,[4] there arose a new demand for tin from the introduction of cannon and fire-arms into warfare, beginning with the latter half of the fourteenth century.[5] There seems to have been at all times a steady demand for the commodity in the markets of the East both prior to and after the fall of Constantinople,[6] a demand due to the custom in those countries of tinning the inside of copper kitchen utensils.[7]

[1] Diod. Sic., v, c. 9.

[2] On this point see Trans. Roy. Geol. Soc. Cornw., iii, 121, citing numerous authorities. Also Reyer, 121; Hunt, 145. The proportion of tin to copper varied from twenty-three to twenty-six parts in one hundred, according to the tone and size of the bell (Trans. Roy. Geol. Soc. Cornw., iii, 122, 123).

[3] In the Greek Church the introduction of bells was slow and came much later (*Ibid.*, iii, 122, 123).

[4] Rogers, i, 599; Hunt, 45.

[5] The proportions of tin and copper in the French artillery before the Revolution were as eleven to one hundred (Trans. Roy. Geol. Soc., iii, 124, n. 16). The use of cannon was not general until the end of the fifteenth and the beginning of the sixteenth century.

[6] See pp. 61, 63. Trans. Roy. Geol. Soc. Cornw., iii, 126.

[7] There developed, of course, other uses for the metal besides those mentioned in the text. Tin salts, for example, became indispensable to dye-houses. In 1577 we find a note concerning the existence of a flourishing business in France in the manufacture of knick-knacks from an alloy of tin and lead (Lans. 24, fol. 50; S. P. Dom. Eliz., ccxliii, 113), while a composition of tin and lead has always served the world as solder (Add. MS. 28079). Finally in 1730 the art of plating iron with tin,

Of the shipment of tin from England in the early Middle Ages we have scarcely any record, except that of the Alexandrian galley which is said to have obtained tin in England in the seventh century.[1] We know that continental trading centres at a later period were distributing points for the product, until the vessels of the Italian and Hanseatic towns learned the way to England. Marseilles has been mentioned as a probable entrepôt for tin during the Roman period, and there is abundant evidence of a considerable traffic there of the same sort throughout the greater part of the Middle Ages. It would seem from the existing records that large shipments of tin regularly left England for Bordeaux. From this port the route lay up the Garonne to Toulouse, and thence overland to Marseilles.[2] The galleys of Marseilles carried the metal to Arles,[3] where part of it was left for further distribution, and then to Pisa,[4] Rome,[5] Sicily,[6] and Naples.[7] By almost the same route a consignment was also sent from England to Narbonne, and from Narbonne to Alexandria.[8]

English tin also found its way to Germany in the days before the German stannaries attained prominence. Cologne from the time of William the Conqueror had stood in close relations with England and in the twelfth and thirteenth centuries probably acted as the centre of the German market. Later, when the home mines came to be a factor in the European tin trade, Cologne gradually ceased to draw its supplies from England and depended solely on the domestic output.[9] But in the meantime Bruges had risen into prominence and by the fourteenth century had become the greatest trading centre for tin in Europe, importing from England and exporting overland through Germany to Venice [10] or through France to Marseilles and the chief Italian marts,[11] thus displacing to a great extent the old route from England to Bordeaux. Bruges became not only the chief entrepôt of the overland tin trade but also the port of supply for the vessels of the Hanse towns and above all for

which had been closely guarded in Bohemia since its discovery there in 1620 (Yarranton, pt. ii, 150, 151) became known throughout Europe (Hunt, 60).

[1] Trans. Roy. Geol. Soc. Cornw., iii, 125. See above, p. 34.

[2] Cal. of Pat., 1429. n. 544; Schaube, 333.

[3] *Ibid.*, 583, 598, 604.

[4] *Ibid.*, 604.

[5] *Ibid.*, 598.

[6] *Ibid.*, 501.

[7] *Ibid.*, 513.

[8] Heyd, i, 422.

[9] Reyer, 125.

[10] Hunt, 45.

[11] Trans. Roy. Geol. Soc. Cornw., iii, 127.

those of the Italian cities.[1] The close trade connection between Bruges and Cornwall is attested by Pegolotti, one of the earliest Italian writers on commerce, who points out the exact correspondence between the weights of Bruges and Cornwall and tells us the expense of the conveyance of tin, namely, "ed lia di spesa a conducere di Cornovaglia a Bruggia, da sol 6 den 8 sterlini il migliazo."[2]

The same author gives some interesting details as to the form in which the tin arrived at Venice. "The tin is imported from Cornwall in England in large slabs of a long square form, each weighing about one and one third cantaro,[3] Barbaresque weight of Majorca, where, and at Venice, they make bundles of tin rods, bound together by rods of tin." Each bundle made up at Majorca was about one and one fourth cantaro, Barbaresque weight of Majorca, and the loss in smelting down the slabs and in casting into rods at Majorca was from one to two per cent, and the expense of making the rods, one soldi, "because of the small Majolichini money, per cantaro, and because every one of the rods made at Venice has on it the stamp of St. Mark, those of Venice get better price than those of Majorca or the Provence,[4] from two or three per cent, although one is as good as the other."

When regular trade communication was opened by the Italians with England, wool and tin were transported by direct sea route to Italy, whence they were distributed among the Levant nations. The precise period for this cannot be assigned. We learn from Pegolotti that in the fourteenth century Cornish tin regularly formed part of the freight of Italian ships to such ports of the Eastern Mediterranean as Alexandria, Acre, Ajazzio in the lesser Armenia, Cyprus, Sabalia, Pera, Caffa, and Tauris. Pegolotti wrote his

[1] Reyer, 125; Hunt, 45. Pegolotti states that Florence and Venice received their tin overland, and Pisa and Genoa by direct sea route in their own ships (Hunt, 46, 47).

[2] Pegolotti, 249. In 1359 we find the Black Prince obtaining letters of protection from the burgomasters of Bruges and Ghent for the export of tin to Flanders (White Bk. of Cornw., ii, Oct., 33 Edw. III). Flanders long retained its importance as a tin market (cf. Guicciardini, 37–39). In 1551 Antwerp is referred to as a tin-distributing point (Cal. Ven. S. P., v, 703). In the seventeenth century a large proportion of the tin went to Holland (Cotton, Titus, B. v., fol. 389; *Aggravii Venetiani*; I. S., *Declaration of Sundry Grievances concerning Tin and Pewter*).

[3] Pegolotti, 130. The "cantaro" was from 140 to 150 pounds avoirdupois.

[4] An indication is here given of the overland route from Bruges through France.

treatise between 1332 and 1345, and certainly the English Channel was navigated as early as this by the Venetians. An instance in 1318 is quoted by Heeren from Guicciardini, and the authorities cited in Rymer date the northern voyages of the Genoese even earlier.[1] Yet in Pegolotti's work no grounds appear for supposing that merchants from either Venice or Genoa had then visited Cornwall. He seems to have written about fifty years after the end of the long struggle between Pisa and Genoa, which, fatal as it was to the naval power of the former, could not have ended her commercial activity. It appears, at any rate, from what he says that some intercourse did exist between Pisa and Cornwall, for in the statement there given of the relative weights and measures of Pisa and other countries we have the following intimation: "Con Cornovaglia d' Inghilterra cantaro l di stagno al peso de Cornovaglia fa in Pisa lib. 140. E costa il migliajo in Cornovaglia da marchi 10 di sterlini." [2]

In this connection may also be noted the participation of the Italians in the tin industry; the first lease of the preëmption of tin was to Antonio Pessaigne,[3] one of the earliest Genoese merchants in England, while other evidence points to the fact that a large share of the tin trade lay in the hands of the Florentine merchants. The connection with the mines was intimate, as the stannary revenues were frequently hypothecated in their favor by the first three Edwards,[4] while we read in 1340 of considerable shipments of tin by them from London.[5]

In the fifteenth century, however, the Venetians, who till then seem to have received their supply from Flanders, began also to make voyages direct to England. Our earliest information upon this point dates from the year 1412, when we read of a decree of the Venetian Senate for the fitting out of four galleys for the Flanders and London voyage, in which occurs an express stipulation that the tin is to be got from London.[6] From that date for more than a century we have ample evidence of the continuity of

[1] Trans. Roy. Geol. Soc. Cornw., iii, 128.

[2] Pegolotti, 130. We see here that the price of a thousand-weight of tin in Cornwall was put at ten marks sterling.

[3] Or Pisane.

[4] See p. 142.

[5] Close, 14 Edw. III, pt. i, m. 22 d.

[6] Cal. Ven. S. P., i, 189. The fleets, as is well known, were under state control.

the direct Venetian tin trade.[1] At the Flemish ports half the ships stopped to take on tin which had been conveyed thither from London and the southern ports by other carriers, while two large galleys made straight for London, where the Venetians had a factory.[2] Each of four ships was in 1485 ordered to bring home eighty thousand pounds of copper and tin, together with other merchandise in amount unspecified.[3] On the return voyage the fleet usually touched at the Barbary cities and did a profitable coasting traffic, but stringent rules forbade the unloading of tin at any place save at Venice itself. From there it was distributed in trading voyages over the Mediterranean countries and the Levant.[4] Besides these state voyages to London, in the later fifteenth and early sixteenth centuries considerable shipments of tin went out from Southampton in Venetian vessels, possibly private fleets and not under government convoy. At the opening of the sixteenth century, however, the periodic journeys of the Flemish fleet had been suspended[5] and the trade in tin, wools, and cloth thrown open to any Venetian.[6] In 1517 the state voyages to Flanders and England were resumed,[7] but the decline of the activities of Venice and the growth of the English marine contributed to bring the carrying trade of the former gradually to an end.[8]

The course of the export tin trade, here imperfectly sketched, does not seem to have been seriously affected by the intermittent attempts, from the thirteenth century onward, to establish tin staples by act of Parliament. Lostwithiel, once natural centre of the trade, was also the lawful staple in the time of Earl Edmund. In later years it was superseded by Bodmin, but in 1314 through a petition in Parliament it once more regained its standing.[9] Thirteen years later we find Truro also made a staple for Cornwall, while Ashburton was made the point of departure for Devon.[10] In

[1] Cal. Ven. S. P., i, 328, 334, 338, 341, 348, 399, 546, 829; ii, 146, 201, 418, 841, 843, 899, 934, 1166; iii, 139, 838; iv, 703.

[2] *Ibid.*, i, 492.

[3] *Ibid.*, i, 399. Apparently the Venetians secured letters of license exempting them from the staple.

[4] *Ibid.*, i, 328, 338, 399, 492.

[5] *Ibid.*, ii, 146.

[6] *Ibid.*; also ii, 418.

[7] *Ibid.*, ii, 841, 1166; iii, 139, 838.

[8] Cf. Cunningham, 3d. ed., ii, 74.

[9] Parl. R., i, 296; Anc. Pet., 315 E, 197.

[10] Close, 1 Edw. III, pt. i, m. 4 d.

1328 all staples were abolished,[1] and until 1376 the export of tin was entirely unregulated.[2] In that year Calais, already the staple for wool and hides, was made to include tin as well,[3] and with the exception of 1390, when for one year tin was exported from Dartmouth only,[4] Calais remained the only legal staple down to 1492. As a matter of fact, however, not only was much tin exported under royal licenses, but large quantities must have been illicitly shipped abroad.[5] An important contraband trade seems to have been carried on from the Channel Islands, as we learn from the petitions of Calais,[6] while other complaints refer to a direct route to Holland and Zeeland.[7]

In England, despite the regulations of the staple, the ports of departure for tin were numerous. London was early prominent[8] and continued to be the chief exporting centre,[9] but shipments went as well from Penzance, Plymouth, Falmouth, Fowey, Penryn, Exeter, and Dartmouth, and others of the Cornish and Devon ports.[10] Occasionally during the fourteenth century Southampton

[1] Stat. 2 Edw. III, c. 9.

[2] No mention of tin occurs in the Statute of Staples in 1353.

[3] Parl. R., ii, 358 b.

[4] Stat. 14 Rich. II, c. 7.

[5] Cf. Pat., 16 Rich. II, pt. iii, m. 16; p. 153.

[6] Parl. R., iv, 53 a, 53 b, 56 a, 56.

[7] *Ibid.*, iii, 67 a, 500 b, 662 a. The policy of granting special licenses to break the statutes of the staple was occasioned by the complaints of exporters and their repeated requests that the tin staple be removed again to Lostwithiel (Parl. R., iii, 295, 296, 319 a, 319). Needless to say, it was not rigorously adhered to but altered frequently at the instance of the men of Calais, who were constantly memorializing the king on the decay of their town (Parl. R., iii, 370 b; iv, 251 a; v, 149 b; Stat. 27 Hen. VI, c. 2).

[8] In 1198 the Chamberlain of London was charged with £379 18s., received in fines from merchants of London for leave to export tin (Madox, i, 531). No statistics for the London tin trade are available for use until the reign of Henry VIII. During his reign it seems that the annual average shipment was about 9000 hundredweight (Schanz, ii, 21, 118).

[9] Noteworthy are the large tin shipments from London not only by the Italian merchant-financiers but by others, such as the Earl of Salisbury (Close, 13 Edw. III, pt. i, n. 38) and the Dortmund merchant, Tidemann van Limberg (Cust. Accts. K. R., bdle. 70, no. 13). On the latter see Daenell, *Blütezeit der deutschen Hanse*, 48, n. 1.

[10] See Schanz, ii, 119–125; Cust. Accts. K. R., bdle. 115, no. 10; Duchy Accts., Excheq. Aug., port. 4; Add. MS. 24744, fol. 222; Close, 11 Edw. II, pt. i, m. 28, 33, concerning shipments from Padstow, Penryn, and other ports. None of these shipments were large. Thus, for instance, at a later date the Fowey shipments in

is mentioned as a base for exports,[1] but its tin trade was sporadic until later, when it rose in importance by becoming the Venetian port of call for the British Isles.

The later trade statistics of Southampton may serve to illustrate the increasing utilization of English shipping for the export of tin. Out of seven hundred and seventy pieces of block tin shipped from that port in 1517 only one hundred and fifteen were by English exporters.[2] In 1539 native merchants exported eighteen thousand-weight, foreigners three,[3] in 1554 and 1555 natives sixty-eight, foreigners four;[4] and although this proportion did not always continue the same, the home shipping never thereafter lost the advantage. In 1576 aliens exported a bare three thousand-weight,[5] while English shippers had one hundred and ten to their credit. The creation of the Levant[6] and East India Companies in the last years of the reign of Elizabeth completed the extinction of the Italian cities as distributers of tin to the East. As early as 1585 the Venetian ambassador at Constantinople had reported the arrival of an English

1586 for one quarter amounted to but 83 thousand-weight (Cust. Accts. K. R., bdle. 172, no. 11). Truro, during the same period, exported 100 thousand-weight; Mountsbay 16 pieces; while Padstow and St. Ives had no trade worth mentioning. It is impossible to say how much of this belonged to coast traffic. Much of the tin was probably taken from Cornish ports to London by the water route, and shipped from London to the continent (cf. Close, 14 Edw. III, pt. i, m. 22 d; White Bk. of Cornw., May–June, 34 Edw. III).

[1] Southampton is first mentioned as exporting tin in 1309 when about one hundred pounds are set down as having left it (Cust. Accts. K. R., bdle. 136, no. 1). In 1324 a larger shipment occurred (*ibid.*, bdle. 136, no. 27), but the trade, small as it was, apparently died out. No shipments occur in the reports for 1339, 1340, or 1383 (*ibid.*, bdle. 137, nos. 10, 11; bdle. 138, no. 10). Suddenly, however, we find it exporting 235 thousand-weight in 1386 (*ibid.*, bdle. 138, no. 16), and 667 pieces in 1391 (*ibid.*, bdle. 138, no. 20). Tin shipments fall to zero in 1396 (*ibid.*, bdle. 138 no. 22), rise again to 553 pieces in 1404 (*ibid.*, bdle. 139, no. 4), and continue with wide fluctuations throughout the fifteenth and the sixteenth centuries. In 1492 the place became the legal staple for tin and lead (Pat., 7 Hen. VII, pt. i), but this could not have been for long, as in 1506 we find Fowey exporting twice as much of the former metal, 148 thousand-weight (Cust. Accts. K. R., bdle. 115, no. 10) as compared with about 71 thousand-weight shipped from Southampton (Add. MS. 6713, fol. 142).

[2] Cust. Accts. K. R., bdle. 142, no. 11.

[3] *Ibid.*, bdle. 143, no. 11.

[4] *Ibid.*, bdle. 145, no. 11.

[5] *Ibid.*, bdle. 146, no. 18.

[6] Cunningham, 3d ed., ii, 75. The Levant Company's exports of tin amounted to one fourth their entire trade (Cotton, Titus B, v, fol. 389).

ship freighted with cloth and tin.[1] Others appeared in quick succession,[2] while note is made of the presence of British merchantmen at Leghorn, Barbary, and the Grecian Archipelago.[3] From the very first, tin had formed one of the articles of traffic of the East India Company,[4] and through its agency was carried into India,[5] Persia,[6] and as far as Japan.[7] The displacement of the foreign carriers in the English tin trade was probably complete by the middle of the seventeenth century.[8]

[1] Cal. Ven. S. P., vii, 267, 329.

[2] *Ibid.*, ix, 40, 96, 941; xi, 348.

[3] *Ibid.*, x, 3, 470, 508.

[4] Cal. S. P. Col., i, 268, 357, 455.

[5] *Ibid.*, i, 455, 682, 1174.

[6] *Ibid.*, ii, 57, 122, 161; iii, 577; iv, 89, 127, 305, 558, 603; v, 95.

[7] *Ibid.*, i, 343, 946, 1030.

[8] The later course of the European supply presents few points of interest. Large amounts were shipped to Holland and thence distributed in Dutch vessels (Cotton, Titus B, v, 389; *Aggravii Venetiani;* Welch, ii, 91). Other lots were disposed of in French ports, the farmers of the preëmption there driving a brisk trade. By the middle of the seventeenth century the countries receiving tin direct from England included almost all the European states (*Aggravii Venetiani; Declaration of Sundry Grievances concerning Tin and Pewter*). The only event of the eighteenth century worth consideration is the sudden rise of the trade in tin to the Indies in 1787, which for a time saved the stannaries from a severe depression (Flower, 20, 21; Trans. Soc. of Arts, x, 250–251; Unwin, *Letters and Remarks,* esp. p. 1, 20; Eng. Min. Almanack, 1849, p. 69).

# CHAPTER III

## EARLY MINING LAW

The story of the Cornish and Devon stannaries, with the creation by usage and by charter of a specially privileged class of operatives, and the persistent interest of and intervention by the Crown in mining operations is not to be taken as an isolated case in the industrial history of Europe. To understand the significance of these facts in their relations with other national industrial activities, a digression will be necessary. We must examine in its main outlines, so far as possible, the origin and development of English mining law in the light of the growth of mineral law elsewhere in Europe. Yet here we have to reckon with the example of some of the nations of antiquity.

Beyond the fact that many of the mines of the oldest nations, as for example those of Assyria, Persia, and Egypt, were state,[1] that is to say, royal property, and worked by slave labor, little can be said about mining among the ancients until we come to Greece and Rome. What we know of the Greeks is that they held all minerals to belong, not to the ground owner, but to the state,[2] — the most familiar example of this being the silver mines at Laurion, which were the property of Athens, the state, however, usually refraining from exploitation and throwing open the mines to all comers in return for an initial payment plus a tax on output.

When the western world came under Roman dominion, these state mines were taken over by the Roman authorities,[3] as were also any mines opened upon the vast tracts of public lands belonging to the commonwealth. Like the "ager publicus," they were leased for a tribute, apparently the tithe of the product, to the *societates publicani*.[4] Apart from this, however, there was in

[1] Cf. Schmoller, Jb., xv, 675.
[2] Arndt, *Bergpolitik*, 28.
[3] Cf. Achenbach, 1869, pp. 16, 22, n. 1; Trans. Roy. Geol. Soc. Cornw., vi, 83; Arndt, 1893, p. 29.
[4] Achenbach, 1869, p. 15; Trans. Roy. Geol. Soc. Cornw., vi, 85.

most provinces no state ownership of private lands. To some extent, of course, the commonwealth was regarded as having the sole ownership of all provincial holdings,[1] in that private lands were merely in "possessio" of individuals and at the will of the state might be burdened with specific or ad valorem taxes. These last were also as a rule farmed by publicans. Of the existence of privately owned mines we are made aware by Strabo and Pliny.[2] Although in their case the "possessio" of the land on which they were located must doubtless first have come into prominence as the source of the product while the surrounding lands were of less importance, although the tax levied upon the output of the mines was thought of chiefly in connection with the mines themselves and not with the soil, to which alone it technically had reference, still there can be found nowhere in Roman law any general principle by virtue of which property in mines was taken from the owners of the soil in which they were found.[3] Mining itself was never taxed, but only the product of mining.

Nevertheless numerous passages may be cited which, if interpreted without reference to the context, might seem to sanction this principle. Ulpian, for example,[4] states that in certain places by customary law a third party might operate quarries without the consent of the proprietor of the soil, in which case, however, the landlord was entitled to public indemnification and was not to be deprived of the ordinary use of the land or of the right to obtain stone for his personal necessities.[5] Similarly, when a well-known passage in the Theodosian code is cited as giving to all comers the right to exploit mines or quarries wherever found,[6] it must be borne in mind that this "freeing" of the mines was to extend to one only of the Roman provinces, to Africa, a soil not susceptible of Roman private ownership. This provision was in 363 and 376 extended still further.[7] Seventeen years later private marble quarrying was

[1] Achenbach, 1869, p. 16.

[2] Trans. Roy. Geol. Soc. Cornw., vi, 85.

[3] Arndt, 1893, p. 28; Trans. Roy. Geol. Soc. Cornw., iv, 76; vi, 86. Achenbach, 1869, pp. 13–15. For a dissenting view see Neuburg, Zeitschrift f. d. ges. Staatsw., vol. 56, p. 61, and vol. 63, p. 378.

[4] Cf. Achenbach, 1869, pp. 17, 19.

[5] *Ibid.*, 1869, p. 18; Trans. Roy. Geol. Soc. Cornw., vi, 248.

[6] Achenbach, p. 19; Cod. Theod., lib. x, tit. 19, *De metallis.*

[7] Achenbach, p. 20.

forbidden,[1] an order which can be understood only with reference to provincial soil or land belonging to the state, since according to the law as laid down in the Pandects the owners and lessees of Italian soil might exploit its minerals themselves.[2] It will readily be seen, therefore, that these and similar [3] provisions adapted to special provinces and peculiar contingencies cannot be adduced as proof that mines were included in the regalia of the Roman state. For the owner of the regalia is the owner of all royal minerals or else the privileged occupant of royal minerals which are looked upon as belonging to no one, while in Roman law one can discover no trace of either condition.

Although the Roman law of mines upon the continent served ostensibly as a foundation for mineral legislation during the Middle Ages, it cannot be said that the old Roman principles were maintained in their real significance. From the foregoing paragraphs it has been seen that while, in its general tenor, Roman law is distinguished by an absence of regalian rights in mines, nevertheless one or two passages in the codes, quite special and exceptional in character, might be construed as a general assumption of regalian rights. In later centuries it was precisely these which were emphasized by the commentators, notably by the Lombard glossographers Azo and Accursius, whose interpretation, for example, of the "De Metallariis" of the Code of Justinian [4] and the above noted passage of the Theodosian code [5] made them applicable to all mines and metals and at all times and places. Although it does not anywhere appear that they inculcated the doctrines of an absolute property in the Crown, they certainly contributed to establish the principle that mines even in private lands were subject to certain public servitudes and were legitimate sources of public revenue.[6]

The establishment of this view as a principle inherent in Roman law supplies us with one clue as to the manner in which mining polity developed in the larger continental states. Of the two broad divisions in the constitutional history of the mining classes of Europe, namely the adjustment of claims between miner and land-

[1] L. 13, Cod. Theod. t. c. Rufino Pl. P.
[2] Achenbach, 1869, p. 20.
[3] *Ibid.*, 22.
[4] Lib. xi, tit. 7, l. 1, 3.
[5] Lib. x, tit. 19, l. 1, 4, 8, 11, 12, 14, 19, 57.
[6] Trans. Roy. Geol. Soc. Cornw., vi, 86; Smirke, 81, 104.

lord and the segregation of the miner by exemptions, privileges, and special duties, the former, as the basic factor in mining history, owes much of its development to the influence of the ancient imperial law, not only in the Holy Roman Empire, but in France, and even as far as Scotland and Scandinavia.

In Germany this interpretation first gained acceptance during the reign of Barbarossa.[1] Previous to that date the legal status of the mines is open to dispute. The mark theory, as applied to mining properties, has been thoroughly discredited,[2] as has also been the hypothesis that ownership was included with that of the soil;[3] and at present the tendency of historians is to assume that almost from the start the mines were the property of territorial princes.[4] It is not important that we should know whether and where the Roman mining system was preserved through the period of the "Völkerwanderung." Perhaps it survived, at least as regards technique and management, in remote districts here and there in the Black Forest or in the Alps. At all events it is supposed that the Germanic princes soon seized the mines or their supposed sites into their own hands, and that then there quickly and spontaneously arose developments such as those which occurred in antiquity. Kings and magnates of all degrees, and especially the bishops, took up the mines and worked them for a time with their serfs,[5] but soon preferred to lease them[6] on terms similar to those on which they parted with their fields. Sometimes the lord reserved in his grants a sort of partnership or share in the mines by virtue of which he claimed the right to a third or other definite portion of the mine itself.[7] More often a percentage of the net proceeds in ore was paid by the miners.

Later, especially in connection with the gradual accretion of authority and power from the hands of the petty princes to the heads of the larger territorial divisions, the dukes, the kings, and

[1] Trans. Roy. Geol. Soc. Cornw., vi, 87; Schmoller, Jb., xv, 674, n.; Gmelin, 220, 241.

[2] Cf. Schmoller, Jb., xv, 680; Inama-Sternegg, iii, bk. iv, 142, 145; Achenbach, 1871, p. 72.

[3] Trans. Roy. Geol. Soc. Cornw., vi, 87; Schmoller, Jb., xv, 674, n.; xxv, 265, 267.

[4] *Ibid.*, Jb., xv, 674, 675.

[5] Inama-Sternegg, ii, bk. iii, 331.

[6] *Ibid.*, iii, bk. iv, 139, 140.

[7] Cf. Smirke, 105–107; Arndt, 1893, p. 33.

the Emperor, we meet with the Roman juristic theory.[1] Where weak lords were in possession of mining rights their superior forced upon them enfeoffment at his hands. Where, on the other hand, the lords were too strong for this course to be taken they were tacitly allowed to share in the regalian privileges.[2] The principles which the Emperor Frederick I sought to establish and which gradually in later centuries came to be generally acknowledged were: first, that mineral rights were essentially disconnected from tenure of the surface; second, that the sovereign was the sole proprietor of mines and alone might grant to individuals the power to work them; third, if the sovereign declined to work a mine it should be divided into shares, of which he should hold a certain number; fourth, in this case also he should appoint the manager and through his agents direct the conduct of the whole concern.

How these principles were applied by the sovereign princes of the Empire is shown by the provisions of a mine charter which the bishop of Trent issued in 1185,[3] granting to all miners the right of tarrying, laboring, and going and coming in the mountains, in the city, and wherever they might wish, freely and without hindrance. Each was entitled to a measured plot of ground for pursuing his discovery of ore,[4] and by three successive proclamations this area was ceremoniously bounded. Demands for these lots made upon the mine master[5] or other local officer of the sovereign lord might not be refused unless by reason of conflicting claims, in which case the first finder was entitled to the ground. The interest of the concessionary was permanent, assignable, and transmissible, but subject to the obligations of continual working, of payments of a fixed proportion of the product, and of

[1] Several investigators, notably Zycha and Inama-Sternegg, are inclined to lay most stress on the immunities granted miners by the landlords. Arndt, on the other hand, goes to the opposite extreme, in ascribing practically all of the mining privileges to the operation of regalian principles (Zeitschrift der Savigny-Stiftung, xxiv, 63, 71 *et seq.*). A review of the most recent literature on the subject of mining history in Germany is found in the Vierteljahrschrift für Social- und Wirtschaftsgeschichte, 1907, pp. 238–292.

[2] Cf. Schmoller, Jb., xv, 675, 676; Inama-Sternegg, iii, bk. iv, 145; Trans. Roy. Geol. Soc. Cornw., vi, 152.

[3] Schmoller, Jb., xv, 677, 678; Gmelin, 220 n. (k); Smirke, 83.

[4] *Ibid.*, 107.

[5] Cf. Trans. Roy. Geol. Soc. Cornw., vi, 155.

a small fixed quarterly rent.[1] Interest in a mine worked under such a grant would be divided into a number of shares prescribed by law, usually 128, of which a single share free of all costs, or four shares subject to costs, were assigned the landowner as compensation.[2] A mine deserted by its occupants was declared vacant after three formal visits of the mine-master and jurates, and the lapse of a year without working *ipso facto* discharged his rights and made the field free.[3] This charter, the first of which we have record, was followed in its main outlines by that of the Hartz silver-miners in 1219,[4] the Iglavians in 1249,[5] and others in Transylvania, Carinthia, and Styria. Everywhere in Germany the territorial lords gave over to the people the right of mining on their own lands or customary holdings, and threw open their demesnes,[6] although they still retained oversight of operations.[7] At first only certain specified areas were declared open, and the occupant might, if he chose, refuse to allow the land to be disturbed.[8] Soon, however, mining was free in all territories.[9]

What the lord bestowed upon the finders of ore was originally far removed from what we now call ownership. It was the right of usufruct, accompanied by the obligation on the part of the miner of constant labor[10] and of the payment of special taxes. The lord, on his part, undertook to protect the miners against third parties,[11] so long as they worked the mine to his advantage day and night. If they ceased, their rights disappeared and the site was leased to others. The share of the product which the lord reserved for his own use amounted originally to a half or a third of the whole, but in course of time this fell by degrees to a tenth and even less, so that from the sixteenth to the eighteenth century the lord's income was probably equal only to that of one of the mine partners.[12] The miners, like the ordinary farm tenants,[13] at first subject to eviction

[1] Smirke, 83.
[2] *Ibid.*, 83, 84, 107.
[3] *Ibid.*, 84.
[4] Inama-Sternegg, iii, bk. iv, 140.
[5] Arndt, 1893, p. 33; Inama-Sternegg, iii, bk. iv, 140; Trans. Roy. Geol. Soc. Cornw., vi, 88, 152–154.
[6] Inama-Sternegg, iii, bk. iv, 146.
[7] *Ibid.*, 48; Arndt, 1893, 32.
[8] Inama-Sternegg, iii, bk. iv, 140.
[9] *Ibid.*, 150.
[10] Schmoller, Jb., xv, 672.
[11] Arndt, 1893, p. 33.
[12] Schmoller, Jb., xv, 671.
[13] Cf. *ibid.*, Jb., xv, 670, 671; Inama-Sternegg, ii, bk. iii, 332.

at caprice, gradually through custom acquired an interest in their holdings which their superiors found difficult to shake off.

In France, in spite of recent efforts to ascribe to the Merovingian and Carolingian dynasties the conception of regalian rights in mines,[1] the general facts underlying the origins of mining law are much as we have seen them in Germany. The probabilities are that in France the idea of the identity of a mine with the land in which it was situated, a conception perhaps originally inherent in Teutonic law, was in full force in the early Middle Ages, only to go down before the example set by the Hohenstaufen, whose ideas upon the regalia served as a model for the French kings.[2] The great lords at the same time advanced parallel claims, but they did not succeed, as in Germany, in enforcing them, since with the development of the state organism the French monarchs were able to draw into their own hands the control of the mines. The few mines operated in French territory had been, as in Germany, absorbed into the hands of local seigneurs, whose occasional ordinances and the customary rules of mining, many of which were brought into the country by immigrant miners from Germany,[3] formed the only mineral law of France until the fifteenth century. This period was ended by a royal edict of 1413,[4] by which Charles VI claimed sovereignty over all mines within the kingdom to the exclusion of all claims to ownership or taxation on the part of any lord, and authorized all miners or others who so desired to search for ore upon payment to the crown of a tithe of the proceeds besides damages to the landlord.[5]

As a result of this ordinance, French mining law should have had a course of development similar to that in Germany, but although Charles VII, Louis XI, Charles VIII, Louis XII, and Francis I confirmed the edict, its provisions seem never to have been fully carried out.[6] The claims of the seigneurs, it is true, do not reappear in their original form, but as landlords they speedily

[1] The proof adduced is of the flimsiest description. For a discussion of this question in detail see Achenbach, 1869, pp. 23–27; Trans. Roy. Geol. Soc. Cornw., vi, 248.

[2] Loisel, i, 286.

[3] Achenbach, 1869, pp. 28, 29.

[4] Louvrex, vii, 386–390; Smirke, 104; Eberstadt, 184.

[5] This had reference to the damage done the surface, and not to the removal of the ore (Achenbach, 1869, p. 30).

[6] *Ibid.*, 1869, p. 31.

reasserted their rights and eventually obtained for them a legal recognition.[1] To this must be added the occasional royal policy of granting mining monopolies,[2] doubtless in the hope that this would stimulate industrial activities to a greater degree than by the opening of mines to all comers. There is, therefore, little in the later history of French mining law which can claim relationship with that of Germany, and its lack of stability, together with the comparative unimportance of the mines themselves, render further comment in this chapter unnecessary.

Of mineral law elsewhere in Europe that only of the Belgic Provinces presents features of special interest. In the coal fields in the principality of Liège and the county of Limbourg,[3] the sovereign prince never succeeded in establishing a property in the mines, which have always been reputed to go with the ownership of the soil.[4] Yet the field was not left wholly to the landlord. When a mine was deserted, any private adventurer who would undertake to drain it acquired a right to appropriate all coal reclaimed at his cost, upon payment to the proprietor of a fixed proportion of the raw product; and this right was so strong that it is a question whether the maker of the "arême" or adit did not become the owner of all the reclaimed coal.[5] Scotland, Scandinavia, and Russia developed their mining polity much later. The two latter countries modelled their laws after those of Germany,[6] while in Scotland the Crown claimed absolute control over all mines,[7] only to grant them out in the sixteenth and seventeenth centuries to a succession of monopolists.

With the conception of mining as an occupation under the immediate supervision of the state, comes the development of the

[1] Louvrex, x, 623; Pigeonneau, i, 421, 422; Achenbach, 1869, p. 33; Trans. Roy. Geol. Soc. Cornw., vi, 240.

[2] Achenbach, 1869, pp. 33–34, 36–39; Trans. Roy. Soc. Geol. Cornw., vi, 241, 242, 249–251.

[3] Discovered in 1189 (Galloway, 17).

[4] Jars, i, 371; Smirke, 86.

[5] Delebecque, i, 141; Smirke, 86. Cf. Delebecque, i, 123, 193, 231–234; Jars, i, 173–402; Louvrex, ii, 191, 228.

[6] Trans. Roy. Geol. Soc. Cornw., vi, 166–172; Swank, 29; Del Mar, 376; Scrivenor, 159–163.

[7] Patrick, introd., lx, lxi, lxiv; Scots Acts, ii, 5, c. 13; iii, 556, c. 31; Register of Privy Council of Scotland, i, 232, 612; Smirke, 85; Treas. Papers, ccxliv, 9.

miners into a specially privileged class of laborers. The prime object for which the lords and the Crown stocked mines with men was the securing to themselves of the very profitable tithes and other payments which could be realized only when mining operations were carried on smoothly and not subject to interruptions from civic or other authorities. To this must be added a second consideration, namely, the difficulty of securing the miners themselves. It has been shown with great clearness by Schmoller that the history of the rise of silver mining in Germany is not merely concerned with mining but also with a colonization movement,[1] and that the miners came as the result of successive immigrations. This is the more credible when one recalls the fact that the mediæval miner was no bondman tied to the soil, but an extremely skilled workman, and, so far as can be judged, quite ready to remove to lands which afforded him the best terms for his labor.

Thus it was that the petty German princes were ready to compete among themselves for the services of as many mine people as could be attracted to their soil. The miners were treated as "hospites" [2] by their feudal lords, and held to their work by the offer of privilege upon privilege.[3] Rights of faggots and water, exemption from ordinary taxes and tolls, from military service, from the obligations of a servile tenure, and from disability and forfeiture by reason of alienage were but a few of the inducements offered. The lords' interests are clearly to be seen in the clauses assuring the miners from interruptions in their work by preventing the seizure of their tools for debt, and, more important still, the institution of a special court for the trial of offenses and suits in which miners were involved, this being held necessary in order to protect them from the local authorities, with whom, in the course of their occupation, they were very apt to conflict.[4]

In Germany, besides all these rights, a still higher honor was offered the mining classes. If a group of mines proved successful, a "mine city" soon arose, as among others Freiburg, Goslar, Iglau,

[1] Schmoller, Jb., xv, 677.

[2] Achenbach, 1871, pp. 29, 30.

[3] Cf. Schmoller, Jb., xv, 678; Trans. Roy. Geol. Soc. Cornw., vi, 155, *et seq.*; Smirke, 107; Louvrex, vii, 386–390; x, 623; Pigeonneau, i, 421, 422.

[4] Cf. Gmelin, 220 n. (k); Smirke, 83, 86, 113; Jars, i, 371; Trans. Roy. Geol. Soc. Cornw., vi, 252, 253; Louvrex, x, 623; Pigeonneau, i, 421, 422; Achenbach, 1869, pp. 34, 35.

Kuttenberg, and Joachimsthal, and the last stage in the freeing of the miners was the raising of their city to the status of a free mine city. The inhabitants, at least such as were bona-fide miners, enjoyed then all the privileges of town law as well as that of the mines, including free brewing and baking, free transportation of goods, the removal of the burdensome gild regulations which hindered the miners in their occupations, and, finally, freedom from taxation and from army service.[1] The city council soon came to possess, side by side with the lord, a wide autonomy and power in mining matters.[2] But when the city population did not coincide with the mine people, there arose from their jurisdictional privileges a union of all those having to do with mining operations which formed a special "Genossenschaft," a jurisdictional body which assembled according to law, possessed its own juries for important cases, but for matters of moment made common cause with the city government as against the lord, the smelters, and the capitalist mine owners.[3]

The degree of relationship between mining law upon the continent and that in England is to some extent a matter of conjecture. That the English kings consistently asserted claims to regalian rights in mines is doubtful; that this was true upon occasions, however, is quite susceptible of proof. Thus in 1262 Henry III issued a writ to the sheriff of Devon, showing information that mines of gold, silver, and copper had been found upon the lands of the King and others, and declaring the right to such mines to be vested in the King and in no one else.[4] In the same year he gave Grenville and Silvester the custody of all newly discovered mines of gold, silver, copper, lead, and all other metals in Devon, the above to answer at the Exchequer for the revenues of the mines, and to have jurisdiction over miners, officers, and mining affairs in general.[5]

[1] Schmoller, Jb., xv, 678.

[2] For example the right which Freiburg in 1294 gained from the lord, "zu setzen alles was unser stadt und unserem bergwerc nutze ist und was wir mit in überein kommen" (Schmoller, Jb., xv, 681).

[3] *Ibid.*, Jb., xv, 681. In Goslar, after the city had won for itself the regalian rights, there stood at the head of the "silvani et montani" a council of six, analogous to the six councillors of the city (*ibid.*, Jb., xv, 683, n.).

[4] Close, 47 Hen. III, m. 15.

[5] Pat., 47 Hen. III, cited by Smirke (Rep. Roy. Inst. Cornw., 1853, p. 14).

In 1289 the King's officers were directed to work his mines of silver, copper, lead, and iron, newly discovered in Ireland.[1] In 1324 Edward II commissioned Wilringworth, a goldsmith, to search for gold in the tin mines of Cornwall and Devon, and to extract, refine, and reduce it to ingots for the King's use, leaving the tinners to dispose of the residuum of tin as accustomed.[2] In 1344 the warden of the King's mines in Devon and Cornwall was directed to associate with certain miners of Wells and Birlond and to examine and test a silver and lead mine in St. Cuthbert parish near Truro, in which the lead was said to be rich in tin.[3]

These and other documents of a similar nature, extending through the reign of Edward IV,[4] assume the right of the Crown to all metals, but are open to the possible objection that they may apply only to mines upon Crown lands or to mines royal in the lands of subjects. This ambiguity is removed by two documents of rather earlier date, one of which is a grant by Richard I to the bishop of Bath of mines of lead wherever he may find them on the episcopal lands in Somerset,[5] and the other a similar grant by Edward I in 1283 to the Carthusian prior and convent of Wytham.[6] A reasonable interpretation would be that the sovereign did claim all metallic mines, and enforced his claim where practicable.[7] But certain it is that this claim was never permanently made good, nor did it rest upon any firm foundation of earlier usage. Nothing inconsistent with the notion that private property in mines accompanied ownership of the soil is to be found either in the notes of mineral property in the Domesday Book[8] or in the charters of the Anglo-Saxon kings. Kemble[9] has declared mines to have been a part of the regalian rights of the latter, but the charters relied on as evidence are inconclusive, as they show merely

[1] Pat., 17 Edw. I, cited by Smirke (Rep. Roy. Inst. Cornw., 1853, p. 15).

[2] Fine R., 18 Edw. III, m. 15.

[3] Close, 18 Edw. III, pt. ii, m. 22.

[4] Cf. Stringer, 17, 20; Pat., 16 Rich. II, pt. ii, m. 7; 15 Edw. IV, pt. i, m. 22.

[5] For the Latin transcript see Dugdale, ii, 289.

[6] Cal. of Pat., 1283, p. 73.

[7] For transgressing the "assize of mines," in 1189, Juel de Espreton accounted for £63 of fines (Pipe, 1 Rich. I, Devon). It would be extremely interesting to know more of this assize, but I have been unable to obtain further information concerning it. A significant fact in connection with the claims of the Crown to the regalia of all mines is the control exercised over mines of every description in the county palatine of Durham by the bishop (Lapsley, 58, 283, 284).

[8] Ellis, i, 135.

[9] Kemble, ii, 69.

that the kings of England before the Conquest owned salt works and mines just as they may at present, and that taxes were sometimes levied on them.[1]

Excluding these exceptional cases, the mediæval English mines, viewed from a legal standpoint, fall into three divisions: first, mines of gold and silver, which have always been consistently regarded as Crown perquisites;[2] second, mines the possession of which went with that of the land on which they were situated; third, districts where mining was carried on under local customs, akin to those prevailing in Germany and other continental countries.

Gold and silver are not minerals common in England, but, judging from the number of documents in the Patent Rolls devoted to the subject, one might infer that the intensity of the search for the precious metals varied inversely with the likelihood of finding them. As far back as the reign of Henry I we find the Crown in receipt of an annual rent from a silver mine near Carlisle;[3] another in Cumberland for a number of years paid a profit to the Angevin kings,[4] while to enumerate all the commissions and prospecting parties which under the Plantagenets were sent out to examine into alleged mines of gold and silver would be too long a task for the limits of the present chapter. Enough has been said already to show that the Cornish and Devon tin mines were a favorite field for royal enterprise in this direction,[5] but search was made also in Somersetshire,[6] Gloucestershire,[7] the northern counties,[8] and even

[1] Kemble, ii, 70, 71. He cites, for instance, charters such as that by which in 689 Osmund of Kent granted to Rochester a ploughland in which was an iron mine (Cod. Dipl., no. 30), or others by which the kings made grants of salt works.

[2] Cf. Bainbridge, 110. The question as to just when a mine was to be considered "royal," together with other points of dissension with regard to mines as between Crown and subject, came to a head in 1568 in the Case of Mines (Plowden, 310–336. See also S. P. Dom. Eliz., xlii, 25, 31, 35, 39, 40), was quieted by the decision, but not definitely decided (Rep. Hist. MSS. Com., Hatfield MSS., pt. i, 345; Stringer, 148; Bushell, *Tracts on Mines*, A; Bainbridge, 113, 114; Pettus, *Fodinae Regales*, 9) until the reign of William and Mary (Stat. 1 Will. & Mar., 1st. sess. c. 30; 5 Will. & Mar. c. 6. See Galloway, 227, 228).

[3] Pipe R., 31 Hen. I (Rec. Com.), p. 142.

[4] Pipe R., *Cumberland, Westmoreland, and Durham*, introd., xxiv, xxv, xxvi.

[5] Fine R., 18 Edw. II, m. 15; Orig. R., 20 Edw. II, ro. 2.

[6] Cal. of Close, 1314, p. 52; 1339, p. 70.

[7] Cal. of Pat., 1462, p. 194; Close, 44 Edw. III, m. 7 d.

[8] Cal. of Pat., 1468, p. 132; 1474, p. 464; S. P. Dom. Jas. I, clxxvii, 45.

Ireland.[1] The best known and perhaps the most productive mines of this sort were the several argentiferous lead mines which the King operated at different points in Devon, at Beer Alston, Beer Ferris, Birlond, and Combe Martin. These mines, which possibly date back to the Roman period,[2] seem to have been actively exploited up to the sixteenth century [3] under direct royal supervision,[4] in charge of a warden [5] appointed by the King. The warden had power to hold a court and try miners for all offences, since the miners had freedom from outside jurisdiction and taxes [6] much as was the case in Germany.

Occasionally also, as in 1338,[7] a general patent empowered land-owners to search for precious metals upon their own estates with the condition that they pay a round tax to the King, besides pledging themselves to bring all gold and silver to the mint for coinage. For the most part, however, mines of silver and gold wherever found were probably leased by the Crown to private individuals.[8] At first the patentees might excavate anywhere save under houses or gardens; then we find that such digging is to be allowed upon payment of damages to the owner of the soil.[9] In later grants the patentee was under further restrictions in that he was not only to make compensation for damages done but must get license from the land-owner[10] before beginning operations. In this way the rights of the Crown, the landlord, and the tenant were gradually protected, but the working of mines became burdensome, and we find a sud-

[1] Cal. of Pat., 1276, p. 161; 1289, p. 322; Cal. of Close, 1338, p. 436; Parl. R., iii, 86; Holinshed, ii, 316.

[2] Watson, *Compendium of British Mining*, 226, 227.

[3] Cf. Westcote, 254; Watson, op. cit., 57, 226; Trans. Devon Assoc., viii, 209–337; Atkinson, 51, 52.

[4] They were occasionally leased (Cal. of Pat., 1461, p. 19; Cal. of Close, 1319, p. 159).

[5] *Ibid.*, 1322, pp. 469, 598; 1327, p. 6; 1330, p. 31; 1332, pp. 448, 516; 1334, p. 164; 1338, p. 559.

[6] Pat., 27 Edw. I, m. 35; Cal. of Pat., 1299, p. 398; 1307, p. 14; 1313, p. 526; 1320, p. 485; 1331, p. 74; Cal. of Close, 1315, p. 244.

[7] Cal. of Pat., 1338, p. 123; Cal. of Close, 1338, p. 528; Parl. R., v, 272.

[8] Cal. of Close, 1314, p. 52; 1338, p. 528; Cal. of Pat., 1338, p. 123; 1462, p. 194; 1468, p. 132; Parl. R., v, 272; S. P. Dom. Jas. I, clxxvii, 45; Parl. R., iii, 86; Close, 44 Edw. III, m. 7 d. For the terms on which the leases took place see Ruding, i, 124.

[9] Close, 44 Edw. III, m. 7 d; Parl. R., v, 272; Calvert, 246, 247.

[10] *Ibid.*, 246, 247.

den cessation of all such grants from the time of Edward IV to the great Elizabethan monopoly of the mines royal.[1]

Mines of the second class, namely those of which the ownership was identified with that of the surface, need not detain us. Mines of coal in particular seem wholly to have belonged to the proprietor of the soil,[2] a fact true not only of England but of the continental countries as well. Nothing is more striking than the way in which the rulers of Germany, of France, and of Belgium[3] omitted to seize possession of dull and semi-valueless minerals such as coal, and occasionally iron, while on the other hand they never once failed to claim all mines of other metals, silver and gold first, and the rest no doubt in an order somewhat corresponding to that of their values. In England a considerable number of mines occurred in royal forests[4] and were of course controlled by the Crown, as were also those upon royal demesne manors and other territories of a similar nature.[5] On private estates, however, the general rule prevailed that the baser metals were the property of the proprietor.[6]

Exception to this rule must be taken in the case of the third group into which, at the beginning of the discussion, we classified certain of the English mines. Scattered about in isolated regions we find mining communities whose laws, customs, and solidarity made them, so to speak, semi-independent states. Except that the miners were obliged to avoid tilled fields and to pay the landlords a certain percentage of the profits from mines upon their estates, their right of search for minerals seems to have been absolute. Yet although enjoying this and most of the other privileges which were identified with mining upon the continent, their connection with the Crown extended to little more than the payment of taxes to the King and a general supervision of the mineral courts by the Crown. Instead of developing in later years into enterprises under state

[1] Calvert, 247.

[2] Galloway, 18, 25, 26, 51, 55; Patrick, xlv. The bishop of Durham, however, had full regalian rights over all mines in the county (Lapsley, 58, 283, 284).

[3] See p. 72.

[4] Cal. of Close, 1276, p. 310; 1282, p. 199; 1294, p. 101; 1320, p. 278; 1328, p. 296; 1332, p. 443; Cal. of Pat., 1385, p. 118; Galloway, 24–26; Bainbridge, 119.

[5] Cal. of Close, 1278, p. 675; 1288, p. 499; 1318, p. 43; 1327, p. 78; Cal. of Pat., 1275, p. 84; 1292, p. 497; 1399, pp. 27, 51; Pat., 17 Edw. I, m. 8; Pipe R., 2 Hen. II, p. 38; 3 Hen. II, p. 90; 4 Hen. II, p. 162.

[6] Cal. of Pat., 1283, p. 65.

subsidy and control, the mines remained from first to last in the fullest sense private property.

The first of these mining districts to which we shall refer is that of the Forest of Dean, next in importance after the stannaries. There coal and iron were probably early mined, but, save for the fact that the King raised revenues from the taxation of numerous forges,[1] and owned others,[2] no information is given us of the district until the year 1282, when six of the ten bailiffs claimed iron ore and sea-coal and had their claims allowed by a commission of inquiry into the privileges claimed by the keepers of the forest bailiwicks.[3] In 1310 one hundred archers and twenty miners were summoned from Dean to Berwick-on-Tweed,[4] and from that date we meet with frequent references to the forest and its inhabitants.[5] We find the customs of the free miners of the forest, both of coal and of iron, confirmed and embodied in a charter ascribed to Edward I.[6] The right of free mining belonged to every inhabitant of the forest, together with the right of access for his ore to the King's highway.[7] The King, as landowner, was in return entitled to a share in the mines,[8] receiving also through a special officer known as the gaveller [9] a penny a week from each mine share, together with the option of certain rights of partnership.[10] As might have been expected, there was a mine court to try cases between the workmen by juries of twelve, twenty-four, and forty-eight miners,[11] while a miners' parliament, an institution found also among the tinners but not met with upon the continent, was held, in later times at least, in the "Speech House" in the wood.[12]

Another centre of mining was Alston Moor in Cumberland, the seat in the Middle Ages of fairly productive lead works. The Alston miners we find early in the thirteenth century given the King's pro-

[1] Cal. of Close, 1228, pp. 74, 141; 1229, pp. 260, 261; Cal. of Pat., 1234, p. 451; 1341, p. 190.

[2] Cal. of Close, 1228, p. 138; 1229, p. 226.

[3] Galloway, 27.

[4] Fourth Rep. Dean Forest Com., p. 4; Rotuli Scotiae, i, 91.

[5] Fourth Rep. Dean Forest Com., pp. 105, 141, 177, 195, 784; Scrivenor, 31, 32. The King himself owned mines and forges there. Cal. of Close, 1228, pp. 141, 138; 1320, p. 278; 1328, p. 296; 1332, p. 443; Cal. of Pat., 1341, p. 190.

[6] Nicholls, 13, 171.

[7] Houghton, pt. ii, c. 13.

[8] *Ibid.*, c. 14.

[9] Rent-collector.

[10] Houghton, pt. ii, c. 14, 15.

[11] *Ibid.*, pt. ii, c. 20, 21.

[12] Nicholls.

tection;[1] later state documents refer in a general way to their "liberties."[2] These, at last, in an exemplification of an inquisition,[3] taken at Penrith in the third year of the reign of Henry V before justices of the King, are in part revealed to us. We find here the usual features. A court of the mines took cognizance of all pleas of felonies, trespasses, injuries, debts, accounts, contracts, and personal actions in respect to the miners and their servants, as well as any others in the moor. The miners elected from their number a coroner and a bailiff, known as the King's sergeant. They were entitled to all fines and amerciaments before the coroner and all waifs and strays found upon their moor. No bailiff of the King nor any other officer might serve a summons or a process within the liberties, unless by default of the miners or their own bailiff to do the same.

Over the Mendip lead-mining district in Somerset the Bishop of Bath and Wells in common with three other lords formerly exercised a sort of mining sovereignty, but under Edward IV, a dispute having arisen between the people and Lord Benfield as to certain rights and customs claimed by the mining folk, a royal inquiry was instituted, and the miners' customs embodied in a code of written law.[4] Liberty of digging ore was allowed any man who first made application to the owner of the soil,[5] and who would agree to pay for the privilege every tenth pound of lead blown at his hearth.[6] Two courts were held each year at the instance of the lord, in which a jury of twelve miners presented infractions of the Mendip law,[7] but as a general rule civil suits and many mine offences were tried either on the manor or in the courts of common law.

In Derbyshire[8] the mineral laws applied, with a few modifica-

[1] Pat., 18 Hen. III, m. 7; 20 Hen. III, m. 13; 21 Hen. III, m. 10; Hunt, 148; Pleadings in Quo Warranto, 20 Edw. I, fol. 117; 2 Coke, 578.

[2] Parl. R., i, 64; Pat., 30 Edw. III, pt. iii, m. 23. Here the liberties are stated as being enjoyed only by those miners who dwell in their "sheles" together, and not dispersed.

[3] Pat., 4 Hen. V, m. 8.

[4] Printed in Houghton (pt. iii). See also Trans. Roy. Geol. Soc. Cornw., vi, 327–333.

[5] Houghton, pt. iii, c. 1. Permission might not be refused.

[6] *Ibid.*, c. 5.

[7] *Ibid.*, c. 8.

[8] For the earlier history of the Derbyshire lead mines see Del Mar, 114, 115; Add. MS. 6681, fol. 203; Pilkington, i, 110–112.

tions and exceptions, only to that part of the county known as the King's Field, the hundreds of High Peak and Wirksworth. It has always been contended that the mining rights of the Crown are confined to such land manors [1] as belonged to the Duke of Lancaster, and that the mining customs have no application to such estates as did not form part of the Duchy. These laws, evidently originating in the infancy of mining, are confined to the working of mines by manual labor only, and hence some of the customs are now obsolete.[2] Upon a person's finding ore he must make a cross in the ground as a mark of possession, at the same time giving notice to the "bar-master," who then attends and receives a measure or dish of ore, the first produce of the mine.[3] This seems to have been the preliminary condition for allowing the miner to proceed in working his "meer" of twenty-nine yards in length of vein, the bar-master at the same time taking possession for the King of the next half-meer, which was usually disposed of, however, to the adjoining meer-holder.[4] Freedom of mining in this way extended to all sorts of meadow lands, meadows, pastures, meers, and marshes,[5] and carried with it right of access to the highway and to water,[6] and the right to take wood from the King's land to timber shafts.[7] There was also a special miners' court to try offences,[8] held by the bar-master,[9] who was also deputed to oversee bounds, detect infractions of the law, and act as general executive head of administration.[10]

Since everywhere else, with the exception of silver and gold mining, the ownership of mines went with that of the surface, the five districts of the Forest of Dean, Derbyshire, the Mendip Hills, Alston Moor, and the stannaries of Devon and Cornwall, represent the extent to which "free mining" as we have seen it on the continent prevailed in England. To explain the existence of these

[1] The lead districts were divided into eight fields or liberties, each with its own code and its own officers. The differences, however, were very slight. Cf. Add. MS. 6681, fol. 197; Pilkington, i, 115.

[2] Eng. Min. Almanack, 1850, p. 220.

[3] Cf. Rep. Hist. MSS. Com., House of Lords MSS., ii (N. S.), 383–386.

[4] Eng. Min. Almanack, 1850, p. 221.

[5] *Compl. Min. Laws Derb.*, pt. i, c. 12.

[6] *Ibid.*, c. 2.

[7] *Ibid.*, c. 11.

[8] *Ibid.*, c. 16.

[9] German "Bergmeister."

[10] *Compl. Min. Laws Derb.*, pt. i, c. 29.

"free-mining" districts, it is a temptation to apply to England the same formula which has been applied to the continent. The mines, originally private property, were at an early stage seized by the nobles, and later, with the growth of the royal power and the introduction from Roman law of the concept of mineral regalia, passed as peculiarly privileged mining regions under the control of the Crown. In support of this, indeed, one has but to mention the mineral taxes which the tinners paid the King as early as 1156 and to quote the stannary charter of 1201 in which John, referring to the mining district as his demesnes, grants the tinners freedom from taxation in true continental style.

But in several important details this explanation fails to cover the entire ground. In the first place, it should be borne in mind that there is nothing in England to correspond to the long and detailed codes of mineral law which the German rulers bestowed upon their miners. A few, and only a few privileges, notably the release of the tinners from the manorial and common law jurisdictions and from ordinary taxation, were bestowed by specific charter from the Crown, and, in so far as they go, make for the truth of the borrowed hypothesis. Aside from these, and with the exception of the later legislation of the stannary parliaments, we may divide the mining laws into two parts, neither of which apparently owed aught to king or noble.

First may be mentioned mining statutes, such as those constantly passed by the parliament of the miners of the Forest of Dean, an institution of unknown origin. Second, — and this is fundamental, — the great mass of English mining law in the Middle Ages represents usage pure and simple, not legislation or grant. Long before the first charter granted to the tinners of Cornwall and Devon, they possessed free customs, not only as diggers of black tin but as smelters and as dealers. The laws of the Derbyshire lead miners were merely investigated, confirmed, and enrolled by Edward I, in 1287.[1] They are expressly stated to have been customs dating back to a time beyond the memory of man, and it is interesting and significant in this connection to find Pliny, in his description of the Britain of the first century A.D., declaring that the lead miners in the interior of the country are governed

[1] Escheat R., 16 Edw. I, no. 34; Add. MS. 6682, fol. 65.

by certain rules of their own making.[1] Much the same may also be said of the so-called "charters" containing the laws of the mines of the Forest of Dean, Alston Moor, and the Mendip Hills. In every case what the Crown did was merely to inspect and codify the usage of centuries, and it seems an open question whether those ancient customs, constantly declared and applied by miners' courts which in some cases may have been almost equally ancient, do not point to a primitive stage in British mining, the tradition of which has perhaps remained more unbroken than in Germany or France.

Another point which requires explanation is the relation of the mines to royal demesne. Admitting, for the sake of argument, that Alston Moor and the Forest of Dean may always have been Crown property, we have remaining the Derbyshire lead mines which until late in the Middle Ages were not on royal demesnes, the Mendip Hills where mining was carried on upon the estates of several noblemen, and the stannaries of Cornwall and Devon which were scattered over the territories of great numbers of local lords.[2] Now, granting for a moment the hypothesis that these mining districts passed through stages of development similar to those which certain writers assume for the mines of Germany and France, and were subject to the regalian rights of their lords, it seems a fair question to ask how it was that under kings like William I and the Angevins, and at a time when Roman law with its imperial concepts was invading all Europe, the mines of Mendip and of Derbyshire were not claimed as subject to the regalian rights of the Crown.[3] How happened it that the tin mines of Cornwall and Devon were paying special taxes to the King's sheriff half a century before John innovated the claim to the stannaries as his

[1] *Nat. Hist.*, xxxiv, 49; Del Mar, 114, 115. A document, valueless as an authority, states that the Derbyshire lead laws date to pre-Conquest times and that William I, in the ninth year of his reign, was induced by the miners to let the law remain unchanged (Add. MS. 6681, fol. 203).

[2] Few of the tin mines in the stannaries were on royal demesne lands, so that no claim can be made that their privileges were all of them the result of royal favors. Not more than three or four of the demesne manors received any toll from tin mines on their lands. For a list of tenants-in-chief in Cornwall cf. D. B., i, 120, 112; Hund. R., i, 53; Journ. Roy. Inst. Cornw., x, 150–169, 175, 375–389. For the later Duchy manors cf. Rymer, iii, 1; Chart. R., 1 Edw. II, no. 24; 11 Edw. III, no. 60, m. 28.

[3] That the Derbyshire lead mines, at least, were in operation during this period is certain (cf. p. 80, n. 8).

demesnes, a claim, by the way, which the barons forced him to renounce? And finally, how did it come about that mining in the Forest of Dean, instead of being open to all comers, was by custom confined to the descendants, inhabiting the forest, of the original free miners?

These are questions obviously impossible to answer with any degree of confidence. I merely desire to point out the possibilities which exist for the origin of the mineral laws of England. Whether, in truth paralleling a continental theory, we may see here simply another instance of the seizure of private property under color of regalian rights, whether the mineral laws in the five English districts are fragments of a customary law which prevailed all over Britain, at a time preceding the right of private property in land, or whether, as the great age and continuity of English mining would seem to indicate, the mines from the beginning were worked under certain special customs demanded by the peculiarities of the mining industry, — these are points which in the present state of our knowledge must be left unsettled.

# CHAPTER IV

## THE STANNARIES AND THE CROWN. ADMINISTRATION AND JUSTICE

WHATEVER may have been their original status, the tinners by the first half of the fourteenth century had already assumed a definite position under the British constitution. By immemorial usage they were, as regards mining, under the protection of their own peculiar laws, in compensation for which they paid special taxes to the Duchy. By virtue of the charters of 1201 and 1305 they had been constituted as two definitely organized corporations, the Stannaries of Cornwall and the Stannaries of Devon, possessing common seals,[1] courts, and executive and possibly legislative bodies.[2]

It is possible that the motives for the separation of the Cornish tinners from those of Devon in matters of administration were based on racial differences. While the Devon tinners apparently sprang from Anglo-Saxon stock, the miners of Cornwall formed a remnant of the Celtic race. In features, manners, traditions, and language, they were more akin to the Welch or Armorican than to the Saxon. Even before the formal separation in 1305 the stannaries and tinners of the two counties had been treated as distinct. De Wrotham, it has been seen, convened two juries of miners, the one for Devon at Exeter, the other for Cornwall at Launceston. The Devon tinners originally paid a tax on each thousand-weight of tin only half as great as that of Cornwall, a distinction maintained until 1837 by the smaller coinage duty imposed on Devon tin.[3] In the various fiscal arrangements entered into by the Crown in the thirteenth century with regard to the stannaries, the mines of Devon and Cornwall had been treated as separate properties,[4] being leased or farmed separately, accounted for separately in the

[1] Convoc. Cornw., 16 Hen. VIII, c. 28; Harl. 6380, fol. 2. The seal of the Cornish tinners has been discovered (Smirke).

[2] See p. 125.

[3] See pp. 132, 149.

[4] Pat., 8 Hen. III, m. 11; Close, 17 John, m. 16.

Pipe Rolls, and as a rule given separate wardens. Yet, if we accept popular tradition, their representatives[1] met until 1305 at intervals of seven or eight years, in a single parliament, and they were subject to the same charter, the confiscation or extension of which, if for any reason occasioned by the tinners of one county, would have affected those of the other in equal degree.

The stannaries of the two counties were, in fact, on an equal footing in their relations to the Crown and to the Duke of Cornwall, whether that title were vested in the Prince of Wales or his successor, or in the Crown itself. The fiscal connection we may profitably defer until the following chapter. We are here concerned, in the main, with the relations which the stannary government and especially the judiciary of the mines held toward the national administration.

At first glance it would seem that by the incorporation of the stannaries into the newly created Duchy of Cornwall the Crown and the nation had abdicated all right of interference. The Prince of Wales, as Duke of Cornwall, formed with his council the fountain head of all stannary administration. He appointed over the stannaries of both counties a single warden and a deputy, or vice-warden, the former of whom stood to the Prince, in many ways, much as the seneschal of a great lordship stood to his lord. As the Prince's representative, he proclaimed throughout the stannaries whatever new laws the former might see fit to enact for the regulation of the mines, and in special cases he might himself promulgate mining laws.[2] In the capacity of executive lieutenant, he appointed such minor officers as the stewards of the several stannary courts, the steward of Dartmoor, four foresters for Dartmoor, the keeper of Lidford castle, and the head bailiffs of the different stannary districts. Through his officers he maintained an oversight of stannary administration, made seizure of smuggled or otherwise forfeited tin, acted in the place of the sheriff as receiver of the goods of felon tinners, and imprisoned stannary malefactors at Lostwithiel or Lidford. He issued the writs for the choosing of representatives for the tinners' convocations, opened these parliaments, both of Corn-

[1] Cf. Pat., 17 John, m. 12; 37 Hen. III, m. 18; Close, 8 Hen. III, m. 14; Rolls Cal. (Rec. Com.), 38 Hen. III, ro. 3.

[2] Cf. Parl. Devon, 16 Eliz., c. 38.

wall and of Devon, acted as their go-between with the Duke of Cornwall, and gave his assent to their legislation,[1] at the same time using the stannary forces for the carrying of such enactments into effect.[2] In times of war he levied troops from among the tinners and acted as their titular commander.[3] In general one may say that the warden was constituted the tinners' spokesman in most of their dealings with outside authorities.[4] In his judicial capacity, he issued most of the writs which, by the stannary custom, were disallowed if from justices of foreign courts. He might also sit as a court of equity and of appeal, midway between the stewards and the Duke, and in this capacity he regulated the conduct of the stewards in their courts and upon occasion commanded the stay of actions which rightfully belonged to other jurisdictions.

Yet although apparently excluded from influence upon stannary affairs by the workings of the customary mining law and the tinners' charters on the one side, and the exercise on the other of the duchy authority through the hands of the warden, the King and Parliament retained the right, as in the case of all private jurisdictions, to overrule or modify, if necessary, stannary law and even stannary charters. Classic examples of the exercise of this right of interference are the interpretation of the stannary charters by Parliament in 1376,[5] the case of Strode, imprisoned in 1512 by the Devon stannary courts for a speech made in Parliament[6] and given redress by an Act,[7] and the Stannary Act of 1641,[8] which, together with some other examples of the same sort, will presently be discussed.

In the enjoyment of special courts, as in the case of so many other privileges, the tinners were merely one of a number of classes which possessed similar rights, although possibly not to so great an extent. Members of the gild merchant in any borough, it will be remembered, were exempt from trial outside the borough courts;

[1] See p. 127.

[2] Cf. the collection of assessments decreed by the stannary parliaments (Harl. 6380, fol. 48, 49).

[3] See p. 167.

[4] Cf. Acts of Privy Council, Eliz., xvii, 328, in which Raleigh, as warden, presents to the Queen the petition of several tinners.

[5] Appendix F.

[6] Trans. Devon Assoc., xi, 300.

[7] Stat. 4 Hen. VIII, c. 8.

[8] Stat. 16 Chas. I, c. 15.

but to the mining classes in general the grant of special judiciaries seems to have come almost as naturally as the right of free mining. The reason is to be found not only in the technical difficulties, abounding in mining law suits, which could not well be solved in an ordinary court, but also in a desire on the part of the King to prevent interruptions of the miners' work by secular courts. The mining classes were under mining law and courts, much as the soldier is subject to military law and courts martial.

The existence of these mine courts was widespread. We see them, in one form or another, wherever free mining customs existed, on the continent, in Germany,[1] France,[2] Belgium,[3] and in other parts of England, Derbyshire,[4] Alston Moor,[5] the Mendip Hills,[6] and the Forest of Dean.[7] We have seen them also under circumstances entirely different, namely in the case of single mines worked upon a great scale with hired labor either by the King[8] or by his lessees. The royal silver mines at Birlond, Beer Ferris, Beer Alston, and Combe Martin, each had its own court, presided over by the keeper of the mine as steward, who by the aid of a jury of miners dispensed justice to the workmen.[9] The latter were exempt from being impleaded or from pleading against their will for trespasses, actions personal, or any pleas in the county courts, courts baron, or the courts of any lords, save before the keeper of the mine and the sheriff of the county.[10] When the King granted mining privileges on a considerable scale, he usually accompanied the grant with the right to buy wood at reasonable prices, to impress

[1] Cf. Smirke, p. 107; Reyer, 79; Inama-Sternegg, ii, 336, n. 5; Trans. Roy. Geol. Soc. Cornw., vi, 160.

[2] Achenbach, 1869, p. 34; Smirke, 104, 112, 113; Trans. Roy. Geol. Soc. Cornw., vi, 252, 257.

[3] Smirke, 86, 113; Jars, i, 391.

[4] Add. MS. 6682, fol. 65; *Min. Laws Derb.*, pt. i, art. 13, 18, 37, pt. ii, art. 40; Houghton, 14, art. 31.

[5] Pat., 4 Hen. V, m. 8; 30 Edw. III, pt. iii, m. 23.

[6] Trans. Roy. Geol. Soc. Cornw., vi, 329–333.

[7] Houghton, pt. ii, art. 21; Parl. R., vi, 347.

[8] The bishop princes of the palatinate of Durham had similar rights over miners in their employ (Curs. Rec., Enrollment of Bishop Hatfield, no. 31, m. 5; Pat. of Bishop Fox, no. 61, m. 4).

[9] Cf. Pat., 27 Edw. I, m. 35; Cal. of Pat., 1299, p. 398; 1313, p. 526.

[10] *Ibid.*, 1307, p. 14; 1313, p. 526; 1331, p. 74.

laborers, to hold courts and exercise jurisdiction, and to have letters of protection for miners and for those who contributed money to mines.[1] The stannary courts are therefore no isolated instance of a mining judiciary. Their comparative wealth of records, however,[2] and the importance which they assumed in the Middle Ages, in their conflicts with the manor and hundred, make them of exceptional interest to the historian of legal institutions.

The clauses in the charter of 1201, which placed criminal and civil jurisdiction over the tinners in the hands of the warden, had resulted in the division of the mining districts into several provinces or "stannaries."[3] They probably grew up from this general grant of jurisdiction which we find to have been usual in other mines[4] and which, perhaps, constituted a mixed personal and local law. The local limits of each of the four Cornish stannaries have never been accurately defined, for they seem to have spread vaguely outward from the aggregation of tin works in certain situations favorable to them. This is indicated by the fact that the name of each stannary designates its original nucleus. Five tracts of stanniferous wastrel with their adjacent vales supplied the ancient stream works of Cornwall. The moor between Launceston and Bodmin, in which the Fowey River has its source, gave its name to the northern stannary of Foweymore, Hensborough Beacon with the tin grounds of Roche, Luxullian, and St. Austell formed that of Blackmore. A smaller district on the north coast, extending inland to Truro, constituted the stannary of Tywarnhail. The stannary or united stannaries of Penwith and Kerrier included two great tracts of waste of which one lies north of Helston-in-Kerrier and the other stretches between Lelant and Land's End.

[1] Pat., 47 Hen. III, m. 12; 15 Edw. IV, pt. i, m. 22. Cf. Cal. of Pat., 1478, Mar. 11.

[2] I have been unable to discover any of the court rolls from the other mining districts of England. A number of the stannary rolls have survived, unfortunately none of a very early date.

[3] The Derbyshire lead districts were also divided into various administrative districts (*Min. Laws Derb.*). The county palatine of Durham was divided into four wards, in each of which was held a tri-weekly tribunal corresponding to the hundred court. Later these courts were continued for the purpose of the sheriff's tourn (Lapsley, 194, 195).

[4] Cf. Pat., 7 Hen. III, m. 12; 27 Edw. I, m. 35; 15 Edw. IV, pt. i, m. 22; 1 Hen. VII, pt. ii, m. 25.

In the neighboring county of Devon, the stannary districts of Chagford, Ashburton, Tavistock, and at a later date of Plympton,[1] each centring about a town of the same name, comprised and encircled the great stanniferous area of Dartmoor.

In each was a court presided over by a steward as the warden's representative. As early as 1243 we find the stannary courts of Devon recorded in the Pipe Rolls.[2] Cornwall, by 1297, contained the stannaries of Blackmore, Penwith and Kerrier, and Tywarnhail,[3] each with its court, but of Foweymore we have no trace until 1342.[4] Gradually there arose a code, partly from prescription and partly, in all likelihood, from enactment by early stannary parliaments, the object of which was to prevent any infringement of the judicial liberties of the mines by either tinner or foreigner. It is unfortunate that we have not the records of the first stannary convocations, with which to trace the successive steps of legal development; we must depend upon the law as defined by the convocations from the sixteenth to the eighteenth centuries. No tinner, it reads, might appear at an assize or nisi prius against another tinner for digging in any man's freehold,[5] or might sue or allow himself to be sued in any foreign court, save for pleas of life, limb, or land,[6] under penalty of a fine.[7] No case determinable in the stannary might be tried elsewhere, violations of this law being punishable whether or no the offender were a tinner.[8] Warrants and writs issued against tinners from foreign courts were not allowed,[9] and officers attempting to serve them were liable to arrest. In this

[1] Cf. Anc. Pet., bdle. 42, no. 2098.

[2] Pipe, 27 Hen. III, Devon.

[3] Min. Accts. Bailiff's Accts., Edm. of Cornwall.

[4] Receiver, 26 Edw. III.

[5] Parl. Devon, 2 Hen. VIII, c. 35.

[6] Cf. Convoc. Cornw., 16 Hen. VIII, c. 26; 22 Jas. I, c. 14; 30 Eliz., c. 10; Parl. Devon, 2 Hen. VIII, c. 26; Harl. MS. 6380, fol. 42, 52; Pearce, *Laws and Customs of the Stannaries*, 139 (Wallin's Case).

[7] Harl. MS. 6380, fol. 42, 46; Convoc. Cornw., 22 Jas. I, c. 7. Cf. Parl. Devon, 2 Hen. VIII, c. 6, 7; *Min. Laws Derb.*, pt. i, art. 13; pt. iii, art. 40; Houghton, 14, art. 31; pt. 2, art. 21 (Dean Forest); Trans. Roy. Geol. Soc. Cornw., vi, 330. Similar rules are found in the gild ordinances; see, for instance, those of the weavers of Kingston-on-Hull, 1564, where "brethren are not to sue each other out of the town" (Lambert, 206).

[8] Cf. Convoc. Cornw., 16 Hen. VIII, c. 15. A similar provision occurs in the Forest of Dean constitution (Houghton, pt. ii, c. 20), and also in Derbyshire (*ibid.*, pt. i, art. 13).

[9] Cf. Add. MS. 6713, fol. 249, c. 68.

category came warrants issued by any justice of the peace,[1] writs of certiorari from the royal courts,[2] and of replevin from any but the warden, and attempts to remove suits from the stannary courts, once they had been begun. Writs of prohibition, habeas corpus, and corpus cum causa were allowed when the matter was one of land, life, or member,[3] but no litigant might procure these under any other circumstances for removing cases from the stannary courts.[4] The use of royal writs of subpœna to sue a tinner out of the stannary for matters there determinable was forbidden, and the writ might be broken with impunity.[5] No appeals were permitted from stannary judgment to foreign courts by writs of error or of certiorari[6] save in certain cases, as for land, life, or limb, where a tinner might be tried in a foreign court,[7] in which case the jury was composed half of tinners. Tinners were immune from jury service save in their own tribunals.[8] Numerous prosecutions recorded on the mediæval court rolls for violations of these regulations not only confirm our views as to their antiquity, but give evidence of their continued vitality.[9]

Since, therefore, in all but the more serious of criminal cases and civil suits involving the possession of land, the tinners were restricted to their own courts, we should expect to find these courts uniting the activities not only of the manorial court baron and the manorial and municipal leet[10] but of the courts of the craft

[1] Add. MS. 6713, fol. 112; Parl. Devon, 2 Hen. VIII, c. 8. Cf. *Compl. Min. Laws Derb.*, pt. i, art. 18, 37; ii, art. 40.

[2] Add. MS. 6713, fol. 129, 130 (Saintallyn *vs.* Treweet).

[3] *Ibid.*, fol. 194; Convoc. Cornw., 12 Chas. I, c. 27 (Add. MS. 6713, fol. 371).

[4] Convoc. Cornw., 22 Jas. I, c. 15.

[5] Add. MS. 6713, fol. 127.

[6] *Ibid.*, fol. 249, c. 68; Coke, iv, 229; D. O. MS. Vol., fol. 337.

[7] See Appendix D.

[8] Convoc. Cornw., 22 Jas. I. c. 17, 48; 12 Chas. I, c. 11 (Add. MS. 6713, fol. 358); Parl. Devon, 2 Hen. VIII, c. 10; Harl. 6380, fol. 43.

[9] Ct. R., bdle. 156, nos. 26, 27; bdle. 157, no. 13; bdle. 159, nos. 1, 27. Cf. Receiver, 11 Hen. VII, the fines recorded for the impleading of tinners in foreign courts.

[10] Sometimes these powers would be enlarged by the inclusion of the right to hold views of frankpledge and trials for breaches of the assize of beer, both of which were ordinarily part of the business of the sheriff's tourn, from which as well as from the shire and hundred courts, the lord's men were exempt. The manorial leet was also used for the selection of manorial officers, and for the declaration of the customs of the manor. The functions of the leet of a municipality have been well brought out by Mr. Hudson, in his work on the Leet Jurisdiction of the City of Norwich (Selden Soc. Publ., No. 5).

gilds,[1] of the hundred and shire court, and the sheriff's tourn. It is not surprising, therefore, that in the course of their existence the stannary courts came into conflict with some or all of those bodies upon whose powers they trenched.[2] The history of the stannary courts must record a continuous struggle between them and others, especially local, manorial, and hundred courts, the latter, doubtless, in seignorial hands.[3]

The exact relations between the tinners and the neighboring manorial lords will probably never be thoroughly understood.[4] To judge from the charter of 1201 and the general course of subsequent history, the two were completely divorced, yet occasionally we meet with indications of the existence of manors which through long usage possessed well defined rights as against the tinners. The manorial court of King's Climesland, one of the duchy manors, put itself on record in 1540 with the declaration that, by time-honored custom, no bounds ought to be cut there by any tinner without the King's[5] license.[6] It would seem also, from an exposition of stannary custom made in 1539 at the Helston stannary court, that tinners living upon the manors of the King or Prince of Wales were brought before the stannary courts not by their bailiffs, but through the agency of the reeve of the manor, who

[1] This met for the punishment of trade offences among gildsmen, and also constituted the ultimate governing body of the craft, electing masters and wardens and promulgating trade regulations.

[2] The German mineral courts probably underwent a similar experience. Cf. Schmoller, Jb., xv, 1021.

[3] Cf. Parl. R., i, 381.

[4] Cf. Schmoller, Jb., xv, 682, n., on the German mines and manors. The statement of the text applies to other localities as well. In the Mendip Hills, the miners' courts took on a closer relationship with the seignorial powers. Two mineral courts were held each year, where justice was done by juries of miners, but it is probable that much of the mining litigation took place elsewhere, in the common law courts or in the courts of the lords themselves, who, it should be added, are said to have summoned and "kept" the courts of the miners (Trans. Roy. Geol. Soc. Cornw., vi, 329, 330, 332, 333).

This was also more or less the case in some parts of the Derbyshire lead districts, as in the lordship of Litten, where the steward of the mineral courts was nominated by the lord of the manor (*Compl. Min. Laws Derb.*, pt. iv, art. 2). It should be recalled that nowhere in England, save in the stannaries, were the miners expressly exempted from pleas of villeins.

[5] Then Duke of Cornwall.

[6] Add. MS. 6713, fol. 123.

divided with the bailiff the fees for the arrest.[1] There must be noted, further, the significant parts played in the execution of stannary justice by the village tithing men and by manorial servants such as the toller,[2] the identity of the bailiff of the stannary of Penwith and Kerrier with the bailiff of the hundred of Kerrier,[3] and the equally interesting fact that the warden used the parish constables of Devon and Cornwall for the collection of stannary assessments from the tinners.[4] More significant still, if genuine, are two stannary laws, probably dead letters from the thirteenth century onwards, which point to the existence at some time of a distinctly seignorial relationship between the manor and the tinner. By the first of these, which is set down in the laws of Devon, we are told that "no tinner is to sue any tinner for any cause but land, life, and limb, save in the stannary court, the court of Lidford, or else in the court of whom he holdeth, after the custom and manner."[5] By the second custom we learn that the lord of the soil has the right to demand that tin dug upon his land shall be washed, dressed, and stamped at his own mill.[6] Of the enforcement of these two customs, probably relics of the period before 1201, it is impossible to find a record, and it may be stated with tolerable confidence that in the unceasing struggle between the seignorial courts and those of the stannaries the former simply endeavored to restrict the privileges of the tinners to the limits laid down in their charters. Among the grievances of the clergy presented to the King for redress in 1237, we find a complaint against miners of tin, lead, and iron for digging on church land,[7] but it is not until the fourteenth century that the call for royal interference seems to have been urgent. Our first official account of the contentions bears the date 1309, when the sheriff of Cornwall was mobbed by the Blackmore tinners on his attempting to levy upon their chattels.[8] A few years later the King was forced to issue a commission of oyer and terminer touching "the men of the commonalty of Devon and Cornwall and the stannary men of the said counties, who in their petitions exhibited before

[1] Add. MS. 6713, fol. 129. [2] See p. 121. [3] D. O. Min. Accts., 21 Edw. III.
[4] Harl. 6380, fol. 48, 49. [5] Parl. Devon, 2 Hen. VIII, c. 7.
[6] Add. MS. 6713, fol. 235. It may be well to add that I have been able to find nothing even approximating this rule in any other stannary document.
[7] *Annales Monasterii de Burton*, ed. Luard, p. 256.
[8] Pat., 3 Edw. II, m. 43 d.

the King and Council have charged each other with the commission of divers trespasses."[1] In the same year the people of Devon had complained that the tinners were digging into and destroying their tilled fields, woods, and gardens,[2] while the tinners on their part asserted that local magnates were impleading them for pleas of serfs.[3]

In 1318 another complaint arose, and another commission of oyer and terminer was appointed, "on complaint of certain men of Devon . . . that the stannary men commit trespasses and assault men of the county in divers hundreds, outside the bounds and limits of the stannary, nor permit themselves to be brought to justice according to the law and custom of the realm, and when the hue and cry is raised against them, they take and beat the King's bailiffs and the bailiffs of others holding liberties there, and leading them into the stannary imprison them in the stannary gaol until a ransom is paid. Furthermore, that they commit many acts of extortion, dig for tin where it has not been the custom to do so and extort money for the privilege of being left undisturbed; that they seize the King's bailiffs sent to those parts to levy his debts, and put them in prison for a ransom; and that they appropriate the tenth part of the refined mineral, which ought to be paid to the lord of the soil." The stannary bailiffs, the petitioners go on to say, are persons of ruffianly character, and "prevent our free tenants and those of others in the different hundreds from presenting themselves at the hundred court to do satisfaction for the breaking of the assize of bread and ale and of hue and cry raised, and when the tithing men with their tithings ought to come to the hundred courts to present all presentable actions according to the law and custom of the realm, they distrain them so that they dare not come, and if they defy them, they put them in gaol and extort a ransom." The warden was said to take money from men of the county who by right ought to be in assizes, juries, and recognitions, "and he extorts from them that they themselves defend the above acts as the liberties of the stannaries, on account of which our hundreds and those of others are impoverished. They also constrain by force that the transgressions and contracts arising outside their

[1] Pat., 8 Edw. II, pt. ii, m. 2 d. [2] Parl. R., i, 297, 312.

[3] *Ibid.*, i, 324. Cf. D. O. MS. Vol., fol. 317, proving that villeins became tinners to get their freedom, and that their lords came to the court and claimed them.

stannaries be pleaded in the stannary courts to the manifest prejudice of the parties." [1]

Whatever the action taken by the Crown in this instance, the struggle continued unabated. In 1320 commissions were issued for the investigation of no less than ten different remonstrances on the part of the people of Devon.[2] In 1333 the King had to warn the warden of the Devon stannaries not to allow the tinners to dig in arable lands or groves to the neglect of the waste.[3] A long list of grievances appears in 1347, in which the Devonshire men complain that "under cover of the King's charters the tinners claim all manner of lands, arable or otherwise, overturn fields and woods, and turn the courses of streams, whereby the land is become wasted and barren. The stannary men and their servants have cognizance of all sorts of pleas for suit of tinners as well outside the stannary as inside, and make attachment and outrageous distress as well outside as inside, claiming the whole country as their stannary." Their request is that the tinners be called upon to show their charters before the Justices of the King's Bench or Common Bench.[4] From other documents it would seem that occasionally the Cornish and Devon folk retaliated in kind by smashing down the miners' works, carrying away the tin which had been dug, and imprisoning the workmen in the local jails.[5]

These and other [6] excerpts are but typical of the long conflict between the tinners and their courts, and the commonalty of the shires with their manorial and hundred jurisdictions. The stannaries were in the end victorious in this struggle, and succeeded in drawing as suitors before the steward's courts of the stannaries, as shown in the continuous records of five centuries, persons of all ranks, civil and ecclesiastical.[7]

[1] Pat., 12 Edw. II, pt. i, m. 15, sched.; Anc. Pet., bdle. 108, no. 5364. Similar complaints were occasioned by the court of the bishop of Exeter (Anc. Pet., bdle. 260, no. 12958).

[2] Parl. R., i, 382.

[3] Close, 7 Edw. III, pt. i, m. 9 d.

[4] Parl. R., ii, 190. In Derbyshire there seems to have been a similar struggle. Cf. Add. MS. 6682, fol. 30, 33.

[5] Cf. Pat., 18 Edw. III, pt. ii, m. 30 d.; 20 Edw. III, pt. ii, m. 15 d.

[6] Anc. Pet., bdle. 107, no. 5324; bdle. 257, no. 12841; White Bk., i, c. 35; Parl. R., ii, 343, 344; Convoc. Cornw., 16 Hen. VIII, c. 21.

[7] *Proc. of Court of Vice-warden*, xi. Cf. White Bk., i, c. 35 (complaint of the prior of Mt. St. Michael).

The chief reason for these disputes seems to have been the lack of definition with which the charters left the powers of the stannary courts. In practice, stannary law covered five subjects. In the first place, it included all rights and interests justly acquired under stannary law and custom in the absolute usufruct of underground soil for the digging of the ore, and also a qualified usufruct of all streams of water whose natural course might run within the surface limits of these rights or whose use might be essential for the pursuit of mining operations; and, conversely, the prevention of injustice by the usurpation of any such rights in violation of stannary law. It secured, in the second place, to the lords of the soil, or to those who might have acquired such underground rights, their due proportion of toll tin. The regulation of all dealings between the miner and the blower or smelter and the enforcement of the assay and the coinage form the third and fourth subjects of stannary law. And finally, it is concerned with adjudicating all matters in dispute between persons engaged in mining operations so as to entitle them to the character either of privileged tinners or of tinners at large,[1] or between such persons and any others not so engaged.[2]

Concerning the rights of tinners, two questions arose from the start. The first concerned the definition of the word "tinner." Did it, as the stannaries claimed, include not only the manual laborers but their employers, the holders of shares in mines, the dealers in tin, and all artisan classes connected with tin mining? Or was it, as insisted by their opponents, to comprise only the working miners, and so long only as they remained at work?[3]

The evidence on this point is not a model of consistency. Our earliest document, the De Wrotham letter,[4] says nothing, of course, with regard to a stannary court, but includes among those classes whose customs are to be respected all diggers of tin, buyers of black tin, first smelters of tin, and merchants of tin of the first smelting. The charter of 1201 is addressed to "all tinners as long

[1] See p. 101.

[2] It may be added here that in later times, when there was enough tin in a mine of copper to furnish a reasonable excuse, the vice-warden extended the jurisdiction of the stannaries over the entire mine.

[3] Cf. Parl. R., ii, 343, 344.

[4] Appendix A.

as they are at work." The charter of 1305 repeats the qualification and apparently adds another confining its scope to the miners on the King's ancient demesnes. The ambiguities of phrasing with which this instrument abounds,[1] and which were eagerly caught up by stannary officials in support of their aggressive campaign against outside jurisdictions, gave rise to many of the complaints which have been cited. These culminated in 1376 with two long petitions introduced into Parliament by the commons of Devon and Cornwall,[2] which with their answers form a landmark in the constitutional history of the stannaries. It was protested that the tinners off the royal demesnes claimed stannary privileges; that not only laboring tinners but their employers enjoyed the freedom of the mines; that the stannary courts were taking cognizance of pleas arising between tinner and foreigner elsewhere than in the former's place of work; that the warden allowed the tinners imprisoned at Lostwithiel and Lidford for felony to run at large; and that in the jail of the stannary he received villeins whom their masters were about to imprison for arrears of accounts, and treated them so well that they refused to return to their lords.

The exposition allowed on these points by Parliament was in some details evasive and in others clear. To the inquiry as to whether in other than the King's demesnes the tinners were free, it contented itself with pointing out that the charters of 1305 permitted the digging of tin in the lands of all parties. For other complaints it appointed a commission of inquiry, whose findings, if ever made, have disappeared.[3] It promised that pleas between tinners and foreigners, arising outside the places in which mining was actually carried on, should not be taken to the stannary courts, and finally

[1] The confusion of legal thought upon the subject is shown by an inquisition of 10 Edward II, taken to find the yearly value of the stannaries, in which it was added "et ad proficua de exitibus dictae stannariae provenientia, viz., quidam redditus nigrorum stannatorum," etc., the word "stannator" being applied not only to the tinner but to the owner of white tin. In Devon the existence of the tax "white rent," levied upon the owners of white tin, whether miners or not, seems to show that the term "tinner" was in practice interpreted broadly.

[2] Appendix F.

[3] Coke (iv, 232) gives the commission, but states that it is impossible to find its returns. It would seem that the exposition itself was, in practice, disregarded by the tinners (Anc. Pet., bdle. 14, no. 656; bdle. 101, no. 5029).

it defined the word "tinner" to comprise only manual laborers in the tin works, and for so long only as they worked there. This exposition, confirmed a few years later by Richard II,[1] remained unchanged in principle for over a century.[2]

After a lapse of one hundred and thirty-one years we find the whole question inadvertently reopened by the charter of pardon of 1507,[3] in which Henry VII forgave the tinners their disregard of the ordinances of Prince Arthur. The persons expressly named in the pardon, and styled without exception "stannatores," were gentlemen bounders, owners of tin works, possessors of blowing-houses, and buyers of black or white tin, — thus indicating a return to a more liberal interpretation of the charters than that of 1376.

In 1524 another event took place which emphasized the importance of the stannary question, namely the issuance of two letters, by the King and by the Marquis of Exeter as lord warden, appointing a commission of five to settle "certain doubtful questions connected with the stannaries." In the first section of their report[4] they decided that no man was to be taken for a tinner, privileged to sue or be sued in the stannary courts, save such as had some portion in tin works or employed some charge in making things requisite for the getting of tin, including, therefore, artisans such as carpenters, smiths, colliers, and blowers. This interpretation again received indirect confirmation a few years later from two statutes passed in 1532[5] and 1536,[6] to prevent injury to Devon and Cornish seaports through silt from the tin works, reciting that if any person be sued by an officer of any of the King's courts of Chancery for pursuing any action under the above statutes, all these suits and every act to be done in any of the stannary courts were to be void. But they provided also that the statutes should not prejudice any of the officers of the stannary, nor their lawful laws and customs, saving only the cases contained in the acts themselves.

Indirect recognition of the applicability of the term "tinner" to others than mere manual laborers is furnished also in the stan-

[1] Harl. 6380, fol. 99; Add. MS. 6713, fol. 196.
[2] Stat. 16 Chas. I, c. 15, preamble.
[3] Pat., 23 Hen. VII, pt. vii, m. 29–31.
[4] Convoc. Cornw., 16 Hen. VIII, c. 1.
[5] 23 Hen. VIII, c. 8. [6] 27 Hen. VIII, c. 23.

nary case of Boscawen *vs.* Chaplin,[1] in which the plaintiff was the owner of extensive tin works and the defendant a dealer in the metal, neither of them being in any sense manual laborers.

The Cornish stannary parliament, held at Lostwithiel in 1588, passed upon the question in comprehensive fashion.[2] It divided all tinners into two classes. In the first were manual laborers, "spaliers" and "pioneers;" these were not to sue or to be drawn into any foreign jurisdiction for the trial of any case whatsoever, save matters concerning land, life, and limb. The other class comprised those gentlemen who had some share in tin works, or who received toll tin as lords or farmers, men who converted black tin into white, or who were necessary for the getting of tin, such as colliers, blowers, carpenters, smiths, tin merchants, and other intermediaries, owners of bounds, makers of miners' tools, and workers and smelters of tin. All these might sue and be sued and impleaded in the stannaries and were free from tolls and tallages, but it was not obligatory for them to refuse to make use of the common law tribunals to settle their differences.

By this time the question of stannary jurisdiction had begun to attract the close attention of lawyers, and from the tinners' complaints to their warden [3] it is evident that a movement was on foot to break down the stannary courts altogether, or at least to reduce them to a shadow of their power.[4] Not only did the Star Chamber, which since the beginning of the sixteenth century had always insisted on its right to hear appeals from the stannary courts,[5] assert

[1] Harl. 6380, fol. 9. See also Trewynard *vs.* Killigrew (Harrison, 55), where both parties to the suit were "esquires," yet the case was referred by the Chancery and Star Chamber to the stannary courts.

[2] Convoc. Cornw., 30 Eliz., c. 7–9.

[3] Raleigh had had trouble as lord warden with the Earl of Bath and the Devon justices in 1589 (S. P. Dom. Eliz., ccxxvii, 8), but had won out, with the coöperation of the Queen (Add. MS. 24746, fol. 92; cf. Harl. 6696, fol. 132; Add. MS. 6713, fol. 113; D. O. MS. Vol., fol. 337 (Trescott *vs.* Richards); Convoc. Cornw., 30 Eliz., c. 13; Acts of P. C., 1552, p. 504).

[4] Cf. *Proceedings of Court of Vice-warden*, p. xi.

[5] Cf. Star Chamber Proc. of Henry VIII, iii, 274; iv, 94–97; ix, 136; x, 222; cclxiv, 7; Star Chamber Pet., bdle. 18, no. 71; bdle. 19, no. 58; bdle. 29, no. 15; bdle. 32, no. 90; Acts of P. C., 1530, p. 364; 1579, p. 295; 1587–88, p. 328; 1591, pp. 11, 71, 72; D. O. MS. Vol., fol. 337. Not all of these petitions and cases were accepted, but certainly a large proportion.

its claims more aggressively, but the Chancery Courts [1] and the Justices of the King's Bench also entertained suits which by right could be heard only in the courts of the tinners.[2] Raleigh was succeeded as warden in 1603 by the Earl of Pembroke, who had scarcely taken office before he was besieged by petitions from the tinners with regard to actions brought against them in the courts of common law and chancery [3] for matters such as cases of trespass, which up to that time had never been sued outside the stannaries. The result of his efforts [4] on behalf of his charges was that the question of the extent of stannary jurisdiction was referred to Chief Justices Fleming and Coke.[5] Their declaration was to the effect that all matters which concerned the stannaries or depended upon the same were to be heard in the stannary courts, while the privileges of the mine were to be free to all blowers and their *bona fide* workers in or about the stannaries so long as they worked there and no longer.

Complaints, however, were still heard from the tinners, that their privileges were being menaced,[6] and the Cornish convocation some years later once more gave out its decision on the question of jurisdiction,[7] defining as privileged tinners, not to be impleaded in foreign courts, all laboring tinners, blowers, owners of blowing-houses, spaliards, adventurers at any charge for the getting of tin, smiths, colliers, or any other person employed in working or washing any tin, or about any utensils for working it.

In 1627 we find the whole subject referred to the decision of a number of eminent jurists,[8] who proceeded to expand and explain the resolutions of the judges in 1608. So far as concerns the comprehensiveness of the word "tinner," little was vouchsafed in the way of additional information. On the one side they placed as privileged tinners all blowers and *bona fide* laborers in the stan-

[1] Cf. Chanc. Proc., Hen. VIII, bdle. 23, no. 5; *ibid.*, series ii, bdle. 79, no. 185; bdle. 165, no. 96. For still earlier examples of Chancery interference, see Early Chanc. Proc., bdle. 17, no. 185; bdle. 68, no. 52; bdle. 18, no. 76; bdle. 65, no. 133.

[2] Cf. Cal. of Pat., 1428, p. 505; 1429, p. 505.

[3] Pearce, 147, 151.

[4] Cf. S. P. Dom. Jas. I, Addenda, Feb. 20, 1605; Pearce, 149, 150.

[5] Appendix G.

[6] Cf. S. P. Dom. Jas. I, lxxviii, 36.

[7] Convoc. Cornw., 22 Jas. I, c. 12–14. In c. 13 is a provision that no tinner is to sue another in a foreign court, abrogated by Stat. 16 Chas. I, c. 15.

[8] Appendix H.

naries; and on the other, such men as jurates of stannary courts, owners, adventurers in tin mines, and others concerned in the stannaries whose personal attendance, however, was not essential to the processes of production.

This left the matter as ambiguous as ever, and in 1632 the King and Privy Council, in conjunction with the judges,[1] once more defined the word "tinner." "Workers in tin, in mine or stream, carriers, washers and blowers, and necessary attendants about the works, ought not to be sued outside of the stannary for any cause arising within the stannary. Other miners that do no hand work, *i. e.* owners of tin lands, owners of bounds, owners of blowing-houses, and their partners, buyers, and sellers of black tin or white tin, before the deliverance, may sue one another or working tinners, or any other, in and for any matter concerning tin or tin works, in the stannary courts. Both these tinners and the workers may sue one another in the stannaries for all causes personal, arising in the stannary and not concerning freehold, life, or member; but a tinner may not sue a foreigner in the stannary for matters personal arising out of the stannary. Of these latter sort of tinners, such only are intended as within some convenient time make profit, or endeavor to make profit out of the coinage."

Four years later[2] we find the parliament of the tinners of Cornwall ignoring both this explanation and that of the year 1627, and referring back to the resolution of 1608. Privileged tinners are "spaliards with pick and shovel, watermen, boll or barrow men, dressers, blowers, and all tinners, laborers and workmen that necessarily attend the getting of tin, or the dressing, blowing, or whitening it, so long as they continue their work, without fraud or covin," and are not to be sued or to sue, outside the stannary courts, save in cases involving life, limb, and freehold. Further, "all the said former privileged tinners, if they shall discontinue their working about tin and tin works, and also all the officers, the owners of tin works in wastrel or several, the adventurers in

[1] Appendix I.

[2] In the same year, however, the court leet at Penzance presented that by stannary custom "all are tinners who take the uses and profits of tin, either by their labors or works, or in their own rights as owners, or in the right of the labor of others, as farmers" (Add. MS. 6713, fol. 279).

tin works, the buyers of black and white tin," and, generally, all others that intermeddle with tin are called tinners-at-large, and have also the liberty to sue and may be sued in the stannary courts [1] "for matters there determinable, and may also sue and be sued at common law at the pleasure of the plaintiff."

One more complaint from the warden,[2] and the Long Parliament attempted to settle matters once and for all by a public statute.[3] This act, *inter alia*, states "that tinners in the counties of Devon and Cornwall had by virtue of their charters enjoyed great liberties which did of right belong to the working tinner and not to any other, or elsewhere working, and were granted to the said tinners for encouragement in their works." It then recites the evils which had arisen from "false or feigned tinners," and says that an endeavor had been made to extend the jurisdiction of the said stannaries, contrary to ancient right and usage and the said charters, out of the places where the tinners did work, no way for the benefit of His Majesty, and that, by the said abuses, great inconveniences did follow.[4] It was enacted that "the said declaration (of 1376) should be duly observed with this, that if any person sued in the stannaries should swear in the court where he should be sued that he was not a tinner,[5] then the defendant should be discharged of such suit unless the plaintiff were a working tinner, and cause of his suit arose within the stannaries, or concerned tin or tin works, and if any person, not being *re vera* and without fraud a working or laboring tinner in or about some tin work, set, or work, within one half-year next before his suit should sue in any of the said courts, or before the warden, vice-warden, or steward of the said stannaries any person or persons that was not a tinner or tinners at the time of the suit commenced, then the defendant should have his action at the common law against the person suing, and recover ten pounds costs, if brought within two years." The sixth section enacts that

[1] Convoc. Cornw., 12 Chas. I, c. 4 (Add. MS. 6713, fol. 355, 356); Add. MS. 6713, fol. 225, c. 6. Abrogated by Stat. 16 Chas. I, c. 15.

[2] S. P. Dom. Chas. I, cccxcii, 41.

[3] 16 Chas. I, c. 15. Cf. Sir John Northcote's Note Book, 104.

[4] Cf. Convoc. Cornw., 30 Eliz., c. 22.

[5] In Devon, in consequence of the abuses of stannary writs by the bailiffs, a law had been passed by the tinners' parliament requiring the registration of every tinner (Parl. Devon, 16 Eliz., c. 12).

it should be lawful for any tinner, if he thought fit, to sue any foreigner at common law, thus removing any danger that the institution of stannary jurisdiction should be absolutely oppressive to the workmen.

Half a century later, we find by an inspection of a code of law under date of 1687 that other persons besides "privileged tinners" are recognized as coming within the stannary jurisdiction,[1] such as lords of the soil, and bounders, also adventurers in tin mines,[2] owners of blowing-houses[3] and buyers and sellers of black and white tin.[4] The term "tinner" is applied to persons who, under the circumstances, could not be supposed to be laboring tinners,[5] since the section which directs of what description of persons the grand jury was to be composed ordains that it be of the best and most sufficient stannators, "to wit, owners of tin land, owners of bounds, and adventurers for tin not being merchants or shopkeepers."[6]

Finally, by the stannary parliament of 1752, owners of blowing and smelting houses, together with their managers and agents and the conductors of their business, were placed under stricter surveillance by the stannary courts and made liable to heavy penalties for any breach of its regulations.[7] Owners and buyers and sellers of black and white tin were to have in these courts remedies for injustice and to the same tribunal were rendered amenable for misconduct;[8] the lords of the soil were protected in their rights, and the rights of bounders were regulated and restrained through the medium of stannary jurisdiction.[9] The rights and interests of adventurers in tin mines were guarded, and mutual differences and dealings of co-adventurers in the same mines were governed and settled in the same way. This interpretation was upheld by the Stannary Acts of 1837 and 1855, by which all adventurers, agents, laborers, in short all connected in any way with mines, either in supplying materials or otherwise, are held to be miners and may sue and be sued in the stannaries.[10]

[1] Add. MSS. 6713, fol. 238 *et seq.*, c. 1–4, 8, 23.
[2] *Ibid.*, c. 5, 6. [3] *Ibid.*, c. 14. [4] *Ibid.*, c. 16, 31.
[5] *Ibid.*, c. 8, 22, 32. [6] *Ibid.*, c. 20.
[7] Convoc. Cornw., 26 Geo. II, c. 1–4, 6, 10, 17. [8] *Ibid.*, c. 8, sec. 1–6.
[9] *Ibid.*, c. 9, 10, 14. [10] 6, 7 Will. IV, c. 106; 18, 19 Vict., c. 32.

Can we harmonize these various interpretations? Little doubt exists that, granted at the outset a customary law of mining, the exemption from outside jurisdiction was purposely confined from the first to those tinners whose personal attendance was deemed essential to mining. As for the others, whether their cases were decided in the manor or in the shire and hundred courts or by the royal justices, it is not given to us to say. But the probabilities of the case lead to the conjecture of a gradually encroaching stannary jurisdiction; the outsiders came to use the stannary courts for their suits and to plead before the warden and his officers to the exclusion of all other tribunals. When we find that as early as the thirteenth century there probably existed many quasi-capitalistic miners who conducted their operations through the medium of hired labor, to say nothing of dealers in tin,[1] when we admit the possibility of disputes on technical questions arising between owners and adventurers and their hired help, which could not be settled save in the stannary courts where the latter were forced to plead, and when we add to this the aggressive spirit which from the fourteenth century characterized the tinners, conditions were evidently existent which justify our conjecture.

To interpret the charter of 1305 one must recollect that, while addressed at the outset to all tinners, it includes in the opening paragraph grants of privileges to specially mentioned bodies. The King, ignoring the charter of 1201 and its confirmation, granted freedom from pleas of serfdom to the working miners on his own estates, with the privilege of being liable solely before the warden or his officers for actions arising in the stannary and not involving land, life, or limb. What follows is addressed to all tinners.[2] To them, including for the first time those not actually engaged in manual labor, it confirms the ancient rights of bounding and of wood and water, and grants the right of pleading in the stannary courts. On the civil side these courts were given cognizance over pleas between tinners and between tinners and foreigners, whenever the case arose within the stannaries, with the proviso that

[1] Appendix A.

[2] The proper interpretation of this general grant of judicial rights is, of course, to make the "stannatores predictos" refer back to "eisdem stannatoribus" a few lines above, who have been confirmed in the prescriptive rights of bounding. They must then include all tinners, both workmen and proprietors.

in any suit in which a tinner was involved, if he wished to put it to an inquest of the country, the jury must consist half of tinners. All criminal cases where the accused was a tinner were dealt with in the ordinary courts, with the concession to the stannaries that the tinners were to be lodged in a special jail. Now the exposition in 1376 referred merely to the franchises granted for the first time by this charter and left untouched the ancient customary rights antedating all written documents. All the questions dealing with the word "stannator" concerned merely the special privileges granted to the tinners on the royal estates. In their case it was defined as including laborers alone, and only so long as they worked. Upon the most important question as to the comprehensiveness of the word as applied to the great mass of tinners outside the royal demesnes, no complaint appeared, and no definition was attempted.

It is not until the period of Tudors that we meet with renewed efforts to settle the conflict of jurisdictions. The charter of pardon granted by Henry VII gave, as we have seen, a wide construction to the term "tinner," but in the report of the 1524 commission there appears a novel and arbitrary mode of interpretation. The old distinction between the tinners on the royal estates and those elsewhere had probably long been dropped, and in its stead arose a theory which attempted to apply to all tinners that contrast between working and non-working tinners brought out in 1376.[1] The main point of the distinction was the claim that working tinners were to be impleaded only in the stannary courts, while others possessed rights of suit at stannary or common law courts at their choice. Granted this proposition, — and it seems to have found acceptance from the start, — the question resolved itself into a discussion as to what tinners belonged in the privileged or working class, and who were merely "tinners at large."

The judges in 1608 decided that the former class comprised blowers and all other laborers while at work. The convocation of 1624 extended the privilege of not being suable in any other than the stannary courts to owners of blowing-houses and to the adventurers at any charge for getting or making tin. It was probably this extension of the privilege which occasioned the second reference to the judges three years later. Their decision was little

[1] Expediency was, of course, the sole excuse for this new interpretation.

more than a reversion to the rule of 1608. The same statement may be made with regard to the rules laid down by the King in Council in 1632, the act of the stannary parliament of 1636, and the Act of Parliament in 1641. An inspection of the two succeeding codes of stannary law, of 1687 and 1752, makes it clear that their provisions are based upon this principle, and an uninterrupted course of usage and practice in conformity with it carries the doctrine down to 1837.

No local limits seem to have been prescribed for the stannary jurisdiction until the charters of 1305. In these, as regards the tinners on the royal estates, exemption was granted from pleading elsewhere than before the stannary courts for pleas arising "infra predictas stannarias" (*i. e.* "quae sunt in dominico nostro"). All others were answerable to the stannary courts for pleas among themselves and between themselves and foreigners concerning trespasses, plaints, and contracts made "in places where they worked, within the stannaries arising." In 1376 this was interpreted to mean that the jurisdiction extended to places where the workmen[1] labored and nowhere else, a decision which, if acquiesced in by the tinners, would have resulted in unending confusion. The charter of 1305 itself, in its clause of preëmption, calls for the coinage of tin in Lostwithiel, Bodmin, Liskeard, Truro, and Helston. Contracts made there for the sale and purchase of tin were unquestionably determinable in the stannary courts. An instance of this has been cited in the case of Boscawen *vs.* Chaplin, where the cause of action arose upon a contract made at Truro for the sale of tin between persons not laborers, which, nevertheless, was tried in a stannary court.

In the discussion before the judges in 1608 the question of jurisdictional limits was again discussed. The warden raised the points that "the place upon the words 'infra stannarias nostras' be declared to extend to the divisions of every stannary court respectively, and not only to the place and to the work," and "that the matters of plea to be determined in that court be declared to com-

[1] The confusion of opinion with regard to stannary rights at that period, as later, is seen in the use of the word "operarii" in this connection. According to the above interpretation of the charter of 1305, the right of stannary pleading was given to all tinners, whether workmen or not.

prehend all manner of suits where one of the parties is a tinner." The judges decided that transitory actions between tinner and tinner, worker and worker might be determined in the stannary courts, even though the cause of action should be collateral to or arise outside the stannaries, or at common law, at the election of the plaintiff. But if in such case only one of the parties were a tinner, the cause of action being transitory and collateral to the stannaries, the defendant might have the case removed to a foreign court. They ruled that the courts of the mines had no jurisdiction over any local cause arising outside of the stannaries, and that the privileges of the workmen in the stannaries did not extend to any local case arising outside the stannaries, whereby any freehold should be demanded, "for that matters of life, members and plea of land were exempted by express words in their charters, and no man can be exempt from justice."

The convocation of 1624 [1] recognized the stannary jurisdiction as embracing the contracts of all persons in the buying or selling of uncoined tin, without limitation or qualification as to their condition, and without reference to the place of the transaction. By the resolutions of the judges in 1627 it was decided, with reference to the extent of the stannaries, that they comprised "every village, tithing and hamlet, and all lands within any of the said villages, tithings and hamlets, wherein any such tin work then was, or at any time thereafter should be settled, found and wrought during the continuance of any such work only, and no longer, and in no other place."

A further step was taken by the Privy Council in 1632. "We cannot yet discern," so runs their decision, "but that the stannaries do extend over the whole county of Cornwall. The exemption of tinners from toll is over the whole county. The power to dig and search for tin is over the whole county, saving under houses, orchards, gardens, etc. The tin wrought in any place of the county must be brought to the coinage. The privilege of exemption or preemption is of tin gotten over the whole county. Fines and amerciaments set in the stannary court are leviable in all parts of the county. Judgments had in the stannary court may be levied over the whole county by process of the stannary. For trespasses in tin works

[1] Convoc. Cornw., 22 Jas. I, c. 5.

process may be executed in the whole county. Watercourses for the tin works or tin mills may be made in any place of the county." As regards Devon the judges declined to express a definite opinion, but let it be understood that in their view, for reasons of expediency, the same rule ought to be applied as in Cornwall.[1]

In the twenty-first section of the laws enumerated by the tinners' parliament in 1636 the stannary jurisdiction is recognized as embracing all dealings in black tin, and gives by action a remedy to the party wronged without any qualification as to the condition of the offending parties, or of those to whom the remedy is given, and without any reference whatever to the place of dealing. But the statute of 1641 returned to the exposition of 1376, "that the words — 'in locis ubi operantur' — be expounded of the village, hamlet and tithing where some tin work is situate, and not elsewhere, and no longer than the same tin work is, or shall be, in working." Its operation, however, was deemed in practice to be confined to the case of laboring tinners, and this construction was adopted in the laws of 1687 and 1752, and so continued until recent times.

It has been seen that, from a legal standpoint, the stannaries were a peculiar jurisdiction under the operation of certain laws, customary and statutory, technical and non-technical, for the administration of which a royal officer, the warden, was responsible. The head of the stannary system was accordingly the King or, after 1338, the Prince of Wales as Duke of Cornwall. Beneath him stood the warden, then the vice-warden, and lastly the lower stannary courts with their stewards and juries of miners. The warden,[2] however, as far back as we can find record, in all but exceptional cases, invariably delegated his judicial powers to his lieutenants, the vice-warden and the stewards, interfering in legal questions only in cases where appeals were made from the vice-warden's verdicts.

[1] A writer of this period argues that there has been really no question but that the stannary jurisdiction extended over all Cornwall, because tin was found there so universally. In Devon, where but a few mines existed, lay most of the difficulty. He argues that expediency and convenience demand that the jurisdiction of the Devon stannaries be extended over the entire country (Hales MS. 83, Lincoln's Inn Lib., fol. 232, 233).

[2] Later styled "lord warden."

The vice-warden's powers were, first, magisterial, for the granting of injunctions,[1] the issue of warrants, and the subpœnas of the peace,[2] replevins,[3] and other writs of similar nature, the prevention by summary process of offences against stannary law, and their summary punishment if perpetrated.[4] Next, as judge of the vice-warden's court, he had original jurisdiction in all matters of equity. The origin of this power remains obscure. We find no mention made in the charter of 1201, of a *locum tenens*, or vice-warden, *eo nomine*, but the word "ballivus" there used is a term of extensive signification and may perhaps be construed to embrace the office of vice-warden. Thus it is written, "stannatores non recedant ab operationibus . . . nisi per summonitionem capitalis custodis stannariarum et ballivi ejus;" and again, "capitalis custos stannariarum et ballivi ejus per eum habeant supra predictos stannatores plenariam potestatem."

The charter of 1305 expressly reserves for the "locum tenens" of the wardens a power, "tenere omnia placita . . . de omnibus transgressionibus, querelis, et contractibus." This expression would imply the confirmation to the vice-warden, as representing the warden, a wide grant either of new or of an existing original jurisdiction, namely, to hold all pleas concerning all trespasses, complaints, and contracts arising in places where tinners worked. We must not forget, however, that the stewards' courts were in existence as early as 1243, and that, therefore, almost from the start they must have divided the jurisdiction over the tinners with the warden and his deputy. As to anything in the charters of John or Edward specifically pointing to the exercise of an equitable jurisdiction by the vice-warden, it has been shown that they contain no phrases which are not to be found in common law writs.[5] On the other hand, the early petitions addressed to the Prince[6] are rarely made to the warden, but to the Prince's proper officer, who-

[1] Harl. 6380, fol. 51. [2] *Ibid.*, fol. 70. [3] *Ibid.*, fol. 43.

[4] For example, he was given power to punish at his discretion tinners detected in selling their ashes to plumbers or pewterers, and upon application he issued orders against the impleading of tinners in foreign courts (Convoc. Cornw., 22 Jas. I, c. 6, 12, 13).

[5] Smirke, 102.

[6] Cf. the case of Abraham the Tinner (White Bk., i, c. 15), or that of the parson of St. Ladoke (Smirke, 25).

ever he might be, nor were they restricted to the subject of mining.[1] It is probable that from these petitions and the orders from the Prince's council there gradually arose an equitable jurisdiction. Similar petitions to great lords and their councils ripened into courts of chancery, notably the case of courts of the President and Council of Wales,[2] and probably only the interposition of Parliament[3] prevented the growth of many local courts of equity. The Prince's council survived those of other subjects,[4] and in the case of the Duchy of Cornwall long usage, the fact that a large part of the petitions related to the stannaries, and the accidental but long-continued union of the wardenship with other high offices of the Duchy[5] coöperated to narrow the Prince's jurisdiction to the same subjects as those embraced by the tinners' charters.

As no records were kept of the proceedings of the vice-warden's court until 1752,[6] it is impossible to say how early it had actual practice in equity. The case of Boscawen *vs.* Chaplin, already mentioned, was evidently dealt with as a special case, as it appears to have been tried before Sir William Godolphin, "under-warden and chief steward," and William Beare, his under-steward, and a jury of twenty-four persons, one half of them tinners and the other half merchants, at a court summoned for the stannary of Blackmore. It would be difficult, therefore, to draw from it any conclusions which could influence the question.[7] An undoubted jurisdiction in equity was exercised by the lord warden in the case of Glanville *vs.* Courtney in 1593,[8] not many years later. Carew in 1602 wrote of the lord warden: "He supplieth the place both of a judge for law and of a chancellor for conscience, and so taketh

[1] Cf. Smirke, 26.
[2] Coke, iv, 242–245.
[3] Stat. 15 Rich. II, c. 12; 16 Rich. II, c. 2.
[4] It was recognized by Parliament (Parl. R., ii, 371).
[5] Notably with that of the steward of the Duchy (Close, 8 Edw. II, m. 7).
[6] In that year the convocation (26 Geo. II, c. 16) directed that the records henceforth be kept.
[7] This case is certainly the only one yet found where the vice-warden presided at the trial of a jury cause. It has another peculiarity, namely, that the jury consisted of twenty-four persons instead of six, the ordinary number in all trials of the steward's court. The question was evidently a new one, and the probability is that it was deemed of so much importance that both of the parties assented to its being tried in the most formal and solemn manner which could be devised.
[8] Smirke, 45. Cf. Acts of P. C., 1593, pp. 266, 342.

hearing of causes, either *in forma juris* or *de jure et equo.* He substituteth some gentleman in the shire, of good calling and discretion, to be his vice-warden." [1] In a dissertation written a few years later we find words to the same effect, and the vice-warden's power in equity was declared by Coke in 1608 to be founded on prescription.[2] From this admission we may infer that a court of chancery by prescription — which could be no other than the vice-warden's court — existed as an integral part of the jurisdiction of the stannaries anterior to the decision of 1608, and that it was a court of original jurisdiction in all matters belonging to a court of chancery to decide; for it would be quite inconsistent with the existence of such a court to suppose that matters properly cognizable there should originate elsewhere, and still less that they should originate in the common law or jury court of the stannaries.

Later stannary laws recognize this as a fact. Thus the twenty-first section of the laws confirmed by the convocation of the Cornish tinners in 1624, and the eighteenth of those in 1636, both of them merely declaratory of custom, state that in cases of disputed rights to the possession of a tin work under bounds the tinners in possession should "continue their possession until verdict rendered against them, and in the meanwhile the work should be sequestered and placed in impartial hands, to answer to the man who should recover the right by legal trial." The objects of this custom could be attained only by application to some other court than that in which the question of right was ultimately to be tried by law and decided by the verdict of a jury, and this other court must necessarily interpose and by an equitable jurisdiction enjoin the claimants not to molest the possessors until the law should have decided the right. It must direct an issue to try the legal right in the steward's court and in the meanwhile it must secure the due administration of the property for the benefit of the party finally successful at law. The last declaration of stannary law, that of Cornwall in 1752, provides that in cases of dispute between adventurers in tin mines as to the limits of their underground rights, the vice-warden, on the application of one of the litigant parties, is to issue his injunction for the staying of any further working of the mine within the contested limits, until these rights should have been de-

[1] *Survey of Cornw.*, ed. 1811, p. 53. [2] Harrison, Appendix I.

cided by a jury in the common law court of the steward.[1] Another clause provides that in case of disputes between adventurers in the same mine or any of their executors as to the expenses of working it, the vice-warden is empowered on petition of the purser of the mine to hear and to decree payment of such expenses as he may find justly due, and in default of payment to decree, in the first instance, the sale of the defendant's portion of tin ore. In case this should be insufficient, he is to direct the sale of the share in the mine itself.[2] The powers given the vice-warden by these clauses are so perfectly analogous in their principle to those of which the judge was possessed by the ancient custom of the stannaries, that they can be considered only as explaining and confirming powers which already existed.

Finally, in his official report to the Prince's council in 1785, Vice-warden Thomas thus described the then court of the lord warden or vice-warden. "This court has a jurisdiction through all the stannaries: the lord warden, or in his absence the vice-warden, is the judge, and causes are heard therein in a summary way on petition in writing, stating the petitioner's case. In the exercise of his functions of original jurisdiction as a judge in equity, it is the duty of the vice-warden to be ready at all times to receive the petitions in writing of all persons relating to subjects cognizable before him as vice-warden, to issue orders in writing in the name of the lord warden, and under the seal of the Duchy of Cornwall, for all persons complained of in such petitions to appear before him at certain times and places within the county of Cornwall, to answer the complaints contained in such petitions, and to hear such complaints and make decrees and orders therein, agreeable to equity, and according to the laws and customs of the stannaries."

The vice-warden's appellate jurisdiction dates back certainly to 1510, and probably much earlier. In that year we see that the course of appeals lay from steward to vice-warden, to warden, and to the Prince's council,[3] and in 1565 this measure was confirmed in the

[1] Convoc. Cornw., 26 Geo. III, c. 9. Cf. also Add. MS. 6682, fol. 507.

[2] Convoc. Cornw., 26 Geo. II, c. 11.

[3] Parl. Devon, 2 Hen. VIII, c. 32. Cf. also Convoc. Cornw., 30 Eliz., c. 36. In the King's Field, in Derbyshire, appeal lay from the barmote court of the mines to the court of the Duchy of Lancaster (Add. MS. 6682, fol. 41). There was no appeal from the miners' verdict in the Forest of Dean (Houghton, pt. ii, art. 22).

case of Trewynnard *vs.* Roskarrock,[1] while subsequent stannary records show that it was maintained.[2]

Somewhat later, however, arose a practice, quite unwarranted by law, of using the vice-warden's court as one of original common law jurisdiction. It would appear that this was exercised chiefly in cases of debts due to merchants and tradesmen for the supply of goods requisite for the working of mines, and to miners for labor there performed. The delays habitually incurred in stannary judgment and execution in cases of debt in the steward's courts were great, and judgments were often to be obtained more expeditiously by a resort to the ordinary common law courts than to those of the stannaries.[3] In affairs of this sort, as appears from the correspondence of Vice-warden Wallis in the early decades of the last century, a jurisdiction had been exercised by the vice-warden's court for a period as far back as there are records of the court, which, as we have noticed, date only from 1752.[4] In confirmation of this statement, we may take the case Rawles *vs.* Usticke, tried before the vice-warden in 1759, in which the question of jurisdiction was not raised in any way, the ground of the subsequent appeal to the lord warden being the personal non-liability of that part of the adventurers against whom the vice-warden had decreed payment of the debt. From this period until the action Hall *vs.* Vivian the authority of the vice-warden to entertain such petitions and to adjudicate in such cases was never brought into question.

That the original common law jurisdiction of the vice-warden was not founded upon a firm basis was indicated with sufficient clearness by the case Trewynnard *vs.* Roskarrock, where the Star Chamber in rendering decision remarked that all actions or suits for redress of wrongs or injuries, the appropriate remedy for which was at common law, must originate in the steward's court. In conformity with this view, the parliament of tinners in 1588 had declared that "every cause that the court will hear should commence in the steward's court and have its due trial there, and that, for default of justice there, the party grieved might appeal."

The improper use of the vice-warden's court appears to have been due largely to the system of adventuring which in the eight-

[1] Coke, iv, 230.
[2] Cf. Convoc. Cornw., 22 Jas. I, c. 40, 42, 45.
[3] Harrison, Appendix I.
[4] *Ibid.*, 93.

eenth and nineteenth centuries was at its heignt in the two counties. The creditors who supplied the mines were made to look to the purser of the company,[1] usually one of the adventurers, and as the purser received all the money arising from the produce of the mine the creditors could usually reach him by petition direct to the vice-warden, while the purser, in turn, had the power to recover against the adventurers in arrears. This right of petitioning the vice-warden to recover costs from adventurers in mines was, so it was claimed, expressly provided for in the eleventh article of the laws sanctioned by the convocation of 1752. Powers were given for the sale of the share of any adventurer and of the tin stuff, if upon an investigation of the merits of the petition a question should arise as to the sum due the petitioning creditor. It would then be incumbent on the vice-warden at the request of either of the parties to direct an issue to the stannary law court to try that point by a jury, and on the return of their verdict to decree accordingly.

The law was silent as to the particular course which the creditor was to pursue for his recovery, and with reference to the different objects it embraced, the law would seem to have a two-fold aspect as to jurisdiction. The first paragraph of the clause was confined to a transaction altogether of a legal character, namely, a contract entered into between an adventurer in a mine, exclusive of his co-adventurers, and a merchant for the supply of goods or material, or a working tinner for the sale of his personal labor. Now the law never directed the mode of recovery but simply restrained the creditor from proceeding against any other of the co-adventurers in the mine; so that it would seem obvious that the mode of proceeding should have followed and conformed to the general law governing stannary jurisdiction in all other cases of debt, for which legal remedy was provided by suit in the steward's court. The other aspect of the case is shown in the two succeeding paragraphs of the law, which related to transactions of an equitable character, namely to the settlement of claims of one of a set of co-partners in a concern against one or more of his partners; and this, as a matter of equity, should have been settled by the vice-warden. The course of procedure, therefore, would have been consistent throughout.

[1] Convoc. Cornw., 2 Jas. II, c. 6 (Add. MS. 6713, fol. 238).

The creditor would proceed against his debtor at common law in the court of the steward. The debtor against whom judgment was obtained by the verdict of a jury in that court would have, as against his co-partners, remedy in equity in the court of the vice-warden.

The practical effect of this unlawful practice of suing for debts directly to the vice-warden was to diminish greatly the activities of the lower courts by withdrawing from them almost all the most important business properly belonging to their jurisdiction, namely, actions brought by merchants and tradesmen for the recovery of debts incurred by the supply of goods or materials, and by tinners for their wages, since, with regard to simple contracts having no connection with the working of mines, recourse had usually been had to the common law tribunals.[1] Matters having continued thus for a considerable period, the jurisdiction of the vice-warden was suddenly called into question and denied by two decisions, in one of which, that in the case of Hall *vs.* Vivian in 1825,[2] he was made to pay heavy damages. As a result both the lord warden, the vice-warden, and the stewards, declined to hold any courts [3] until their respective jurisdictions should be settled, and continued in this determination until the passage of the Stannaries Act of 1837.

This Act, with the later stannary legislation embodied in successive statutes in 1839, 1848, 1855, 1862, 1869, 1875, and 1897,[4] has taken the bold step of abolishing the steward's courts in their entirety, and of vesting the common law as well as the equitable jurisdiction of the tinners in the vice-warden. The latter, who under the new laws must be a barrister, holds his court, which now embraces mining matters of all sorts in both counties, at Truro at least once in three months, and in his capacity as judge of equity

[1] The statute 16 Chas. I, c. 15, had allowed the tinners to sue one another outside the stannary courts.

[2] Not reported, but quoted in substance by Smirke (37).

[3] Delabeche, 619. Cf. Parl. Papers, 1831–1832, xxxv, 407. No stewards held courts in the stannaries from Jan. 1, 1824 to June 30, 1828. Only one case was heard from 1828 to the end of 1831. The vice-warden's court was suspended from 1822 until September, 1828, while only twelve cases were heard from 1828 to 1832.

[4] 2–3 Vict., c. 58; 11–12 Vict., c. 83, sec. 7; 18–19 Vict., c. 32; 25–26 Vict., c. 89, sec. 68; 32–33 Vict., c. 19. Cf. Parl. Papers, 1868–1869, v, 193, 207; 38–39 Vict., c. 66, sched.; 59–60 Vict., c. 45.

may receive appeals upon any grounds upon which appeals are granted by the courts at Westminster. From the common law side also of the vice-warden's court appeals may be taken by writ of certiorari to the court of the King's Bench; [1] and, on paper, a litigant may appeal from the vice-warden to that now purely ornamental officer, the lord warden of the stannaries, who in fact, however, always refers the matter to the justices.[2]

For ordinary purposes resort was usually had to the common law courts of the stannaries, those held by the stewards of the eight stannary districts. Ordinarily the stewards were appointees of the warden,[3] so that the latter in reality was the creator of all stannary courts immediately beneath his own equitable and appellate jurisdiction. What little is known of their origin has been dealt with in a preceding chapter; certainly the presence of entries of the profits of stannary courts in Devon in one of the Pipe Rolls of Henry III[4] points to the existence of the stewards' court as early as 1243. Carew speaks of them as follows: "The tinners of the whole shire are divided into four quarters; to each of these is assigned by the lord warden a steward, who keepeth his court once in every three weeks. They are termed stannary courts, and hold plea of whatsoever action of debt or trespass whereto any one dealing with black or white tin, either as plaintiff or defendant, is a party. Their manner of trial consisteth in the verdict given by a jury of six tinners, according to which the steward pronounceth judgment." [5] A later writer says of them: "They are courts of record for trying civil actions between tinners, or between tinners and any other persons, arising within the stannaries, and for recording proclamations of new tin bounds, and giving judgment thereon. Each of these courts has jurisdiction throughout the respective stannary in which

[1] 6–7 Will. IV, c. 106, sec. 42.

[2] Journ. Stat. Soc., li, 504.

[3] Smirke, 97; Add. MS. 6682, fol. 507; S. P. Dom. Jas. I, xxviii, 130. In the Forest of Dean the mineral courts were presided over by the constable of St. Briavels, assisted by the gaveller and the castle clerk (Houghton, pt. ii, art. 20).

In Derbyshire, the courts were held by the "bar-master" or steward, nominated in the case of the King's Field by the Crown, elsewhere probably by the lord of the manor in which the mines were situated (cf. *Compl. Min. Laws Derb.*, pt. iv, art. 2). Every three weeks a court was held, and twice a year a "great court" or leet (cf. *Compl. Min. Laws Derb.*, pt. iv, art. 2; pt. i, art. 16; pt. ii, art. 18, 25; Houghton, ii, art. xxv).

[4] Pipe, 27 Hen. III, Devon.

[5] Carew, ed. 1811, p. 58.

it is held, and the causes are tried therein before the stewards of the respective courts (who are the judges appointed by the lord warden), and a jury consisting of six tinners. The duty of the stewards of the four law courts of the stannaries, which are held from three weeks to three weeks,[1] is to hear all causes and subjects cognizable in their courts that are brought before them, and to record the verdicts of the juries therein, and to give judgment in such cases according to the laws and customs of the stannaries; and they are to take care of the records, plaints, entries, proclamations of tin bounds, and other proceedings in their respective stannary courts for two years after the determination of the several suits and causes to which the records relate, and afterwards to deliver them over to the vice-warden to be preserved and kept in such place in the stannaries as the lord warden or vice-warden shall direct."

The stewards as well as the vice-warden, and in most cases concurrently with him, were invested by stannary law with the powers of magistrates within their respective districts, for the prevention by summary process of offences against the laws, these powers being likewise wholly distinct from their judicial functions as judges in their respective courts.[2] With respect to the legitimate exercise and the limits of their judicial functions, we may merely refer once again to the various interpretations of the charter of 1305, which from time to time were in vogue, with the remark that by the principles laid down in 1608 by Coke and Fleming we may in general assume that it was within the competency of the stewards', or common law courts of the stannaries (with those exceptions specified in the charter), to take cognizance of, and to redress every matter concerning the stannaries, which would by the common law of the land in ordinary cases be properly cognizable and remediable by the ordinary English common law jurisdiction. The substitution of this peculiar jurisdiction in the place of the general jurisdiction of the country must be supposed to have been complete for all the purposes for which it was substituted, and the system as effective for all these purposes as the one which it was made to represent and exclude.[3]

[1] Convoc. Cornw., 22 Jas. I, c. 25. [2] *Ibid.*, c. 12.

[3] Harl. 6380 contains various forms of procedure used in the stewards' courts in the time of Henry VIII and Elizabeth.

In addition to their magisterial functions in the prevention of offences against the stannary laws, and of these judicial functions at the ordinary stannary courts for the trial of actions at common law relating to stannary affairs, it was the duty of the stewards to hold special courts at the request of litigants, when necessary, for the trial of rights in tin works.[1] These courts, which must be distinguished from the special juries sometimes convened by the steward to declare the custom, if doubt existed,[2] were held for trial of tin works within the tin bounds or at the mine itself and are rarely entered under the name,[3] although probably the inquisition and findings of juries respecting trespass in such works, of which entries are frequent, may be referable to adjourned courts so held in the works themselves.[4] It was the steward's duty also to hold so-called "customary courts." These, according to ancient custom, were always held "the morrow after certain fairs within each stannary, for the benefit of such as do attend the fairs and courts," and were always to be kept at the place where the fair was held, or else at the market town nearest to the same place within the stannary, and no further off.[5] The steward was also to preside at the courts leet within the respective stannaries.

If we omit from consideration the special and the customary courts, we shall find the legal business of the lower courts transacted in thirteen sessions held each year in each of the eight districts,[6] of which two, one held in the spring and the other in the fall, went by the name of "great" or "law courts." In all thirteen, however, the cases both criminal and civil were decided, if required,[7]

[1] Convoc. Cornw., 16 Hen. VIII, c. 30.

[2] Cf. Smirke, 68.

[3] Ct. R., bdle. 156, no. 21.

[4] Smirke, 96.

[5] Convoc. Cornw., 22 Jas. I, c. 18, 27. These probably were similar to the "courts of pie-powder."

[6] This is by actual count upon the court rolls, although the stannary law prescribed a court every three weeks, cf. Parl. Devon, 16 Eliz., c. 29; Convoc. Cornw., 12 Chas. I, c. 22 (Add. MS. 6713, fol. 230); 22 Jas. I, c. 27. A tri-weekly barmote court was held in the Derbyshire mines by a bar-master and a miners' jury (Houghton, pt. ii, art. 25; *Compl. Min. Laws Derb.*, pt. iv, art. 2; pt. i, art. 16).

[7] Cf. Ct. R., bdle. 156, no. 27 (Blackmore, 3 Rich. II), "Ballivus presentat quod Thomas Robertus Gilmyn levavit hutescium juste super Johannem Gentil, quia entravit in domum suam et cepit et asportavit diversa bona valencia 100s. Et defendens dicit quod non est culpabilis, et *petit quod inqueritur*."

by a jury of tinners [1] returned by the court bailiffs. Previous to the remodelling of the courts in 1837, and save in cases where a special jury was struck,[2] the practice was to try by six jurors,[3] but by the stannary codes of Devon and Cornwall in 1552 and 1524 respectively the number is set down as four.[4] As to whether the latter number was used in earlier centuries, the court rolls are too brief to give any information, but that trials as early as the thirteenth century were conducted on the jury principle the wording of the charter of 1305 would seem to furnish satisfactory proof.[5]

The pleas entered in the courts consisted of all sorts of actions, personal and otherwise. Many of them related to trespasses for taking tin and entering into "opera stannaria" or tin works.[6] As soon as the practice was established of entering bounds upon the court rolls,[7] we find them described as "opus stanni" or "opera stannaria," and there seems little doubt that the court exercised, as might have been expected, the power of adjudicating upon this species of property. There are also, down to a late date, numerous

[1] Cf. Parl. Devon, 25 Hen. VIII, c. 8. One is tempted to follow the analogy of the manorial courts and those of the shire and hundred, by making the introduction of trial by jury in the stannary courts date to the thirteenth century (Maitland, *Select Pleas*, lxv–lxvi). This may have been the case, but we have no means of proving it.

[2] Harrison, 170; Add. MS. 6713, fol. 104; Add. MS. 6682, fol. 507; Smirke, 68; Harl. 6380, fol. 16. In 1596, at Marazion, a special court was held before the vice-warden and the steward and a jury of 25 veteran tinners, chosen from 72 summoned by the bailiff to declare the custom of the stannary on a certain obscure point of law arising in the case of Clies *vs.* Nanspan. In 1357, again, special juries of 12 were impanelled in Blackmore, Foweymore, and Tywarnhail stannaries, for an inquisition concerning the alleged deterioration of Fowey as a port, by reason of the silt from the tin mines (Ct. R., bdle. 156, no. 26).

[3] Cf. Add. MS. 6713, fol. 130; Smirke, 99. In Devon the ordinary jury in civil suits involving over £5 was 24 (Parl. Devon, 6 Edw. VI, c. 6), and if less than £5 four, except in cases of trespass (*ibid.*, c. 14). Twelve was the usual number of jurors in the mineral courts of Derbyshire (*Compl. Min. Laws Derb.*, pt. iv, art. 2). In the Forest of Dean the juries were from 12 to 48 (Houghton, pt. ii, art. 22).

[4] Parl. Devon, 6 Edw. VI, c. 14; Convoc. Cornw., 16 Hen. VIII, c. 3.

[5] The clause, namely, which provides for the choosing of a jury, half of tinners, in case the tinner impleaded for an offence occurring outside the stannaries willed to throw himself on the country. "If for matters touching the stannaries wholly, let inquest be made as has been usual."

[6] Ct. R., bdle. 156, no. 27.

[7] See p. 161, n. 1. Cf. Convoc. Cornw., 30 Eliz., c. 15; Vinogradoff, 372, 374.

entries of hue and cry levied in respect to trespasses upon tinners,[1] and presentments by bailiffs of unjust levyings of the hue and cry were a common cause of amercement.[2] This process with regard to mere trespasses was not peculiar to the stannaries but prevailed also in the Forest of Dean, and was enjoined against disturbers of the staple.[3] Cases of debt and contract,[4] assault and battery,[5] thefts of money and ore,[6] and the diverting of a miner's watercourse,[7] and cases peculiar to the stannaries, such as the smelting of impure tin,[8] formed a large proportion of the legal business transacted at both the leet and the court of general session.[9] Another class of cases, appearing early in the records, serves in a way to indicate the separation of the tinners from the ordinary courts of the country; such cases, for example, as the trespassing with swine and geese on a neighbor's cornfield,[10] cutting another's timber,[11] infractions of the assize of beer,[12] baking of unwholesome bread,[13] and, shortly after the Black Death, evasion of the Statute of Laborers.[14] Instances also are not lacking of an entire tithing being fined for failure to repair its roads.[15]

Offenders might be presented for trial in one of several different

[1] Ct. R., bdle. 156, no. 27; bdle. 159, no. 16. Ordinarily such cases would be dealt with at the sheriff's tourn or in the manor.

[2] "Hutescium injuste levatum" was an offence cognizable by the leet at common law (Fleta, fol. 113). This process was also prevalent in forest law (Coke, iv, 294). It would seem that the presentment for trespass did not preclude suit for damages by the injured party.

[3] 27 Edw. III, c. 2, sec. 4.

[4] Ct. R., bdle. 168, no. 5; bdle. 156, no. 27; bdle. 157, no. 7.

[5] *Ibid.*, bdle. 156, no. 27; bdle. 157, nos. 7, 13.

[6] *Ibid.*, bdle. 164, no. 34 (Tynwarnhail, 49 Edw. III). Cf. also *ibid.*, bdle. 157, no. 7; bdle. 156, no. 27.

[7] *Ibid.*, bdle. 157, no. 13. Cf. Early Chanc. Proc., bdle. 68, no. 52.

[8] Ct. R., bdle. 165, no. 38 (Ashburton, 6 Hen. VI).

[9] Cf. the sheriff's tourn (Pollock and Maitland, i, 546).

[10] Ct. R., bdle. 159, no. 1 (Foweymore, 3 Hen. IV).

[11] *Ibid.*, bdle. 156, no. 27 (Blackmore, 3 Rich. II).

[12] *Ibid.*, bdle. 156, no. 27. The assize of beer was a source of great profit to the manorial courts (Maitland, *Select Pleas*, xxxvii, xxxviii).

[13] Harl. 6380, fol. 67, 68; Add. MS. 6713, fol. 71. This would naturally have fallen to the manorial courts or to the sheriff's tourn (Pollock and Maitland, i, 546).

[14] Ct. R., bdle. 161, no. 81 (Penwith and Kerrier, 29 Edw. III).

[15] Decennarius et tota decenna de Tremodret et Tregarrek in misericordia quia non reparavit malam viam apud Roche (Ct. R., bdle. 157, no. 13).

ways. Ordinarily this was the duty of the bailiffs or court officers, but in the exercise of this function we find them assisted by the village tithing men,[1] who, in a way, served as a link between the stannaries and the hundred and shire courts, besides emphasizing the analogy of the leet to the sheriff's tourn.[2] As regards civil suits, the stannary laws inform us that in any case in which a bailiff was involved, the return was made to the court by the tithing man of the place where the venire arose,[3] and the same procedure held good in case the bailiff was challenged in a suit between party and party on the ground of favoritism.[4] Presentments were made also by the toller, who in this light seems to have stood as a connecting link between the stannaries and the manor. He first appears as the servant of the landlord who received from the tinners their toll.[5] Next we find him, no doubt in his master's interests, intervening in the bailiff's absence to arrest ore in dispute between party and party, and to commit it to impartial persons until the contention was settled.[6] From that it was but a short step for him to be empowered to make returns for civil suits when neither bailiff nor tithing man was able to act.[7] Repeated instances also can be found of presentments by the toller of offenders, in the same nature as those brought by the tithing men and bailiffs.[8]

The two leet courts held each year with view of frank-pledge[9] contain several novel features. In the stannary of Blackmore, as far back as our earliest stannary court rolls can carry us,[10] the tithing men of eight tithings in the hundred of Powder appear with their tithings to do suit and to present criminals, acting in this capacity as ex-officio bailiffs.[11] Another feature was the present-

[1] Ct. R., bdle. 156, no. 27. [2] Maitland, *Select Pleas*, xxx, xxxvii.

[3] Convoc. Cornw., 12 Chas. I, c. 28 (Add. MS. 6713, fol. 231); Add. MS. 6713, fol. 298, c. 63; fol. 246, c. 52.

[4] Add. MS. 6713, fol. 279 (Penwith and Kerrier custom).

[5] See p. 160. [6] Harl. 6380, fol. 39.

[7] Add. MS. 6713, fol. 279, c. 52; fol. 298, c. 63. Two tollers made the return in a civil suit, and were named by the parties at variance.

[8] Ct. R., bdle. 161, no. 81 (Penwith and Kerrier, 30 Edw. III).

[9] Add. MS. 24746, fol. 122; Ct. R., bdle. 156, no. 2. [10] 1355.

[11] Cf. *ibid.*, bdle. 157, no. 13 (Blackmore, 4 Hen. VII); Convoc. Cornw., 12 Chas. I, c. 28; 2 Jas. II, c. 28. "By ancient custom the tithing men are sworn officers of the court, and shall and may warn any person within the stannary to appear in court ex-officio, and without a writ of summons." These eight tithings were, of course,

ment of criminals by a grand jury of twenty-four tinners. Whether the composition of this body was, in early history, subject to special regulations is a matter of doubt. Toward the end of the sixteenth century, at any rate, it had taken on a semi-aristocratic character. In the Cornish parliament of 1588, it was declared that the grand jury ought to consist of the most discreet tinners every year to be entered on the court books, and a century later the regulations are still more explicit. None but the best and most sufficient were eligible, including such men as were owners of tin lands, owners of bounds, and adventurers for tin who were neither merchants nor shopkeepers.[1]

The functions of the leet, however, were not ended with the presentment of criminals and the ordinary legal business of a court, but in many respects resembled those of the manorial halimote, or the court of a gild, inasmuch as it seems to have been the common meeting-place of all tinners and the administrative centre of the stannary machinery. It met, as has been stated, twice each year, and in all likelihood was attended, theoretically at least, by every tinner in the district.[2] No direct proof can be offered on this score, but circumstances all point in that direction. The election in Devon of the jurates of the parliament by a full court,[3] and the petition which the Cornish tinners later sent up to the Crown for the privilege of a similar practice,[4] are evidence that such was the case; while again, if we examine the scanty records in the court

amenable equally with tinners to the Blackmore court. We see them fined for not repairing their roads and bridges (Ct. R., bdle. 157, no. 13).

[1] Convoc. Cornw., 2 Jas. II, c. 20. Cf. *Compl. Min. Laws Derb.*, pt. iii, art. 28. "We say that able fit men if they be not miners, if they have parts and be maintainers of mines and known by the barmaster or his deputy to understand well the custom of the mines, ought to serve for jurors, especially in the difficult and weighty matters and causes." A long charge from the steward preceded the presentment (Harl. 6380, fol. 67; Add. MS. 6713, fol. 71), in which the jury was charged to declare a long list of offenders, including such manorial offences as the breaking of the assize of bread and ale.

[2] Cf. Smirke, 98. As to how long actual attendance upon the leet was maintained, and at what period the power began to pass from the hands of the mass of the miners, the records tell us nothing. In Saxony the old miners' assemblies had fallen into disuse by the later Middle Ages (Schmoller, Jb., xv, 1015) and in Tyrol by 1490–1556 (*ibid.*, 1016). In some parts of Austria, however, they survived until a later date (*ibid.*, 1016).

[3] Cf. Trans. Devon Assoc., viii, 321.

[4] See p. 129.

rolls, we find in a roll of 1638 for Penwith and Kerrier, at the head of the membrane, a long list of tinners who have been summoned and who have not appeared for suit, "therefore let them pay a fine."[1] In a Foweymore roll we are made aware of the defection of certain tollers and owners of blowing-houses, who owe suit and have not come, — "*idcerto*, let them be in mercy, every man."[2] Blowers and owners of blowing-houses are fined for default in attendance at a leet at Tywarnhail in 1638.[3] Colliers, also, as well as tollers, blowers, and owners of blowing-houses, are referred to in the statutes and declared customs as owing suit at each law court.[3]

The nature of the special business there transacted was such as called for all possible publicity. There it was that the tinners chose such officers as they had within their power to elect.[4] What the manner of procedure had been in earlier times it is difficult to say, but by the seventeenth century the power of choice had become vested in the grand jury. Thus from the records of the Cornish parliament of 1636 it appears that they designated a receiver for the stannary common funds,[5] while, at the same time and place, the receiver for the time being accounted for his charge before the steward and jury and turned it over to his successor.[6] At the law day also a jury of twelve decided on the measures to be used against tinners who refused to pay their stannary assessments;[7] four of the "most discreet" tinners assisted the steward in the assessment of fines and amercements;[8] and the grand jury nominated a few petty officers of a quasi-manorial character, such as supervisors

[1] Ct. R., bdle. 156, no. 21.

[2] *Ibid.*, bdle. 159, no. 16 (Foweymore, 20 Edw. IV). Et presentant defaltas Johannis Olyver, tollator de Trewyntmore, Johannis Halya, tollator de Wallys, tollator de Redehill, tollator de Traddemore . . . flator de Glynmille, Willielmi Martyn flator de Tolbanek, qui sectam debent ad hunc diem et non venerunt.

[3] Add. MS. 6713, fol. 292, c. 37 (Penwith and Kerrier customs); fol. 242, c. 23.

[4] Cf. the law court of the gild merchant as described by Gross (*Gild Merchant*, 1890, ii, 105).

[5] Composed of half the profits of the stannary court, together with the proceeds of various assessments laid on the tinners by the stannary parliaments. Cf. Add. MS. 6713, fol. 379; Harl. 6380, fol. 49; Convoc. Cornw., 16 Hen. VIII, c. 6.

[6] Convoc. Cornw., 12 Chas. I, c. 32; Add. MS. 6713, fol. 233.

[7] Parl. Devon, 2 Hen. VIII, c. 23; Convoc. Cornw., 30 Eliz., c. 27.

[8] A similar custom prevailed at the sheriff's tourn (Pollock and Maitland, i, 546).

of roads, and port reeves,[1] and by virtue of an act of the stannary parliament of 1624 might choose court bailiffs in case their own parliamentary representatives neglected to do so.[2]

The leet served also to register the initiation oaths of stannary officials. In open court appeared the head bailiff and in presence of the jury took solemn oath to fill his office justly and well.[3] So it was also with the supervisors of blowing-houses, the paysor, and the assay master and his deputies, who weighed and tested the tin at the coinage towns.[4] The owners of blowing-houses appeared in court and registered their house marks in the steward's book, presenting also their blowers for the oath of faithful service.[5] In the Michaelmas session occurred the ceremony of verifying the gallon and footfate measures, used in the apportionment of black tin, a duty which was performed by the head bailiff with the aid of the stannary jury,[6] or perhaps of a few impanelled solely for that occasion.[7] There should be mentioned, in addition to all this, the proclamation by the stewards of royal ordinances or statutes of the realm which dealt with the stannaries, and the occasional presentment by the grand jury of the more important customs in use in that particular stannary.[8] It seems to have been from this last function that there arose the institution referred to as the stannary parliament.

Few subjects in stannary history are more obscure in origin than the tinners' parliaments. The presence of similar bodies in the mines

[1] Ct. R., bdle. 156, no. 21 (Tywarnhail, 14 Chas. I). "Item presentant Edrum Whitta et Richardum Shole fore supervisores viarum et gardiandos portuum infra stannarias predictas pro anno sequenti." It is hard to account for this one instance of such a choice, except on the assumption that in this particular district the tinners comprised practically all the population.

[2] Convoc. Cornw., 22 Jas. I, c. 33.

[3] *Ibid.*, c. 32.

[4] *Ibid.*, c. 13; 12 Chas. I, c. 12. On this occasion the warden or vice-warden was supposed to be present. Cf. also Add. MS. 6713, fol. 272, c. 30.

[5] Harl. 6380, fol. 32; Convoc. Cornw., 16 Hen. VIII, c. 17, 18; 12 Chas. I, c. 7, 8; 22 Jas. I, c. 2, 3; Parl. Devon, 2 Hen. VIII, c. 13; Ct. R., bdle. 156, no. 21.

[6] Cf. Convoc. Cornw., 2 Jas. II, c. 10 (Add. MS. 6713, fol. 266); Add. MS. 6713, fol. 113; fol. 243, c. 33; Convoc. Cornw., 22 Jas. I, c. 25.

[7] *Ibid.*, 12 Chas. I, c. 20 (Add. MS. 6713, fol. 229).

[8] Cf. Add. MS. 24746, fol. 122; 6713, fol. 279; Ct. R., bdle. 156, no. 21 (Blackmore leet, 15 Chas. I).

of the Forest of Dean,[1] and the exasperating way in which all direct documentary evidence before the time of Henry VIII has disappeared, tempts, in the absence of proof, to speculation. In both Cornwall and Devon the parliaments were probably an expansion of, and an offshoot from, the grand juries or the special courts in the stannary judiciary, which, as we have seen, were called upon occasionally to declare the customs of the mines and which often prefixed to their presentments of criminals a confirmation of existing stannary law. It is stated in the older local histories, and frequently copied in later writings, that until 1305 the tinners of Devon and Cornwall met in one parliament every seven or eight years, on Hingston Hill near Callington,[2] after which date the Devon stannary parliament was kept on Crockerntorre, and one for Cornwall at various towns in the latter county.[3] All that can be stated from existing evidence, however, is that the records of the Devon parliaments go back only to 1510,[4] while those for Cornwall refer back to 1588,[5] by which latter date the parliament or convocation of the stannaries of Cornwall was assembled in accordance with the articles of the Charter of Pardon.[6]

This document was the outcome of the cupidity of Henry VII, who in consequence of real or assumed violations of the stannary

[1] Nicholls, 45, 47, 49, 54, chap. iv. This body was known as the Mine Law Court with a jury of 48 free miners, who by 1685 were half coal and half iron miners. It is to be carefully distinguished from the court of the mine, and was a parliament in all but name, meeting at irregular periods as did the parliaments of Cornwall and Devon. The Speech House, where it met, still stands in the Forest. The lead miners of Derbyshire and of the Mendip Hills do not seem to have possessed parliaments, but the Great Barmote Court of the Derbyshire miners really performed the functions of a parliament, together with those of courts. Whether or no it was chosen by popular ballot is doubtful (*Compl. Min. Laws Derb.*, pt. iv; pt. i, art. 40).

[2] Cf. Carew, ed. 1811, p. 17; Notes and Queries, 3S, v, 374; Journ. of Science, i, 283.

[3] Journ. of Science, i, 283; Add. MS. 6682, fol. 507.

[4] A Devon parliament is said to have been held in 1494 (Trans. Devon Assoc., viii, 319), but I can find no documentary proof of it.

[5] If other records existed, they were probably destroyed during the Civil War when Lostwithiel and Luxullian were partially burned. The presentments of 16 Henry VIII were not those of a convocation, but of commissioners appointed to inquire into and settle certain doubtful customs of the stannaries in Cornwall (Harrison, 34).

[6] Pat., 23 Hen. VII, pt. vii, m. 29–31; Add. MS. 6713, fol. 168–180.

laws on the part of the Cornish tinners, especially of the series of regulations initiated by Prince Arthur,[1] had declared the stannary charter forfeit, and restored it in 1507 only upon payment of a fine of £1100.[2] To the original instrument he added a grant of new powers to the parliaments. The especial provisions of the document which deal with the legislative powers of the stannaries applied merely to Cornwall, and declared not only the powers of the convocation but the manner in which its members were to be chosen. Upon the calling of a convocation by the Duke, through the lord warden,[3] twenty-four "stannators" were nominated, six by the mayor and council of each of the four towns of Lostwithiel, Truro, Launceston, and Helston, representing the stannaries of Blackmore, Tywarnhail, Foweymore, and Penwith and Kerrier; and the convocation so constituted had power to allow or disallow "any statute, act, ordinance, provision, restraint, or proclamation" thereafter "to be made by the King, his heirs, or successors, or the Prince of Wales, Dukes of Cornwall, or their Council for the time being," which should be "to the prejudice of any tinner, or other person having to do with black or white tin." In these provisions the parliament of Devon had no share, but it was hardly conceivable that what was granted to the one parliament could long be withheld from the other, and there is good reason to believe that in practice Devon, as well as Cornwall, passed judgment upon whatever arrangements the King or Parliament made concerning the stannaries.[4]

The manner, however, in which the Devon tinners' parliament was chosen was quite different from that in the neighboring county. Each of the four stannary districts of Chagford, Tavistock, Ashburton, and Plympton elected twenty-four representatives or "jurates" in a special court held for the purpose, in which all classes of tinners, including bounders, owners of tin works and of tin, adventurers, and laborers, and all others concerned in tin or tin works had a voice,[5] and the ninety-six representatives so chosen

[1] Add. MS. 6713, fol. 101–104. [2] *Ibid.*, fol. 136.

[3] Cf. S. P. Dom. Eliz., clxxxv, 45.

[4] For instance, the preëmption contracts in 1703 and 1710 were referred to the Devon Great Court as well as to the Cornish convocation for ratification.

[5] Convoc. Cornw., 30 Eliz., c. 30; Trans. Devon Assoc., viii, 320.

met at Crockerntorre [1] on the open hillside for the transaction of business. Of these assemblies we know but little. Journals of the parliament, if ever kept, have long since perished, and our sole information consists of the customs reaffirmed and the law codes passed, together with what indirect evidence is furnished by a knowledge of the procedure in Cornwall.[2]

There the order of the day scarcely differed from that of the House of Commons. The convocation, meeting at Truro or Lostwithiel behind closed doors, listened to the reading of a set speech from the lord warden, and then elected a speaker who, having been approved by the warden and having appointed the necessary clerks and doorkeepers, proceeded to open the session. From this both the lord warden and the vice-warden were excluded.[3] Usually the first piece of legislation was the ratification of all that had been enacted in the previous sessions, and this was followed by fresh legislation either to emphasize or amend already existing laws or to put into force new ones. The "stannators" were summoned to the convocation "to consult, enquire, and take into deliberation to resolve upon such orders as in their judgments shall be thought expedient for the redressing and amending of any inconveniences or abuses within the stannaries, and to reduce things in question and doubtful (touching liberties and customs) to a certainty." [4] Sometimes a contract for the preëmption of tin was urged on the King's behalf, but in most cases when this matter arose a special session was summoned, which dealt with it to the exclusion of almost all other business.[5] By an old custom, whatever was enacted must be signed by the "stannators," [6] the warden, or vice-warden, and the Duke of Cornwall, but in 1636 this was modified to allow of

[1] Westcote, 76. Polwhele (*Hist. of Cornw.*, bk. i, c. 2, p. 29) believes that Crockerntorre was the seat of one of the assemblies of the ancient British.

[2] The journals of several of the Cornish stannary parliaments are extant. Cf. Add. MS. 6713, fol. 417 *et seq.*

[3] Convoc. Cornw., 12 Chas. I, c. 1.

[4] Pearce, 21. "No one stannary can make any laws, but it must be by the stannary parliament of Cornwall." (Add. MS. 6713, fol. 113, Helston Court, 8 Hen. VIII).

[5] As in 1703 and 1710 (Add. MS. 6713, fol. 417 *et seq.*, 458 *et seq.*)

[6] Convoc. Cornw., 30 Eliz., c. 5. This, however, is not prescribed in the Charter of Pardon, and was therefore probably the rule in the convocations of earlier times.

a law being signed by merely sixteen instead of the entire twenty-four as formerly.[1] The session, which might extend with frequent adjournments over several weeks,[2] usually broke up with a vote of thanks to the lord warden and another to the Duke.

The question as to whether either parliament really represented all classes of tinners requires to be answered with some discrimination. In Devonshire the greater number of jurates and the democratic manner in which they were chosen would seem to indicate that the parliament, in earlier years at least, enjoyed to a considerable extent the support of the masses. In Cornwall the case was quite different, due largely to the fact that the nomination of representatives was the privilege of the mayors and councils of the stannary towns. How they exercised their power for the first century and a half we do not know, but in 1687, in consequence of the failure of the convocation to ratify a royal contract for the preëmption, we find the lord warden suggesting to the King the possibility of so returning members from the stannary towns "that they will consist of sober, loyal persons;"[3] while ten years later we hear complaint that the mayors of the stannary towns who returned convocators were the bitterest enemies the tinner had.[4]

Doubts may be raised as to whether, even where appointments were free from bias, the Cornish parliaments ever represented any class but the large mine owners or tin dealers. Membership seems to have been prized and to have been eagerly sought after, possibly with some sinister motive of ulterior gain, for during the seventeenth and eighteenth centuries a large amount of underhand dealing in stannary affairs seems to have been practiced. The writer of an account of a Cornish parliament which met in 1750[5] remarks that "during the time of an election, gentlemen think it worth while to come down into the county who were never seen in it at any other time, and to neglect their business and the pursuit of their affairs for three months together, without having or pretending to have the least knowledge of tin or stannary matters."

[1] Convoc. Cornw., 12 Chas. I, c. 1.

[2] Cf. Journ. Roy. Inst. Cornw., xvi, pt. ii, p. 296.

[3] Treas. Papers, ii, 10.

[4] *Tinners' Grievance.*

[5] *A Statement of the Proceedings of the Convocation for the Stannary of Cornwall,* 1750, by a Cornishman.

The representatives seem to have been chosen largely from the chief families of Cornwall, and a perusal of the lists of members reveals but few who were not baronets, knights, esquires, or gentlemen.[1] Further evidence to the same effect is supplied by the origin of the body of men known as "assistants."

As early as 1588 the tinners' parliament had petitioned the Queen that their members might be doubled and the additional six from each stannary be chosen, not by the mayors and the council, but by the stannary courts as in Devon.[2] The request was not granted, but by the year 1674 we find the members each nominating for himself an "assistant," and the latter summoned by the vice-warden to consult with the convocation, the idea being expressly stated that by this means the latter would be kept better informed of the feeling among tinners of the lower classes.[3] It is highly questionable whether this was ever the effect. The assistants as well as their principals seem to have been of the gentry,[4] and whatever

[1] Thus for 30 Eliz., two knights, seven esquires, two gentlemen, eight yeomen, and five untitled (Add. MS. 6713, fol. 195); for 22 Jas. I, ten esquires, and fourteen gentlemen (*ibid.*, fol. 207); for 12 Chas. I, one knight, fifteen esquires, eight gentlemen (*ibid.*, fol. 223, 224); for 25 Chas. II, three baronets, two knights, eighteen esquires (*ibid.*, fol. 377, 378); for 2 Jas. II, three baronets, two knights, eighteen esquires, and one gentleman (*ibid.*, fol. 259); for 2 Anne, twenty-two esquires, and two untitled (*ibid.*, fol. 417); for 9 Anne, two baronets, and twenty-two esquires (*ibid.*, fol. 458, 459). On the other hand, we have for Devon, in 2 Hen. VIII, ninety-six untitled (Pearce, 189, 190); in 25 Hen. VIII, ninety-six untitled (*ibid.*, 207); in 6 Edw. VI, ninety-six untitled (*ibid.*, 217, 218); in 16 Eliz., one esquire, twelve gentlemen, and eighty-three untitled (*ibid.*, 240); in 4 Jas. II, three baronets, thirty-nine esquires, forty-three gentlemen, and nine untitled (Bodl. MS. Add., c. 85).

The aristocratic nature of the Cornish parliament at the end of the seventeenth century may be seen by a bill introduced into Parliament in 1679 to remodel the stannary parliament. It proposes, among other things, to restrict the franchise for election of stannators so as to exclude day laborers or receivers of wages. No person might be elected who did not hold a freehold of £400 per annum, or, *if a tinner*, of £2000. The bill failed. It was evidently inspired by the refusal of the Cornish convocation in 1674 to agree with the King as to the preëmption (Rep. Hist. MSS. Com., ix, pt. ii, 93, 94).

[2] Convoc. Cornw., 30 Eliz., c. 5.

Oddly enough, we find the Parliament of the Dean mines suffering from the reverse cause, namely, that the gentlemen, as time went on, refused to attend (Nicholls, 54).

[3] The practice is referred to as one in use previous to 1674 (Add. MS. 6713, fol. 392).

[4] Thus in 1710 there were eighteen gentlemen, one esquire, and five untitled (*ibid.*, fol. 468, 469).

their station their function was apparently confined to the approval of such propositions as chanced to be submitted for their opinion. They were allowed no votes; they were not even present at the sessions, but were herded in an apartment by themselves and called in only upon occasions, nor does it appear that they ever ventured to dissent from any bill or contract upon which their advice was asked.[1]

As to the actual effectiveness of the opposition which the parliaments of either county could make to regal or princely demands, the meagreness of the records does not permit of a satisfactory answer. The Domestic State Papers and the records of the Cornish convocations, so far as they go, reveal two or three instances in which the royal will was thwarted. Thus in the reign of Charles II we find the appointment of Penzance as a coinage town nullified for some time by the refusal of the convocation of Cornwall to extend to it the ordinary laws of the coinage.[2] In 1674, again, the tinners were at loggerheads with the King on account of their persistent refusal to delegate their contracting powers to a select committee, who were to be summoned to Whitehall and overawed into signing a contract for the preëmption.[3] On the whole, however, the relations of the two bodies with the Crown seem to have been friendly enough. There would in any case be but little cause for friction, since the flurries between stannaries and royalty were usually conflicts of courts, while the proclamations and statutes which the parliaments were called upon to ratify were few and seem seldom to have been unreasonable. As time advanced, the occasions for the calling of parliaments grew less and less frequent, so that with the last session as far away as 1752 there seems little probability that they will be revived.[4]

[1] Add. MS. 6713, fol. 423, 426, 427.

[2] Pearce, 103.

[3] Add. MS. 6713, fol. 378, 379.

[4] Of the number of recorded sessions, some have already been indicated. The full list, so far as I have been able to ascertain it, is as follows. For Devon six sessions, all at Crockerntorre, viz., 1510 (2 Hen. VIII), 1532 (33 Hen. VIII), 1552 (6 Edw. VI), 1574 (16 Eliz.), 1688 (4 Jas. II, see Bodl. MSS. Add. c. 85, fol. 51, 52), and 1703 (2 Anne; Bodl. MS. Add. c. 85, fol. 55–58).

For Cornwall nine sessions were held at various coinage towns of the county, viz., 1588 (30 Eliz.), 1624 (22 Jas. I), 1636 (12 Chas. I), 1662 (14 Chas. II; see Add. MS. 18748, fol. 89, 90; 6713, fol. 336), 1674 (25 Chas. II; Add. MS. 6713, fol. 377 *et seq.*), 1687 (2 Jas. II), 1688 (Treas. Papers, ii, 58), 1703 (2 Anne), 1710 (9 Anne), 1750 (23 Geo. II; cf. Journ. Roy. Inst. Cornw., xvi, pt. ii, pp. 304, 305), and 1752 (26 Geo. II).

# CHAPTER V

## THE STANNARIES AND THE CROWN. TAXATION AND REVENUE

THE importance of the stannaries as a fiscal asset was, of course, from earliest times the dominant factor which made for royal control; and the increase of stannary revenue is the keynote of the policy pursued toward the tinners by the Crown. It was only because they knew that the King could be counted upon to block all attempts which threatened even remotely to cause mining to slacken, that the tinners dared defy for centuries the courts of the manor, the hundred, and the shire. It is especially significant that of all the commissions appointed by the Crown upon petition from the commonalty of the shires to investigate the aggressions of the tinners, the report of not one is extant, while we may entertain a reasonable suspicion that that of 1376 was deliberately quashed.

Although by their charters of 1305 the tinners were relieved of the ordinary forms of taxation, it must not be imagined that they got off scot free. Like the inhabitants of the royal demesnes,[1] they were liable to taxation of a special nature at the will of the Crown and without the mediation of Parliament, although it must be said in all fairness that as a rule the taxation of the tinners was not subject to caprice, but was the result of gradual usage, hardened into law.[2] It cannot be denied that considerable sums were drawn

[1] Vinogradoff, 92.

[2] This seems also to have been the case in other English mining districts. In Derbyshire the mine taxes were "lot" or toll lead, the thirteenth dish going to the lord of the soil (Houghton, pt. vii, art. 12; *Compl. Min. Laws Derb.*, pt. iii, art. 12), and "cope," or 6d. per load of nine dishes, paid to the lord, perhaps in lieu of his claim to the smelting of all lead ores (Farey, i, 384). In the Forest of Dean, the King received, through his gaveller, a penny per dole at the time of the first entry (Houghton, pt. ii, art. 15), and the miners of "beneath the wood" (Middle Dean, Little Dean, and River Dean) paid each week twelve charges of ore to the King's forges, receiving for it a penny per charge (Houghton, pt. ii, art. 19). "Lott lead," in the Mendip district, was the tenth pound of lead blown at the hearth, and was paid the lord of the soil (Trans. Roy. Geol. Soc. Cornw., vi, 329). In Dean and Mendip, however, the miners were subject to ordinary taxation as well.

from the stannaries by the titular dukes of Cornwall, but on the other hand taxation must have lain more lightly on the tinner than on other classes, else we should not hear so often of commissions of oyer and terminer sent down into the west to determine who are true and who are "false" tinners and to ferret out the wealthy men of the counties who have made themselves tinners simply to be free from tallages and subsidies.[1] The taxes which afforded the greatest revenue, the coinage and the preëmption, were probably paid in the end by the consumer of tin rather than its producer, since England had in the Middle Ages an almost complete monopoly of this very important product. If this did not advantage the tinners, it was probably for the reason that they were in the hands of the dealers, who tended to absorb the lion's share of the profits. It should be remembered, however, that until the year 1305 the stannaries did not enjoy exemption from ordinary rates, and so virtually paid double duty.

The oldest of the stannary taxes was a duty of 30d. per thousandweight in Devon and 5s. in Cornwall,[2] levied upon tin of the first smelting, that is to say, at the preliminary fusion of the metal at the mine itself. This tax the sheriff farmed annually for a lump sum, roughly readjusted every few years to correspond to its increased productiveness as the output of the mines became greater. For some years the farm was but £16 13s. 4d.[3] It was soon raised to £20;[4] then to £23 6s. 8d.;[5] in 1189 to one hundred marks;[6] and two years later to £80,[7] at which figure it remained for some time. Entries appear also, which seem to show that others than the sheriff[8] sometimes farmed a portion of this tax, while now and then, more frequently during the later years of Henry's reign, the mines were farmed by the sheriffs in conjunction with several associates,[9] as in the years 1174 to 1176,[10] when William Bulzun,

[1] Lay Subs. R., bdle. 95, nos. 12, 22; Pat., 12 Edw. III, pt. i, m. 23 d; 16 Edw. III, pt. iii, m. 2 d; 17 Edw. III, pt. i, m. 40 d; pt. ii, m. 5 d, 32 d; Close, 11 Edw. III, pt. ii, m. 20.

[2] App. A. I have found no explanation for this difference in taxation, which was later perpetuated in the coinage duties.

[3] Pipe R., 2–6 Hen. II, Devon; App. L.

[4] *Ibid.*, 7 Hen. II, Devon; App. L.

[5] *Ibid.*, 9 Hen. II, Devon; App. L.

[6] *Ibid.*, 15 Hen. II, Devon; App. L.

[7] *Ibid.*, 17 Hen. II, Devon; App. L.

[8] *Ibid.*, 15, 23 Hen. II, Devon.

[9] *Ibid.*, 1 Rich. I, Cornw.

[10] *Ibid.*, 16–21 Hen. II, Devon.

Alan Furnell, and Juel de Espreton shared the farm between them.

Aside from this tax, the King occasionally turned an honest penny by trading in tin on his own account. The Lion-Heart took in 1195 a modest "flyer" to the extent of £90,[1] in tin purchased on his behalf by his collectors in the western counties. Evidently finding the investment profitable, he repeated the operation two years afterward on a larger scale,[2] and his example was followed later by John.[3] In addition to these sources of revenue, the Crown received toll tin from the mines on the royal manors[4] and probably considerable sums from the sale of licenses to export,[5] to say nothing of customs duties upon the tin so shipped.

Such in brief was the fiscal side of the stannaries down to 1198, when an entirely new system was introduced. It was in this year that the old "sheriff economy" was displaced,[6] and Justiciar Hubert Walter sent William de Wrotham into the western counties as warden of the stannaries. The innovations which the latter introduced have already been cursorily mentioned. His first business was to reform the weights used to weigh the tin at its first smelting, and for this purpose he accepted the findings of two juries, one from the county court of Devon and the other from that of Cornwall. This done, he appointed two men to take charge of the corrected weights and to them was added a clerk, appointed by the King as a check upon the warden and his subordinates. At the same time an ordinance was issued forbidding the purchase, sale, or removal of tin of the first smelting from the stannaries or from the place appointed for its weighing and stamping, until it had passed through the hands of the keepers and clerks "of the weigh and stamp of the farm."

A second ordinance forbade the retention of any tin of the first smelting for more than two weeks unless it had been properly

[1] £90 9s. 4d. (Pipe R., 7 Rich. I, Cornw.).

[2] *Ibid.*, 9 Rich. I, Cornw. He cleared £352 6s. 10d. on this transaction.

[3] *Ibid.*, 1 and 4 John, Cornw.

[4] These items do not appear in the Pipe Rolls for the reason that they were accounted for by the manorial bailiffs and were probably not entered separately in any of the early rolls.

[5] Madox, i, 531; Close, 5 Hen. III, m. 9; Pat., 5 Hen. III, m. 8.

[6] It was in force in 1197.

weighed and stamped. This operation was performed by the three commissioners above mentioned, who not only weighed the tin and with a blow from the official hammer stamped upon each block the royal arms, but also kept a record in triplicate of each piece, together with its weight, the name of the owner, and the amount of duty paid. These records were probably turned over to the warden as a check upon the amount of the tax, which, collected upon the spot by the officers of the first smelting, was delivered to him.

The other innovations concerned the collection of a new duty of one mark in each county upon each thousand-weight of tin, as it issued from the second and more careful smelting.[1] The same preliminary treatment took place as before; the weights were verified by juries from the shires, and ordinances issued to prevent the holding back of tin from the second smelting. This was done at certain towns designated by the warden. Exeter in Devon and Bodmin in Cornwall were appointed permanently and, as they were also the county seats, the weighing and stamping must have been carried on in the public halls. Other places were named from year to year by the warden and the necessary halls hired in each at the King's charge.

No second smelting might be made save in the presence of the keepers of the weight and stamp of the second smelting. These were three as before, two appointed by the warden and one by the King,[2] with the proviso that the warden's appointees should be men of substantial property, no doubt in view of the large amounts which they handled as collectors. The three keepers travelled from town to town, carrying with them the weights and the stamping hammer, the latter securely sealed while not in use. Triplicate parchment rolls recorded the amounts of tin, the tax on each, and the name of the owners; but with regard to the disposal of the money a new method was adopted. The old "farm" went straight from the keepers to the warden, who had leased it for a fixed rent. The new tax, or the "marks of the new farm" as it was commonly called, was not leased. The tinners of the second smelting paid the money

[1] No one was to keep beyond thirteen weeks any tin of the first smelting weighed and stamped; it must be put into the second smelting and the tax paid.

[2] He received the King's wage and wore his livery.

to two royal treasurers appointed in each town, and it was forbidden to remove the tin from the hall until the mark had been accounted for. The treasurers made out tallies and chirographs similar to those of the keepers, on whom they acted as checks, and the money with all the chirographs was turned over to the warden,[1] who accounted for it at the Exchequer.

From the elaborate precautions taken by the warden and the King against peculation it goes without saying that the new tax produced a considerable yield. In 1199 it stood at £600. With the decline in the production of tin the taxes fell off for some years, but in 1212 the tax amounted to £668, and to £799 in 1214.[2] Omitting the figures for the tax on the first smelting or for toll tin, these amounts were fairly important at a time when the combined revenues of Cornwall and Devon, without the stannaries, was not more than £500 per annum.[3]

During the next few years the fiscal machinery doubtless remained much as it had been established by De Wrotham. As might be expected, it is referred to only incidentally in the charter of 1201. The officers of the second smelting, now designated as "weighers," received no salary, but were exempt from aids and tallages in their home vills, and the like privilege was granted to the treasurer of the marks. Some new officials were appointed at about this time. Head bailiffs chosen by the King[4] for each of the eight stannary districts, officers in many ways analogous to the sheriffs of the counties, acted as treasurers for the King's share of the profits of the courts and for the small poll taxes imposed in the several stannaries. Stewards appear in the stannary courts, which must have come into being in 1201 under the warden's plenary jurisdiction, although they could not have been upon a firm footing until the middle of the century.

The warden himself took on new responsibilities. He continued to account for the stannary revenues at the Exchequer, to issue ordinances, appoint officers, and exercise criminal and civil juris-

[1] Mention is made of one other official, the clerk of the chirographs, who wore the King's livery and received his wage.

[2] App. M.

[3] Pipe R.

[4] White Bk., 32 Edw. III, c. 93 d; Pat., 11 Edw. III, pt. i, m. 39; 9 Rich. II, pt. i, m. 7; 1 Hen. VI, pt. iii, m. 10; pt. ii, m. 14; pt. iv, m. 5; Parl. R., vi, 610 b.

diction, and, beginning about the year 1220, he, or rather they — as the office of warden seems to have been put into commission — appear once more as the lessees of the stannary revenues. In that year we find John, son of Richard, and Stephen le Croy, made "bailiffs"[1] of the stannary and stamp of Cornwall, farming its profits[2] at an annual rent of a thousand marks,[3] with the power of granting licenses for the transportation of tin.[4] This lease was held until 1225, when they were superseded by the King's brother Richard.[5] In 1220 the Devon stannaries were granted to Waleran the Teuton, as warden,[6] at a rent of two hundred marks,[7] and in 1235 to Richard de Tragford[8] for ninety marks, not including an annual tithe of £10, due the Bishop of Exeter. In 1254 William de Englefield, sheriff of Devon, became warden of the stannaries there,[9] and in 1278 the wardenship was bestowed upon Edmund of Cornwall. At about the same period began the practice of turning over the revenues of the mines to creditors of the King or to hangers-on of the court. Isabella, the queen-dowager, with her husband the Count of Marche, received the stannaries of Devon in 1217,[10] and in 1224[11] Berengaria, the widow of Richard I, for a short time held all the stannaries in part payment of a debt.[12] Richard, the King's brother, in 1231 received not only the stannaries but the entire county of Cornwall.[13] In 1253 the stannaries of Devon were granted to Adimus Wymer,[14] and in 1304 those of Cornwall to William Servator.[15]

This confused history of grants, resumptions, and re-grants from 1231 to 1300, taken in connection with the passing of Cornwall into the hands of Richard Plantagenet and his son Edmund, helps to explain why no entries for the Cornish stannaries appear in the Pipe Rolls for the above mentioned period, and why, in consequence,

[1] Pat., 5 Hen. III, m. 4, 8; Close, 5 Hen. III, m. 8.
[2] *Ibid.*, 9 Hen. III, m. 4; 10 Hen. III, m. 27.
[3] Pat., 5 Hen. III, m. 8; Fine R., 5 Hen. III.
[4] Pat., 5 Hen. III, m. 8; Close, 5 Hen. III, m. 9.
[5] *Ibid.*, 9 Hen. III, m. 7, 9.
[6] *Ibid.*, 4 Hen. III, m. 16; 6 Hen. III, m. 6.
[7] Fine R., 5 Hen. III, m. 7; 6 Hen. III, m. 2.
[8] Pat., 19 Hen. III, m. 16.
[9] Orig. R., 38 Hen. III, ro. 3.
[10] Pat., 1 Hen. III, m. 5; Close, 1 Hen. III, m. 23.
[11] Pat., 8 Hen. III, m. 11; Close, 8 Hen. III, m. 14.
[12] Rymer, i, 243; Pat., 4 Hen. III, m. 3.
[13] Chart. R., 15 Hen. III, m. 4.
[14] Pat., 37 Hen. III, m. 18.
[15] Orig. R., 32 Edw. I, ro. 7.

the fiscal history at this point must remain obscure. In 1243, however, the Devon stannaries were accounted for to the King, and the Pipe Roll entries exhibit to a marked degree the changes which had been wrought in the fiscal arrangements. The ancient "farm" disappears and its place is filled by a tax of 2s. 2d. per thousandweight, called "profits of the small stamp." The "marks of the King" are retained in their original amount under the name of "profits of the great stamp." A further tax is designated as "black rent," and the stannary courts account for four shillings in "profits and perquisites." [1]

Pipe Roll entries for the Devon stannaries were resumed in 1289.[2] By this time a new tax had been imposed, a "white rent," and the stamp duties consolidated into a single charge of 18¾d. per hundredweight, an experiment suspended three years later,[3] and permanently adopted only in 1302.[4] Stannary revenues from Devon were accounted for during these years by a "clerk warden," who received a small yearly salary from the King,[5] or, as was usually the case, from the warden for the time being to whom the King had leased the mines. There is good reason also to believe that the profits of the stannary courts, together with the black rent and the white rent, were paid in by the stannary bailiffs, and that at the coinage towns the same procedure was gone through, and under the same officials, as in the days of De Wrotham.

From 1231 to 1300 Cornwall remained under the earls Richard and Edmund, and as but one stannary roll of these years has survived, our information is meagre. The fiscal arrangements probably followed much the same course as those in Devon. In 1302 we find a few new taxes, with no indication of changes in the system of collection. Besides the coinage duties,[6] by this time increased and

[1] Pipe R., 27 Hen. III, Devon.

[2] *Ibid.*, 20 Edw. I, Devon. In 1263, 1264, and 1271, the Devon stannaries were accounted for in lump sums of £43 3s., £60 16s. 4d., and £39 18s. 2d. (Pipe R., 48, 53 Hen. III).

[3] *Ibid.*, 23 Edw. I, Devon.

[4] Duchy Accts., Excheq. Aug., port. 5.

[5] £6 1s. 8d. (Pipe R., 27 Hen. III; 48 Hen. III; 55 Hen. III; 15 Edw. I; 20 Edw. I; 23 Edw. I; 27 Edw. I; 34 Edw. I; Accts. Excheq. K. R., bdle. 260, no. 15; Cal. of Pat., 1304, p. 326).

[6] No traces appear of the separation of this tax into profits of a large and a small stamp.

rolled into a single tax of four shillings on each hundred-weight, the Cornish stannaries paid "tribulage," "dublet," and the "fine of tinners," and also accounted for the profits of their courts. If the arrangements for collection had undergone any change, it was more likely to be an expansion than an alteration of De Wrotham's scheme. The consolidation of the coinage duties did away with the necessity for the three keepers of the first smelting, but with this exception the fiscal machinery probably increased in complexity rather than diminished.

On the whole, then, we may say that the seventy-five years immediately preceding the charters of 1305 were a period of steady but quiet development. The old stampage or coinage duties were consolidated and increased, new taxes were added, small in amount and obscure in origin. The machinery of taxation in both counties was maintained upon the old lines, but developed in order to meet new conditions. The stannaries themselves, after having been successively under lease, grant, or direct royal control, were resumed by Edward I, and all taxes paid into the Exchequer.

Little nominal change in the forms of stannary taxation is apparent after the opening years of the fourteenth century, the only point perhaps worthy of note being the disconnection of the warden from the fiscal side of the administration. Until the year 1338 it had been he who had accounted for the receipts at the Exchequer,[1] but with the inclusion of the stannaries in the newly created duchy of Cornwall this duty passed into the hands of the receiver of the duchy, while the collectors' accounts were scrutinized by the auditor.[2]

First among the various items of revenue which should be mentioned before the taxes properly so called, we may note the "pleas and perquisites of the stannaries," by which is meant the profits of the eight stannary courts. In no case were the totals other than small,[3] not merely because the King or Prince was entitled only to a share in the proceeds of some of the heavier fines but also by reason

[1] Thomas de la Hyde, the "steward of Cornwall," who accounted for the stannary revenues in 1300 (Pipe R., 29 Edw. I, Cornw.) was probably also the warden.

[2] S. P. Dom., Eliz., cvi, 55; Lans. 19, fol. 99.

[3] In 1450 the four for Cornwall totalled the exceptionally low amount of £3 (Receiver, 29 Hen. VI). See App. S, T.

of the fact that a large part of the courts' business was furnished by civil suits. The first recorded payment of court profits is that of the Devon stannaries in 1243, but in all probability it began in both counties as soon as the stannary courts were established.

"Dublet" [1] was a small local tax levied upon five tithings in Penwith and Kerrier, which brought in the annual fixed sum of 11s. 8d. Of this nothing can be said beyond the curious fact that it was collected and returned by the bailiff, not of the stannary, but of the hundred of the same name.[2]

Equally curious and difficult of explanation is the so-called "fine of tin," which appears first in 1342.[3] From the earliest date the amount was usually fixed at 65s. 8d. in the stannary of Blackmore, and 42s. in the hundred of Pyder; but by whom it was paid is not known.[4]

Occasionally receipts appear from the sale of tin which, by reason of having been sold uncoined or having been fraudulently marked, had been declared forfeit to the Prince. Three instances only have been found of such seizures, the first in 1579,[5] when £5 4s. were realized from the sale, the second in 1580,[6] when the amount from this source reached the sum of £76, and the third in 1607,[7] when it totalled £11 9s. 8d. It is improbable that these were the sole years in which tin was seized, and the accounts for this item may well have been rendered to a different person or have been joined with that of some other source of revenue.

Among these minor sources of income, mention may here be made of the blowing-houses which the Black Prince owned at Lostwithiel,[8] of waifs and strays,[9] of the small sums which Edward

[1] Possible derivation: d'oblée (doblata), *i. e.* an offering, or "oblata" (in lieu of tribulage?). See Smirke, 47.

[2] It first appears in 1302 (Pipe R., 30 Edw. I, Cornw.), and last in 1507 (D. O. Ministers' Accts., 22 Hen. VIII).

[3] D. O. Ministers' Accts., 16 Edw. III. App. R.

[4] This fine of tin seems to have been distinct from another payment of the same name, which was made by certain freeholders of the assessionable manors and has been explained as an acknowledgment of the fact that the Crown retained a control over the minerals in the soil. Cf. Concanen, Rowe *vs.* Brenton, pp. xiv, 60, 112–319.

[5] Receiver, 21 Eliz.

[6] *Ibid.*, 22 Eliz.

[7] *Ibid.*, 4 Jas. I.

[8] White Bk., 32 Edw. III, c. 89.

[9] Cf. Add. MS. 6713, fol. 305, c. 91.

IV realized from his shares in several of the mines themselves,[1] and of the toll tin derived from the Duchy manors. To these should be added the coinage of tin halfpennies and farthings by James II and by William and Mary, who profited largely by the high seigniorage.[2]

The tax known as "tribulage," or "shovel money," which represents the only attempt at a capitation tax in the Cornish stannaries, was paid in two stannaries only, Penwith and Kerrier, and, after 1342,[3] Blackmore. It was collected solely from laboring tinners. The first rate was that of one halfpenny per head in Penwith and Kerrier,[4] which in 1301,[5] with a population of three hundred and

[1] Accts. Excheq. K. R., bdle. 266, no. 2. I have not mentioned in the list of miscellaneous receipts the forced loans raised by the Crown upon tin which the merchants were about to ship (Close, 13 Edw. III, pt. ii, m. 13, 19), or the right often exercised of compelling exporters to take out licenses (cf. Pat., 22 Hen. VII, pt. i; 6 Edw. IV, pt. i, m. 11; pt. ii, m. 8; Parl. R., v, 603 b; Cust. R., bdle. 115, no. 10; S. P. Dom. Jas. I, xxviii, 96; lxxxi, 19). These come rather under the head of general export duties.

[2] Cal. S. P. Dom., 1651, p. 313–315; S. P. Dom. Chas. II, xxxvii, 19; ccxxx, 75; Treas. Papers, ii, 23, 26, 37, 42, 45, 48, 49, 50; iv, 11; vii, 73; xvii, 67; lxxxiii, 138; lxxxvi, 102; John Collins, *A Plea for the Bringing in of Irish Cattle*.

The plan seems to have been debated by Parliament in 1651 (Cal. S. P. Dom., 1651, 313–315), and then taken up by Charles II immediately upon his restoration, at a time when every effort was being made to exploit the stannaries for the royal purse (S. P. Dom., Chas. II, xxxvii, 19). Laid aside for the moment, as well as in 1666 when the proposal was renewed by the farmers of the preëmption (S. P. Dom. Chas. II, ccxxx, 75), it was put into operation by James II shortly after his succession. What little is known of the transaction appears from a few reports in the State Papers, from which it would seem that commissions for the new issue were granted on Lady Day, 1684, and ran through the reign of James II, and for six years of that of William and Mary. The coinage was performed by commissioners, and copper and tin purchased at the enormous rate of 20d. per pound were turned into halfpence and farthings at a considerable seigniorage (Treas. Papers, lxxxvi, 102). In all £10,000 worth of tin was purchased, and £65,000 worth of coins issued (Treas. Papers, xxviii, 1). The new coins did not prove a success save as a means of raising revenue. Large shipments were made to the army in Ireland (Treas. Papers, vii, 73), and when that means proved insufficient to exhaust the output, it was urged that they be dumped upon the American plantations. By 1703 the coinage of copper pence had been farmed, and as it was impossible for the Crown by its authority to keep the price of tin at so high a level that the tin farthings would not depreciate, the farmers were subject to considerable loss, since by the terms of their contract they must receive tin for copper by tale (Treas. Papers, lxxxiv, 138; lxxxvi, 102).

[3] D. O. Mins. Accts., 16 Edw. III. App. O.

[4] Mins. Accts., Bailiff's Acct. of Edm. of Cornw. App. O.

[5] Pipe R., 29 Edw. I, Cornw. App. O.

twenty-eight laboring tinners, produced 13s. 8d. in tribulage. In 1349 the yield was 20s., but the manner in which the Black Death decimated the stannaries made it necessary to raise the rate permanently to twopence, although that of Blackmore remained unaltered. The record of tribulage payments [1] is necessarily incomplete from the fact that the duty was collected and paid to the receiver by the head bailiffs, and often included in their accounts with the perquisites of the stannary courts. In all stannaries the wide fluctuations [2] of the total sums, so far as they go, act as a barometer of mining activities. For a few years previous to the Black Death the amount was 20s. in Penwith and Kerrier and about half that amount in Blackmore. In 1350 only 6s. 8d. was gathered in the former, but in 1417 the receipts had risen to £13 6s. 8d. In Blackmore by 1369 the amount obtained was only 10d. Beginning with the sixteenth century the tax was frequently farmed at a fixed rental,[3] but by 1645 it had ceased to appear as part of the Duchy income.

What tribulage was to the Cornish tinners, black and white rent signified to those of Devon. Black rent, the earlier of the two, was also the shorter-lived. It consisted of a levy of twopence for every black tinner, that is to say, for every digger of tin ore in Devon. It first appears in the Pipe Roll for 1243,[4] when we find it assessed upon one hundred and forty-nine miners. Like tribulage, it fluctuated as the population of the mines varied, — the last entry, that of 1301,[5] showing something over £3 10s.

White rent in its original form consisted of two pounds of white or smelted tin, paid by each "maker of white tin," by which is to be understood those tinners who brought the ore to the smelter and received the finished product.[6] The stannary bailiffs, however, accounted for these payments in terms of money, according to current tin quotations. In 1292 the account was rendered at the rate of 10s. per hundred-weight, in 1293, 9s., in 1294, 8s.[7] How

[1] See App. O.

[2] Lending color to the belief that much of the mining population was transitory, and that tinning was in many cases a by-occupation.

[3] Cf. Pat., 5 Hen. VIII, pt. ii, m. 12; White Bk., 32 Edw. III, c. 93 d. App. O.

[4] Pipe R., 27 Hen. III, Devon. App. N.

[5] *Ibid.*, 29 Edw. I, Devon. App. N.

[6] Not, as Smirke interprets it, the smelter of tin.

[7] App. Q.

long this system of payment in kind lasted we do not learn; but in 1400 we find the rent commuted to 8d. the head, and in this form it survived until the Commonwealth. White rent was always small. The payments in 1288 amounted to but £2 8s. 2d., and the most ever obtained was in 1518, when five hundred and seven white tinners responded with almost £17.

An exceedingly fertile source of revenue was afforded the Crown by the royal right of the preëmption of all tin mined in the two counties.[1] This right, asserted in the charters of 1305 and later celebrated under the Stuarts as the tin monopoly, dates back to the twelfth century. It was exercised not, as in the case of the right of purveyance, for the satisfaction of the King's household necessities, but systematically as a business enterprise. We have seen this done by Richard I in 1195 and 1197, and by John in 1199 and 1202. Afterwards, the stannaries were turned over in return for an annual rent to a succession of petty wardens who, besides the right of granting licenses for exporting tin, perhaps exercised the right of preëmption as well. When the mines came under the power of the earls of Cornwall, the preëmption was doubtless habitually exercised,[2] and probably formed an important source of Richard of Cornwall's great wealth.

After 1305, however, the attempts which the King made upon the tin were met by a vigorous, and on the whole successful, opposition. The first instance in point occurred in 1312, in connection with the financial arrangements between Edward II and his creditors, the Florentine Bardi.[3] In that year Antonio Pessaigne, a

[1] Preëmption was constantly exercised by the German princes with respect to silver and copper ores (Schmoller, Jb. xv, 670, 689; Inama-Sternegg, ii, 335), and also with respect to tin (Reyer, 85). In Derbyshire the preëmption of lead was the King's prerogative (Rep. Hist. MSS. Com., House of Lords MSS., ii, N. S., 383–386; Cal. Home Office Papers, i, 434; Treas. Papers, xliv, 25; S. P. Dom. Chas. I, cccx, 11; cccxli, 129; ccclxxvii, 5; cccc, 11; Add. MS. 6682, fol. 69).

[2] Pat., 8 Edw. II, pt. i, m. 12, 29 d; Min. Accts., Bailiff's Acct. of Edm. of Cornw.

[3] Cal. of Pat., 1314, Nov. 27. The loans which these financiers made to the first three Edwards are too well known to need recounting. The security on which the loans were based, however, is not well understood. The profits of Ireland (Cal. of Pat., 1299, Oct. 31), the license to purchase and change pollards and crockards (*ibid.*, 1300, Mar. 14), and finally the mines royal, were successively turned over for exploitation. When the silver mines were pledged to the Bardi, the usual form was for the latter to rent and work them outright (Cal. of Pat., 1308, p. 137; 1301,

Genoese, obtained a lien on the coinage dues [1] in Cornwall and Devon and a commission by virtue of which he was authorized to buy, on the King's behalf, all tin that might be coined.[2] Of the success of this move we have no knowledge, but two years later the King made over to Antonio his rights of preëmption.[3] These the Italian exercised for two years, and by his treatment of the tinners aroused intense hostility in the stannaries. The tinners complained that Antonio forced them to bring all their tin at their own expense to Lostwithiel, weighed it there with weights of his own, and finally paid them about one half the market price.[4] The result was that the miners, despite royal commissions,[5] sold their tin to whomsoever they pleased, attacked and maltreated the patentee's factors,[6] and finally in 1316 procured from the King the revocation of the patent.[7]

A few months later, however, Edward granted a five-year lease to Abyndon, his butler, and Pecok, his yeoman.[8] They were met by the tinners with the same reception as that given to Antonius [9] and came to the same fate, the tinners finally prevailing upon the King to withdraw the grant.[10] In 1336 John of Eltham, the King's brother, was given the preëmption of forty thousand-weight of tin.[11] No resistance seems to have been offered by the tinners, and accordingly Edward in 1338 sent his agents Suthorp and Moveron to buy up all tin for his own use.[12] This move the tinners met by an attitude of passive resistance and the stoppage of all mining

p. 625; 1310, p. 234; Cal. of Close, 1309, p. 95), and there is still extant a form of an indenture drawn up between the King and these Italians, setting forth the terms under which the mines of Devon were to be worked (L. T. R. Mem. R., 69 c, m. 34. See p. 192).

[1] Orig. R., 6 Edw. II, ro. 4.

[2] Close, 6 Edw. II, m. 25.

[3] Pat., 8 Edw. II, pt. i, m. 12, 29 d.

[4] App. E.

[5] Pat., 7 Edw. II, pt. ii, m. 10; 8 Edw. II, pt. i, m. 28 d, 29 d; Close, 8 Edw. II, m. 7.

[6] Pat., 9 Edw. II, pt. i, m. 25 sched. d.

[7] Pat., 10 Edw. II, pt. i, m. 4.

[8] Pat., 10 Edw. II, pt. ii, m. 24, 28. They also secured the issues of the coinage after Antonius had been satisfied of his debts (Close, 10 Edw. II, m. 12; 14 Edw. II, m. 23; Orig. R., 10 Edw. II, ro. 17).

[9] Pat., 10 Edw. II, pt. ii, m. 17 d.

[10] Close, 14 Edw. II, m. 23.

[11] Close, 10 Edw. III, m. 15 d. He was warden of the Devon stannaries (Close, 14 Edw. III, pt. i, m. 50) and, as Earl of Cornwall, was a large dealer in tin on his own account (Accts. Excheq. K. R., bdle. 262, no. 26).

[12] Close, 12 Edw. III, pt. ii, m. 25.

operations. Within a month the King not only countermanded his orders, but restored the tin already taken.[1] But the claims of the Crown were in no wise renounced. In 1347 the preëmption was successfully exercised, against the protests of tinners and merchants alike,[2] by the Black Prince as Duke of Cornwall. Of the patentee, a German merchant called Tideman of Limberg, we read that he was to have the first purchase of all tin as well as the office of receiver of the Duchy at a rent of thirty-five hundred marks.[3] The last instance during this period of the exercise of the preëmption was in 1367, when, as we know from the account rolls, the Prince of Wales bought the product of the mines at 20s. per hundredweight and resold it to the merchants at an advance of 6s. 8d.[4]

For two centuries the preëmption was apparently never exercised,[5] and when in the sixteenth and seventeenth centuries it was again put into operation,[6] its revival was actuated not entirely by the fiscal motive in the main predominant in the earlier exercise of the claim. This, indeed, played an important part in the successive Stuart tin monopolies, and the revenue accruing to the Crown in this way, amounting to £2000 in 1601 [7] and rising by degrees to not less than £12000 in 1628,[8] exceeded even that produced by the coinage duty. But a second and weighty argument for the revived use

[1] Close, 12 Edw. III, pt. ii, m. 20. He "saved his face" by the excuse that Cornwall and Devon had granted him a subsidy, so that the preëmption was no longer necessary.

[2] Parl. R., ii, 168, 180, 203; S. P. Dom. Jas. I, xxiii, 57.

[3] Pat., 17 Edw. III, pt. ii, m. 3; S. P. Dom. Jas. I, xxiii, 57; Cust. R., bdle. 70, no. 13.

[4] Accts. Excheq. K. R., bdle. 263, no. 15. The former seems a reasonable price to have given the tinners. The ordinary market price was nearer one mark (Pat., 14 Rich. II, pt. ii, m. 22).

[5] The advisability of exercising the preëmption was discussed under Henry VIII, but no action was taken (Lans. 24, fol. 50).

[6] It is interesting to note that in this age of monopolies, not even the coal mines escaped royal attention. In 1638 Charles I incorporated a company of coal buyers to purchase all coals exported from Sunderland, Newcastle, Blyth, and Berwick, paying the King 1s. custom per chaldron. This agreement was hardly made before it was broken into by the Civil War (Galloway, 140–141). The proposal was revived under Charles II ( S. P. Dom. Chas. II, cxix, 24), but nothing was done to carry it into effect. Similarly James I frequently asserted his rights over the Forest of Dean coal and iron mines (Galloway, 206–208).

[7] S. P. Dom. Eliz., cclxxxvi, 26.

[8] Receiver, 9 Chas. I.

of the preëmption was that in this way only could the tinners, ground down by the merchant dealers, find a permanent and equitable price for their metal. The discussion of this motive as well as of the economic consequences of the tin monopolies may best be deferred until a later chapter. For the present it will suffice to examine briefly the preëmption of tin as it was exercised during and after the reign of Edward VI.

In the closing months of the latter's reign the preëmption was leased to Gilbert Brokehouse,[1] and although his patent lasted but a few months, the fact that it initiated a new phase of royal intervention and furnished an example for those that were to follow gives it a fair claim to our attention. By the terms of the agreement he was permitted, in consideration of an annual rent of three thousand marks, to purchase at the ordinary market rate all the tin coined. For three days he must be ready to sell to all comers at a profit of one halfpenny on each pound weight, but at the end of that time he might sell or transport his tin as he pleased. Bad judgment and slowness in paying the miners cost him whatever respect he may have enjoyed from them, and upon their petition Queen Mary suspended his patent in December of the same year.[2]

For nearly fifty years after this fiasco the question of the formation of a tin monopoly was repeatedly discussed by the Privy Council,[3] but the plan was never carried into effect. If we except

[1] Pat., 7 Edw. VI, pt. iii.

[2] He seems to have been unable to get sufficient cash to start the purchases, and so attempted to impress the tinners by displaying a number of sacks, presumably full of silver but actually of lead. He hoped to realize in advance on his sales to merchants, and thus deal with the tinners without a cash reserve. The cheat, however, was discovered. (S. P. Dom. Mary, iv, 5; Lans. 76, fol. 34; D. O. MS. Vol., fol. 92.)

[3] In 1565 a proposition emanated from Warden Godolphin to buy off the Queen's prerogative (S. P. Dom. Eliz., xxxvi, 46). A patent was granted in 1577 to a Mr. Halford but revoked at the last moment (Lans. 24, fol. 47), and another in the same year to Sir George Carey was also revoked (Lans. 24, fol. 44, 47, 48, 50). In the closing years of the century the matter was urged more and more strongly. An anonymous writer, under date of 1591, proposed that the Queen take the preëmption into her own hands (S. P. Dom. Eliz., ccxl, 131). In 1595 Lord Oxford and Lord Treasurer Buckhurst competed furiously for the patent (S. P. Dom. Eliz., ccli, 71, 120, 121; cclii, 49, 52; ccliii, 46, 48), and the result of their rivalry was that the Queen sent an agent, Middleton, into the West to discover the real conditions prevailing among the tinners (S. P. Dom. Eliz., ccliii, 52).

the short and doubtful exercise[1] of the preëmption by Queen Elizabeth in 1599 or 1600 and by Raleigh in the following year, the next instance is that of Brigham and Wemmes, who received a patent for the preëmption in 1601.[2] They came into difficulties, however, on the one hand with the London pewterers and on the other with the Levant Company, which at that time carried on much of its trade in tin and was unwilling to endure competition. To ruin the patentees, therefore, the company for eighteen months refused to deal in tin, and as much of the output went usually to Levant ports the two preëmptors were put to considerable loss.[3] In the domestic market they were still less fortunate. At the time of the granting of the patent the tinners had been dependent for their sales upon a coterie of Londoners, among whom were several pewterers. By purchasing all the available tin about London and alternately inflating and depressing the market, the London merchants made it impossible for the tin farmers to sell at a profit, and having at the same time captured the pewterers' organization, they prevailed upon the King in 1603 to cancel the patent.[4]

The next to enter the field was James himself, who by dint largely of forced loans[5] raised £24,000, sufficient for the purchase of a year's output at the comparatively generous price of 59s. per hundredweight.[6] As his object was to force a sharp corner in tin and then to sell at his own price, the King made no attempt at first to dispose of his stock of metal, but accumulated some millions of pounds'

[1] Doubt exists as to whether or no Elizabeth exercised the right of preëmption on her own behalf. If she did, as may be argued from several allusions in the State Papers (S. P. Dom. Eliz., cclxxiii, 74; cclxx, 123; cclxxvi, 18; Carew, ed. 1811, p. 56), it was probably in 1599 or 1600, but in 1599, according to other evidence, the preëmption was granted to Sir Bevis Bulmer (S. P. Dom. Eliz., cclxxiii, 9), and in 1601 we learn that it was in the hands of Raleigh (D'Ewes, 647). The whole subject is extremely obscure.

[2] For the terms of the agreement, see S. P. Dom. Eliz., cclxxvi, 26; Jas. I, xxiii, 57.

[3] S. P. Dom. Jas. I, vi, 79. As early as 1602 the patentees were in financial straits and had been forced to ask for a state loan of £20,000, on security of some of their redundant tin (S. P. Dom. Eliz., cclxxxvi, 26).

[4] *Ibid.*, Jas. I, ii, 4; ix, 75; xxiii, 57; Welch, ii, 37–39. See p. 49.

[5] *Ibid.*, vi, 84; ix, 12 (Warrant Bk., i, 135); xii, 80; xv, 75. These loans, made on security of tin, were apparently punctually repaid. For the manner of purchase, cf. S. P. Dom. Jas. I, ix, 19; xii, 85; Addenda, xxxvii, 20.

[6] S. P. Dom. Jas. I, vi, 80; xxiii, 57; Add. MS. 36767, fol. 258.

weight in the Tower and at various seaports.[1] By 1606, in spite of the efforts of the London speculators, he had succeeded in stiffening the market price sufficiently to be able to sell several lots to the pewterers and others at high rates;[2] but the disposal of his stock took so long and the necessity for regular purchases absorbed so much of his capital that in the following year he was glad to relinquish the preëmption to private hands.

The new farmers formed a syndicate headed by Sir Thomas Bludder, and until 1642 with the exception of a few years, this company, with gradual changes in membership as old members dropped out and new ones were admitted, handled under various patents the entire output of the mines. Sir Thomas was to take over at cost the King's own stock[3] and pay the tinners 56s. for each hundred-weight purchased, besides a rent to the King of £2000 a year, in return for which he and his fellows received a twenty-one year lease of the preëmption and transportation of tin as well as of the transportation of pewter.[4] But the heavy outlay occasioned by the taking over of the King's tin[5] crippled the patentees at the start, and when in order to recoup themselves they raised general prices, they forced many of the retail dealers out of business. Thus they ended by spoiling the market and injuring their own fortunes beyond repair.[6] Within six months over £20,000 worth of tin lay on their hands, and in barely a year's time their patent was called in and a new one issued to Ralph Freeman, his two brothers, Adrian Moore, and John Eldred.[7] This in turn was wrecked by the London pewterers, who in the following October induced the King to violate the terms of the agreement by granting them the sole right of casting tin into bars, as well as an annual surplus at reduced rates.[8]

[1] S. P. Dom. Jas. I, viii, 137; Add. MS. 36767, fol. 67.

[2] Add. MS. 36767, fol. 67, 258.

[3] Stowe, 143, fol. 144.

[4] Receiver, 1608. Unlike the patent of Brigham and Wemmes no loan need be made to the tinners. Cf. also S. P. Dom. Jas. I, xxxii, 81; xxxvii, 87; Warrant Bk., ii, 81.

[5] James had over a million pounds of tin which, at £28 per thousand-weight, meant the investment of £28,000, to say nothing of the continual call for money to buy tin from the tinners, in all, about £45,000 (S. P. Dom. Jas. I, Oct. 1614).

[6] S. P. Dom. Jas. I, xxxiv, 65.

[7] *Ibid.*, xxxvii, 87; Receiver, 1609; S. P. Dom. Jas. I; Indirect Warrant Bk., p. 57.

[8] Add. MS. 36767, fol. 219, 228; S. P. Dom. Jas. I, xxxviii, 50; Warrant Bk., ii, 77.

It would be of little interest or profit to follow in close detail the history of the later tin contracts, the terms upon which they were granted, and the preëmptor's relations with the tinners on the one hand and with the pewterers and other consumers on the other.[1] We find that in 1645 the King, in the financial straits of the Civil War, preëmpted for himself large consignments of tin, which were shipped to Ostend to be sold for the benefit of the royal party.[2] In the following year the stannaries were almost ruined by the carrying of the war into Cornwall, but Parliament, which had succeeded in holding the western counties, farmed out what little was raised to members of its own party.[3] From approximately 1650 to 1660 the preëmption fell into disuse,[4] but immediately upon the Restoration it was revived and passed in quick succession through the hands of two companies,[5] neither of which was able to use it to much profit. In spite of various subsequent attempts to organize

[1] The following manuscripts are sources for the preëmption down to the Civil War: Receiver, 7, 11, 13, 14 Jas. I; 9 Chas. I; Lans. 1215, fol. 226–230; Hargrave, 321, fol. 689–693; Cotton, Titus, B, v, fol. 389–393; S. P. Dom. Jas. I, lxxxi, 24; cix, 6; cxviii, 83, 84; S. P. Dom. Jas. I, Grant Bk., p. 112, 135, 309, 338; S. P. Dom. Chas. I, xxxviii, 81; cxlvi, 69; cclxxix, 62; cclxxxv, 12; cclxxxvi, 32, 66; cccv, 67; cccxiv, 65; cccxvii, 107; cccxxii, 1; cccxxvi, 28, 60; ccclxxvi, 77; ccccvi, 83, 102; ccccxiii, 31; ccccxx, 38; ccccxxiv, 101; ccccxlvi, 44; ccccli, 84; cccclv, 23; cccclxx, 81; diii, 85; Coll. Sign Man. Chas. I, xi, no. 12; Coll. Proc. Chas. I, no. 168; S. P. Dom. Chas. II, xxx, 69; liii, 41; lxxiii, 47; Entry Bk., xiii, 238; Treas. Papers, ccviii, 30; Welch, ii, 67, 68, 71, 76, 77, 81, 86, 106, 109. For the average expenses of the preëmptors, see Cotton, Titus, B, v, 391; S. P. Dom. Jas. I, cxviii, 83.

[2] S. P. Dom. Chas. I, dvi, 78 *et seq.;* dvii, 3, 31, 54, 71, 108.

[3] D. O. Audit Accts., 1646.

[4] *Tinners' Grievance;* Treas. Papers, ii, 44; Add. MS. 6713, fol. 381.

[5] Receiver, 13 Chas. II; S. P. Dom. Chas. II, xxix, 93; xcix, 12; ci, 17; cxxvii, 9, 80, 125; clxxiv, 142; clxxv, 44; ccxxx, 75; ccxciii, 185; Treas. Papers, ii, 44; Cal. Treas. Papers, i, 99, 104, 211, 302; Welch, ii, 128, 131, 132. The king in 1662 attempted to exercise the preëmption in his own behalf, and even called a parliament of the tinners of Cornwall in order to discuss terms but for some reason the scheme was abandoned (cf. Add. MS. 6713, fol. 336; S. P. Dom. Chas. II, lvii, 88; lvii, 109; lxvii, 118; lxxxi, 8). In 1665 Charles seems once more to have conceived the idea of a venture in tin, although at that time he had leased the preëmption to Richard Ford. Five hundred tons were taken from the preëmptors, paid for by tallies, and shipped to Ostend. This incident, in all probability, was one of the factors that caused Ford's downfall in 1668 (cf. S. P. Dom. Chas. II, cxxviii, 63–65; cxxix, 16; ccxix, 80; cclvi, 21). The Earl of Bath, as lord warden, received a grant of the lease both of preëmption and of coinage duties (S. P. Dom. Chas. II, lxx, 47; Entry Bk., xiii, 327), but seems never to have used his privilege.

a lease,[1] the preëmption was allowed to lie dormant for a number of years. It was put into operation once again by Anne, from 1703 until 1717,[2] but thereafter it disappears from history. The several later efforts on the part of company promoters [3] proved ineffectual to revive an unpopular and long moribund enterprise.

Passing now to the tin coinage duties, we have seen that by the year 1305 they had become entirely dissociated from the process of smelting, and shortly before that date had been fixed at 4s. per hundred-weight in Cornwall and 18¾d. in Devon.[4] These taxes were assessed and paid at the several coinage towns in the two counties, established by the charters of 1305 and altered occasionally by royal decree,[5] usually in order to keep pace with the shifting of the centre of the tin industry. In practice it often happened that only one or two of the towns in Cornwall would be utilized, since the tinners, with a free choice, naturally preferred the town nearest their mines. Of the five towns appointed for Cornwall in 1305, only two, Lostwithiel and Truro, appear four years earlier to have been used for coinage purposes, while in the two centuries following the issue of the charters they remained the sole places of coinage.[6]

The stampage itself is a matter of no little interest. For two hundred years it took place, as a rule,[7] at but two periods in the

[1] Treas. Papers, ii, 12, 16, 17, 41–43, 45, 46, 48–50, 57, 58; iii, 1, 13; viii, 32; ix, 1; S. P. Dom. Chas. II, ccxxiii, 186; H. O. Let. Bk., ii, 87; Add. MS. 6713, fol. 353 *et seq.*, 385, 387.

[2] Under two patents (Treas. Papers, ccviii, 30; Stowe, cxliii, 56–59; Add. MS. 6713, fol. 437–442. Cf. Treas. Papers, cxi, 26; ccxvi, 37). For the disposal of Anne's tin, see Treas. Papers, xcii, 109; cxxxi, 28; xcv, 88; xcix, 97; cxxii, 59; cxxxix, 21, 56; cxxxvii, 97; cxxxvi, 38; cxlvii, 37; clxxiv, 16; clxxxv, 16; clxxxvi, 26; cxci, 16; clxxxvii, 15; cxcix, 24; cc, 9; ccvi, 49; ccxv, 63; ccxvi, 37; ccxxi, 5; ccxxv, 7; ccxlvi, 55.

[3] Lans. 1215, fol. 230; Hunt, 41; Gent. Mag., i, 487.

[4] One document only (Cal. Treas. Papers, i, 61) refers to an alternative of eight pounds of tin per hundred-weight.

[5] Thus, in 1328, Plympton was made a coinage town for Devon (Pat., 2 Edw. III, pt. i, m. 27), and Penzance for Cornwall in 1663 and 1670 (S. P. Dom. Chas. II, lxxvi, 68, 89). Cf. also Anc. Pet., 315 E 168.

[6] Information on this point is taken from the Receiver's Rolls in the Duchy Office.

[7] The earlier coinages of the fourteenth century were quite irregular. Cf. Acct. Excheq. K. R., bdle. 260, no. 20; bdle. 261, nos. 8, 17; bdle. 263, no. 26.

year,[1] Midsummer and Michaelmas. These terms were fixed by the sovereign,[2] or, after 1338, by the Duke of Cornwall, but the minor arrangements with regard to the number of days to be spent in each town and the order in which the towns were to be visited were decided by the officers of the coinage for the two counties,[3] or later by the tinners in their parliaments.[4] There were three chief coinage officials, the receiver,[5] the controller, and a steward. The controller and receiver were royal nominees[6] and represented solely the interests of the Crown. The steward possibly represented in some degree the interests of the tinners as against those of the Duchy, and as the regular presiding officer of the tinners' court he appeared, of course, only at the coinage held within his own stannary.

As the day drew near, the controller and receiver journeyed from town to town,[7] carrying in a sealed bag the stamping hammer and the weights.[8] At the towns they were met by other functionaries, the weigher,[9] and the assay master,[10] together with the requisite number of local porters.[11] Thither also came the tinners, despatch-

[1] Cf. Cal. of Pat., 1304, p. 326; Convoc. Cornw., 16 Hen. VIII, c. 32; 12 Chas. I, c. 32 (Add. MS. 6713, fol. 232). They were increased to four in the seventeenth century (Receiver's Views, 1660–1700; Treas. Papers, ix, 1; cxxii, 17; ii, 55).

[2] Cf. Pat., 9 Edw. III, pt. 1, m. 8; Convoc. Cornw., 16 Hen. VIII, c. 31.

[3] It is hard to say whether or no there were not, in early times, two sets of officials, one for each county (cf. Pat., 7 Edw. III, pt. ii, m. 2; 13 Rich. II, pt. i. m. 30; Close, 8 Edw. III, m. 27). One set appears in later years (Parl. R., v. 533, 587; Pat., 1 Edw. IV, pt. i, m. 4).

[4] Convoc. Cornw., 2 Jas. II, c. 29, 33.

[5] S. P. Dom. Eliz., xlvi, 54; Accts. Excheq. K. R., bdle, 263, no. 13.

[6] Close, 8 Edw. III, m. 27; Pat., 21 Edw. III, pt. iii, m. 14; 13 Rich. II, pt. i, m. 30; 1 Edw. IV, pt. i, m. 4; Cal. of Pat., 1423, p. 85; Parl. R., v, 533, 587; S. P. Dom. Chas. II, xxix, 98.

[7] S. P. Dom. Eliz., xlvi, 54.

[8] This was the special charge of the controller, but the bag was sealed, when not in use, with the seals of all three officials (S. P. Dom. Eliz., cvi, 55).

[9] Also known as the "troner" or "paysor." He was appointed by the receiver (S. P. Dom. Eliz., cvi, 55), but in earlier times by the King (Pat., 12 Rich. II, pt. i, m. 3; 1 Hen. IV, pt. viii, m. 34; 1 Edw. IV, pt. ii, m. 21).

[10] Appointed by the King (S. P. Dom. Chas. II, Entry Bk., 2 Chas. II, xxx, 89; S. P. Dom. Chas. II, Docquet 23, no. 297; 24, no. 55; Treas. Papers, ii, 44; Carew, ed. 1811, p. 45, n.) His salary was £200 per annum in later times, and he often resigned active service to a deputy (Treas. Papers, ii, 44).

[11] Appointed by the receiver (S. P. Dom. Eliz., cvi, 55).

ing their metal in advance by packhorses or carts, while from London and the southern ports came the would-be purchasers, including country chapmen, London dealers, pewterers' factors, and a sprinkling of Italian and Flemish traders. At noon on the first day of the coinage, all assembled by the coinage hall, a warehouse-like structure near the market-place, in which the tin had previously been stored in anticipation of the official inspection and public sale. An open space was roped off at the front, the King's beam was brought out and rectified by the controller and weigher,[1] the weights were solemnly unsealed and handed to the weigher,[2] the assay-master made ready his hammer and chisels, and the steward, controller, and receiver took their seats facing the beam. When all was in readiness, the porters brought out the blocks one at a time and placed them upon the scales. Each had been stamped with the private mark of the owner, and as the steward carried with him a register of these marks no difficulty occurred in identification. The weight of each was shouted out by the weigher,[3] and taken down by the three officials. The blocks, on leaving the scales, were taken in hand by the assay-master, who chiselled a small piece from a corner of each[4] and rapidly assayed it to make sure that the metal was of the proper quality. If so, the controller with a blow from the hammer struck upon the block the Duchy arms; but should it be below standard, it was "tared"[5] by the assay-master, that is to say, a relative figure was placed by him upon each block at which it might be sold below the market price for tin of standard quality.[6] If hopelessly corrupt, it was set aside for re-melting.[7]

When the tin had been weighed and stamped, it was returned to the coinage hall and a bill made out by the clerk[8] for each tin owner, setting forth the number of his blocks, their weights, their fineness, whether "soft and merchantable tin" or "hard and corrupt," and finally the amount of coinage due upon the whole.

[1] Convoc. Cornw., 2 Jas. II, c. 11.
[2] Cf. Close, 18 Edw. II, m. 2.
[3] S. P. Dom. Eliz., cvi, 54, 55.
[4] Treas. Papers, ii, 44; Convoc. Cornw., 22 Jas. I, c. 12; *Tin Duties*, 8. The piece chiselled off used to be his perquisite. Later it was redeemed by the owner for 2d.
[5] *Tinners' Grievance;* Lans. 18, fol. 52.
[6] S. P. Dom. Eliz., cvi, 54.
[7] Carew, ed. 1811, p. 45, n.
[8] *Tinners' Grievance*; Lans. 18, fol. 52.

These served as vouchers for the tin, and during the sixteenth and seventeenth centuries, if not earlier, tin bills were bought and sold [1] much as Scotch pig iron warrants are at the present day. The tin itself was kept at the hall under control of the weigher [2] until the owner had paid the requisite duties.[3] Ordinarily a few days sufficed,[4] but occasionally tin was left standing for a longer period,[5] and as no charge was made for storage the dilatoriness of the tin owners led ultimately to the enactment of rules checking the practice. In 1636 it was declared that all tin coined at Michaelmas must be paid for within ten days.[6] In 1687, however, we find more leeway allowed. All tin lying unpaid for after a year and a day might be assessed by a jury drawn from the stannary court and sold to the first comer, the amount of the tax and dues deducted and the residue given the owner.[7]

The above arrangements were subject to variations. When the stannary revenues were pledged to a creditor of the King or granted to a favorite, or when the preëmption of tin was farmed, it was usually provided that the patentee should be allowed to be present in person or by deputy to see that accounts were truly kept. In this case each tinner received at the coinage a voucher for the proper sum due for his tin by the terms of the contract, and this bill he presented for payment at the patentee's local office, of which one would be established at each coinage town.[8] When the preemption was farmed together with the coinage, very often the controller, receiver, and assay-master were appointed by the lessee.[9]

To meet the difficulty continually arising that the ordinary coinages were insufficient to deal with all the tin produced, there existed the institution known as the "post coinages," which consisted of one or more supplementary coinages held by special war-

[1] Hargrave 321, fol. 41. Cf. Harl. 6380, fol. 39. Ownership of a tin bill implied ownership of the tin (Harl. 6380, fol. 9).

[2] Add. MS. 6713, fol. 254, c. 90.

[3] S. P. Dom. Eliz., cvi, 54, 55.

[4] Convoc. Cornw., 12 Chas. I, c. 32 (Add. MS. 6713, fol. 232).

[5] Receiver's Rolls.

[6] Convoc. Cornw., 12 Chas. I, c. 32 (Add. MS. 6713, fol. 232).

[7] *Ibid.*, fol. 254, c. 90.

[8] Convoc. Cornw., 12 Chas. I, c. 32 (Add. MS. 6713, fol. 232).

[9] Treas. Papers, ii, 44.

rant.[1] Tin stamped in this way paid an additional charge of 4d.[2] on each hundred-weight, known as the "fine of tinners" or "post-groats." It seems to have been levied only in Cornwall, and naturally fluctuated widely in amount from year to year, at times falling to zero, and again rising to over £100, since it was dependent largely upon the necessities of the tinners and their ability to hold their tin for another six months.[3]

The coinage revenues, however, depended not only upon the production of tin but upon the efficiency of the laws which compelled it to be brought to the coinage. The tax was heavy, and the value of the metal and the ease with which it could be got out of the country led from early times to the systematic shipping of tin from smelting-house direct to purchaser. As early as 1198 De Wrotham had prohibited such illicit sales,[4] and had compelled masters to take oath not to receive it on shipboard without the warden's license. Smuggling, nevertheless, seems to have proceeded with unabated activity for centuries.[5] It was not confined to any one class of the community, but was indulged in by merchants,[6] sailors,[7] blowers and miners,[8] without distinction.

Many were the methods taken to avoid payment. Occasionally the stamp was counterfeited,[9] but more often the tin was run into small bars[10] and sold either to wandering chapmen[11] or to sailors from the coast.[12] The houses of the Cornish seaports had their cellars

[1] Add. MS. 6713, fol. 117. But in cases where it was rendered necessary by lack of water for washing the ores or by too great an abundance of water, the royal warrant was not needed (Convoc. Cornw., 16 Hen. VIII, c. 31).

[2] In 1517, 1s. (Receiver, 9 Hen VIII), and in 1518, 8d. (*ibid.*, 10 Hen. VIII).

[3] App. R.

[4] App. A.

[5] Close, 7 Edw. II, m. 10; 8 Edw. II, m. 7; Pat., 7 Edw. II, pt. ii, m. 10 d; 8 Edw. II, m. 28 d; 10 Edw. II, pt. 11, m. 17 d; 16 Edw. III, pt. iii, m. 14 d; 17 Edw. III, pt. i, m. 43 d, 38 d; 18 Edw. III, pt. ii, m. 30 d; 21 Edw. III, pt. i, m. 25 d; pt. ii, m. 9 d; 1 Hen. IV, pt. viii, m. 14 d; S. P. Dom. Eliz., ccxliii, 113; Jas. I, clxxxvii, 26.

[6] Close, 7 Edw. II, m. 10; Pat., 8 Edw. II, pt. i, m. 28 d; Treas. Papers, xvii, 10; Hargrave 321, fol. 41; *Tinners' Grievance*.

[7] Pat. 7 Edw. II, pt. ii, m. 10 d; Coll. Proc., Chas. I, no. 164.

[8] S. P. Dom. Jas. I, clxxxvii, 26; Treas. Papers, ii, 10, 44; xi, 10.

[9] Pat., 7 Edw. II, pt. ii, m. 10 d; 8 Edw. II, pt. i, m. 28 d; Close, 8 Edw. II, m. 7.

[10] S. P. Dom. Eliz., ccxliii, 113; Jas. I, clxxvii, 26; Will. & Mary, xxxviii, 75; Treas. Papers, ii, 44. The bars were known as "pocket tin."

[11] S. P. Dom. Jas. I, clxxxiii, 26.

[12] S. P. Dom. Chas. I, Proc. Chas. I, no. 164.

fitted up with furnaces and kettles, and here a brisk business was done melting down uncoined blocks and running the metal into small wooden moulds, in order to escape detection.[1] Others did not trouble themselves about melting the blocks but during the night rushed them on board ships which, ostensibly coming for slate and stone,[2] carried the tin to Holland, where it was sold at a high profit.[3]

To cope with this illicit commerce, a long code of stannary law had been put in force. All tin must be blown before Michaelmas.[4] No tin might be sold until coined upon pain of forfeiture.[5] No coinage might take place save in the regular towns.[6] No black tin of Cornwall ought to be blown, or white tin of Cornwall stamped, in Devon.[7] Other statutes prescribed the route which tin should take after being whitened. It must not be moved from the blowing or smelting house to any place except to the regular coinage towns;[8] it must not be carried by night;[9] it must be carried by the shortest route and within a reasonable time.[10] At the same time laws were passed to regulate the trading in tin. No one should receive or buy any black tin, tin stuff, or leavings of blowing-houses from any suspicious person not known to be an adventurer for tin, a maker of black tin, or an owner of a blowing-house, or receive any tin from a tinner otherwise than openly at a wash, or before at least one sufficient witness. An attempt was even made to guard the tin before it was removed from the blowing-house. To each owner of a blowing-house was assigned a mark, registered at the nearest stannary court in the steward's book, and this he must place upon every piece [11] and account to the stannary exchequer for every separate block of tin that left his establishment.[12] He must employ no man without first presenting him before the court and seeing that he

[1] Treas. Papers, ii, 10; S. P. Dom. Will. & Mary, xxxviii, 75.

[2] S. P. Dom. Jas. I, clxxxvii, 26; Close, 8 Edw. II, m. 7.

[3] *Tinners' Grievance.* Cf. Penzance in the seventeenth century (Treas. Papers, ii, 10, 44).

[4] Convoc. Cornw., 2 Jas. II, c. 7; Parl. Devon, 2 Hen. VIII, c. 4.

[5] Convoc. Cornw., 30 Eliz., c. 44; Parl. Devon, 2 Hen VIII, c. 29; 6 Edw. VI, c. 5.

[6] Add. MS. 6713, fol. 249, c. 66. [7] *Ibid.*, fol. 248, c. 60.

[8] Convoc. Cornw., 22 Jas. I, c. 4. [9] *Ibid.*, 2 Jas. II, c. 17.

[10] *Ibid.*, 16 Hen. VIII, c. 14; 2 Jas. II, c. 17.

[11] Add. MS. 6713, fol. 101–103; Convoc. Cornw., 2 Jas. II, c. 22; Parl. Devon, 2 Hen. VIII, c. 13.

[12] Add. Ms. 6713, fol. 101–103.

there take an oath to execute his duties with due regard for the laws against impure tin.[1] Similarly every tinner who brought black tin to the blowing-house had to register a private mark at the Lostwithiel Exchequer, so that tin captured from smugglers might be properly identified.[2]

In the later Stuart period a determined effort was made to put down smuggling through the appointment of a supervisor of blowing-houses.[3] One man at first attempted to fill the office, but when it was seen to be too large an undertaking, four under-supervisors were appointed, who divided the inspection between them.[4] What their duties were we know largely from the correspondence of the energetic Mr. George Treweek, who held the post of supervisor-general in the reign of James II.[5] The supervisor was continually running about, visiting the blowing-houses, examining into the records of the day's work, and taking note of the amounts blown in each. He was to see that no small moulds were in use, that the workmen had been properly vouched for, and that the house-mark and the tin owner's mark were on each block. On the coinage day he compared his records of the tin blown in each house with those which the blowing-house proprietor had turned into the steward's court, ferreting out irregularities and seizing all forfeited tin. He must also be active along the coast, to intercept tin that might be carried down to creeks or inlets and hoisted aboard suspicious luggers. He must be prepared also to hunt down suspected tin even to London, and claim it on proof of its falsity. It was no very easy position to fill, and all the more difficult when, as Mr. Treweek bitterly complained, the officers were systematically obstructed in the performance of their duties and on flimsy pretexts haled before the stannary courts for punishment.

That much tin was sold uncoined is undeniable, and to a large extent this factor vitiates whatever official statistics may have to say concerning the production. But with regard to the sums actually

[1] Convoc. Cornw., 16 Hen. VIII, c. 17; 22 Jas. I, c. 2, 3; 2 Jas. II, c. 35; Harl. 6380, fol. 32.

[2] Add. MS. 6713, fol. 102; Convoc. Cornw., 16 Hen. VIII, c. 18; Add. MS. 6713, fol. 241, c. 22.

[3] He first appears in 1660 (Cal. Treas. Papers, i, 13).

[4] Treas. Papers, xi, 10. Each had about seven or eight houses to supervise.

[5] *Ibid.*, ii, 10.

paid in as coinage dues, the figures are clear. Until the preëmption was set to farm, the coinage money was far and away the largest item of revenue which the stannaries afforded. Over £1600 was in this way accounted for in 1303, and the amount rose and fell in wide fluctuations according to the output of tin. In 1355, shortly after the Black Death, less than £1000 was realized from Cornwall and nothing from Devon, whereas seven years before Cornwall alone had paid £2500. From an average of about £2000 a year in the reigns of the first Stuarts the tax dwindled away during the Civil War, and during the Commonwealth was replaced by an excise which lasted until 1660.[1] From 1660 began a steady rise, bringing it in 1700 to £6380, in 1710 to £9600, and in 1750 to £10000.[2]

Post groats first appear in 1302 and for a few years averaged £5 8s. From 1306 to 1507 the entry ceases to appear. It was never a very certain tax until after the Restoration, when the rising production of tin crowded the coinages to repletion, causing the post groats in 1675 to rise as high as £124.[3]

Long before the tin duties were abolished they had become an anachronism and a most intolerable nuisance to producers. Of the original items of revenue the coinage, the prëemption, the fine of tinners, toll tin, and the profits of the courts emerged unscathed from the Civil War; the rest, although not formally abolished, had at one time or another tacitly ceased to be collected. But the machinery of collection remained, a mere excuse for fees and perquisites. There were fees to cutters, fees to weighers, fees to porters, and fees to scalesmen, fees for the use of the beam, drink money, gift money, house money, and dinners to the coinage officials,[4] amounting in all to 1s. 3d. per hundred-weight over and above the regular taxes.[5] The necessity for reform was denied by scarcely any one,[6] and finally in 1838 the whole system, with the exception of toll, was quietly swept away in favor of a small excise duty levied at the smelting-house.[7]

[1] *Tract on Tin Duties.*

[2] For the figures of the revenue from coinage dues see App. K.

[3] App. R.

[4] *Tin Duties*, 9.

[5] *Ibid.*, 7.

[6] Hansard, xliv, 1002–1005.

[7] Stat. 1 & 2 Vict., c. 120.

# CHAPTER VI

## PRIVILEGES AND TRADE RULES OF THE TINNERS

It must be admitted at the outset that so far as concerned the enjoyment of special privileges from the Crown the tinners were treated no differently from many other classes of workers, such, for example, as the moneyers, or the tenants of the royal demesnes.[1] Particular favor, however, was usually shown the mining classes, a fact for which analogies may readily be gathered from what has been said already concerning mining law in England and elsewhere. Thus, for example, we know that the "King's miners," in his mines at Tisdale, Irresdale, and elsewhere, near Carlisle, received letters of protection against outside interference, which the sheriff of York was expected to enforce,[2] while Henry III acted in similar fashion towards the miners of Northumberland[3] and York. In 1320 a commission of oyer and terminer was issued to John de Treiagu, Henry de Bukerel, and Simon Balde, "touching all trespasses committed against miners in the King's service in the mines of Devon."[4] The fact that similar patronage was accorded to miners not in the King's service, whose obligation to the Crown was limited to the taxation on output, adds still greater emphasis to this benevolent attitude of the Crown toward the industry.

None were favored more than were the mediæval tinners. As far back as 1336, for example, occurs a case in point. The tinners complained to Edward III that they were unable to make any sales

[1] "The King's manor was treated as a franchise isolated from the surrounding hundred and shire. Its tenants are not bound to attend the county court or hundred moot, and are not assessed for danegeld or common amerciaments or murder fine, are exempted from the jurisdiction of the King's sheriff, do not serve on juries and assizes before the King's justices, are free from all market tolls and customs house tolls. They do not get taxed with the country at large, and on the other hand are liable to be tallaged by the King without the consent of Parliament, and by virtue of his private right" (Vinogradoff, 92).

[2] Cal. of Pat., 1219, p. 224. [3] *Ibid.*, 1222, p. 339; 1223, p. 336; 1229, p. 249.

[4] *Ibid.*, 1320, p. 485.

of tin in Bristol, owing to letters patent issued to the London girdlers whereby the garnishing of girdles with false work of lead, pewter, or tin was forbidden. The answer was that inasmuch as this measure was prejudicial to the stannaries and to the King's revenue, and to John of Eltham, earl of Cornwall, the Bristol men and others might consider themselves relieved from the pressure of the prohibition, and might work tin at pleasure.[1] Tinners' wages, again, were not subject to tithes,[2] a provision, however, which in later centuries seems to have been disregarded.[3] The stannary weights were not to be prejudiced by Henry VII's Act for the standardizing of weights and measures.[4] Henry VIII's statute concerning artificers and laborers did not apply to tinners,[5] but Elizabeth's statute of 1563 did not exempt them. Warrants of the stannary court were free from the duties on stamped vellum, parchment, or paper, instituted in 1698,[6] and even as late as the eighteenth century we find a drawback granted on duties on coals used in fire engines for the drainage of Cornish mines of tin and copper.[7]

Of the ancient privileges of the tinners that of bounding was probably the oldest and certainly the most important, containing as it did the very essence of the free mining systems which have been described.[8] In the earliest stannary charter we find it merely confirmed as an ancient custom, and it is impossible to set a date at which it originated. The actual process consisted in marking the angles of the area desired to be enclosed by small holes and heaps of turf, or else by poles erected at each corner, with a furze bush

[1] Pat., 10 Edw. III, pt. i, m. 20 d.

[2] Convoc. Cornw., 22 Jas. I, c. 16. This was not the case in the Derbyshire lead mines (Add. MS. 6682, fol. 38).

[3] Convoc. Cornw., 22 Jas. I, c. 18.

[4] Stat. 11 Hen. VII, c. 4, sec. 6.

[5] Stat. 6 Hen. VIII, c. 3, sec. 6.

[6] Stat. 9 & 10 Will. III, c. 25, sec. 45.

[7] Stat. 14 Geo. II, c. 41, sec. 3.

[8] Cf. Trans. Roy. Geol. Soc. Cornw., vi, 239 (Edict of Charles VI of France). In France, however, a reasonable price had to be paid for the land bounded. By the edict of Louis XI in 1471 the landowners were to have an indemnity (*ibid.*, 240); and the right of searching for ore was limited to miners and to the King's officials (Smirke, 112). The landowners, furthermore, had preference over all others in working mines upon their lands (*ibid.*, 112, 113). The Parliament of Paris subsequently limited the general right of search to desert and uncultivated places (*ibid.*, 113; Trans. Roy. Geol. Soc. Cornw., vi, 230). For mining rights in France under Henry IV see Achenbach, 1869, pp. 35–37. For Germany see Smirke, 107; Jars, iii, 486, 505, art. i.

at the top.[1] The privilege differed in extent as between the two counties. In Devon we find the parliament of tinners in 1510 reaffirming their "ancient right to dig tin in any place in Devon where found,[2] and also to carry water to the works," and a fine of £40 was to be inflicted upon any one who obstructed the tinners in the exercise of this right.[3] Furthermore, the Devon tinners paid no toll tin,[4] save when their works were on meadow lands, good pasture, and lands commonly used for tillage, when they gave up each tenth bowl of ore.[5] In Cornwall all wastrel land and also any several and enclosed land anciently bounded and assured for wastrel by the delivery of toll tin to the owner before hedges were set up, together with so much of the Duchy several and enclosed land in the demesne manors as had anciently been bounded, might never be taken up by tinners in the course of their work.[6] To pitch bounds in land not coming under the above categories the permission of the owner must first be obtained,[7] and it was a rule of the stannary judiciary that action might be brought against any tinner who subverted trees and woods in any private grounds.[8] Further restrictions forbade the

[1] Journ. of Science, i, 285; Harl. 6380, fol. 27.

[2] This was also the case for iron mining in the Forest of Dean (Houghton, pt. ii, art. 4), where, if the miner were denied this right, the King's gaveller interposed in his behalf (*ibid.*, pt. ii, art. 13). In the High Peak, Derbyshire, it was lawful to dig all sorts of grounds, but if arable land or meadows were broken into but not worked according to the custom of the mines, the owners might fill in the works (*Compl. Min. Laws Derb.*, pt. i, art. 12).

[3] Parl. Devon, 2 Hen. VIII, c. 2 (cf. also *ibid.*, 16 Eliz., c. 20). Under this statute arose the well-known Strode case (Stat. 4 Hen. VIII, c. 8; Trans. Devon Assoc., xi, 300).

[4] Toll tin should be distinguished from "farm tin" paid the bounder if he leased his claim to others (Bainbridge, 144; Convoc. Cornw., 25 Chas. II, c. 4).

[5] Parl. Devon, 16 Eliz., c. 17. This custom, remarks Smirke, is untenable in law, but Devon contains but few tin works and so the question is of small practical importance (Eng. Min. Almanack, 1849, p. 157).

[6] App. D; Convoc. Cornw., 12 Chas. I, c. 2 (Add. MS. 6713, fol. 335; 25 Chas. II, c. 4; Add. MS. 6713, fol. 236, c. 4.

[7] Convoc. Cornw., 16 Hen. VIII, c. 25; Add. MS. 6713, fol. 112. In Germany, according to Schmoller (Jb. xv, 679–680), when it came to the question of the exploitation of minerals in tilled fields, the occupant, in spite of regalian rights, refused consent until the destructive stream-work methods had given way to lode mining which did little damage to the surface, and that usually in land of little value. This the peasant would usually give for a compensation.

[8] Add. MS. 6713, fol. 249, c. 63.

sinking of shafts within twenty-four feet of the highway, or so situated as to choke the rivers and havens with their refuse.[1]

Toll tin, or the landlord's dues, consisted of a certain fixed proportion of the ore raised from a mine, a proportion which might vary with local custom [2] but in general consisted of the fifteenth dish.[3] The toller, as the lord's agent, in earlier days received his master's share at the periodic washing of the ore at a mine, of which warning must be given the lord some days in advance.[4] Instead of toll tin, the landlord, apparently at his choice, might be given a certain share or "dole" in the enterprise itself,[5] a custom which has its analogies elsewhere in England,[6] as also in Germany; [7] or in some places, if he liked, the landlord might place one workman [8] in the mine for every fourteen hired by the adventurer. A practice of somewhat later date allowed the latter, in lieu of toll, to contract with the landlord to employ a certain number of men and boys annually and to pay the landlord an agreed rate for each one so employed.[9]

[1] Convoc. Cornw., 16 Hen. VIII, c. 33; 2 Jas. II, c. 7; Parl. Devon, 16 Eliz., c. 1–7.

[2] Convoc. Cornw., 12 Chas. I, c. 4 (Add. MS. 6713, fol. 224).

[3] Convoc. Cornw., 25 Chas. II, c. 4; Add. MS. 6713, fol. 235, c. 2. The tenth dish was given by the Mendip lead miners (Trans. Roy. Geol. Soc. Cornw., vi, 329). In France the King received a royalty of one tenth. The landowner received one tenth or one twentieth, and at a later period one fortieth (Trans. Roy. Geol. Soc. Cornw., vi, 248–249; Achenbach, 1869, p. 737).

[4] Add. MS. 6713, fol. 236, c. 6. Little toll tin seems to have been realized upon the Duchy demesne manors, and that little was almost always farmed out for a lump sum (Duchy Accts. Excheq. Aug., port. 2; Add. MS. 24746, fol. 353; S. P. Dom. Jas. I, Grant Bk. 134; Chas. II, xxix, 98, Docquet 25, no. 95; Pat., 1 Hen. VIII, pt. i, m. 20).

[5] Add. MS. 6713, fol. 235, c. 3.

[6] In the Forest of Dean (Houghton, pt. ii, art. 14), and in Derbyshire (*Compl. Min. Laws Derb.*, pt. iii, art. 1).

[7] In Germany the custom went even farther, at least in Saxony and Bohemia, for besides the seven "lanen" comprised in each allotment others were measured along the vein for numerous public personages. Thus in Freiburg there was one for the margrave, one for his wife, and also for the chief household officials, the marshal, the steward, and the chamberlain, the city council, and the mine master (Schmoller, Jb., xv, 694). In the Iglavian law, there was also a King's "lane," and sometimes one for the landowner (*ibid.*, xv, 695; Smirke, 82). As a rule, by the thirteenth century these were leased to the operators of the regular mines (Schmoller, Jb., xv, 696). For later practice in Prussia, cf. Trans. Roy. Geol. Soc. Cornw., vi, 158, 161.

[8] Carew, ed. 1811, p. 44.

[9] Pryce, 132.

When the tinner had settled upon his holding and had pitched his bounds, it was necessary for him to have them registered upon the steward's book at the nearest stannary court[1] and proclaimed at three successive sessions.[2] If the bounder's claim were not successfully resisted, he became in due course entitled to his writ of possession, and his right became absolute.[3] In Cornwall the tin bound so acquired was a chattel real,[4] and devolved on the legal per-

[1] Parl. Devon, 16 Eliz., c. 35, 36; Ct. R., bdle. 101, no. 18; Harl. 6380, fol. 31; Add. MS. 6713, fol. 103. Smirke (p. 101) advances arguments to show that this practice of proclaiming bounds originated in 1495 with Prince Arthur. His reasons are as follows. First, in the court rolls of 17 Henry VII we find bounds entered as above, but not until then; second, an express declaration of the tinners of Penwith and Kerrier, in a court held in 1616, is to the effect that proclamations are not as old as the bounds themselves; third, the fact that both in Arthur's ordinances and in the articles confirmed by him but adopted by a parliament in Devon, 10 Henry VII, the practice of proclaiming bounds in court was instituted.

[2] Convoc. Cornw., 30 Eliz., c. 15; Harl. 6380, fol. 31.

[3] In Derbyshire, upon a man's finding a vein of ore, the law required him to fix a cross in the ground as a mark of possession, giving notice at the same time to the barmaster, an officer "chosen by the miners and merchants to be an impartial person between the lord of the field and the miners, and between the miners and merchants" (*Compl. Min. Laws Derb.*, pt. iii, art. 1). The barmaster, having received the first produce of the mine, *i. e.* a measure of ore, delivers to the miner "two meers" of ground in the same vein, each meer being twenty-nine yards long and fourteen broad. The barmaster then took possession of the next half-meer for the lord of the field, *i. e.* usually the King (*ibid.*, pt. iii, art. 1. See also English's Min. Almanack, 1850, p. 220, 221; Farey, i. 357). The mining laws of the Mendip Hills in this connection make an interesting provision. Any man intending to be a miner must get leave of the lord of the soil, or his lead reeve or bailiff, which license, however, by custom might not be denied. After having once been licensed in this manner, no further permission was ever necessary during his lifetime. (Trans. Roy. Geol. Soc. Cornw., vi, 328) The length of the miner's "pitch" on the vein was determined by the distance to which he could throw an axe. ("Every man who begins his pitch shall have his axe throw two ways after ye rake or chine," *ibid.*, 329, 331). This having been effected, the miner had twenty-four hours allowed him "to make the same lawful with a pair of stillings, sufficient beams and forks."

In the Liège coal region, when a man desired to drain an abandoned mine, he must go through certain formalities of proclamations and notices (Delebecque, iii, 123, 193, 231–244).

The German method, so far as any one method may be said to have existed, consisted in the enfeoffment of the would-be miner at the hands of the mine master or lord's officer, after which the bounds were proclaimed at three successive mine courts, as in the stannaries (Schmoller, Jb. xv, 693; Jars, iii, 505, art. iii). For later Prussian and Saxon practice, see Trans. Roy. Geol. Soc. Cornw., vi, 157, 160.

[4] Convoc. Cornw., 2 Jas. II, c. 1, 4. So it was in the Forest of Dean (Houghton,

sonal representative of the bounder. In Devon, on the other hand, the tin bound was real property, and devolved upon the tinner's heir.[1]

With the privilege of pitching bounds went the obligation of keeping the marks in repair. Ordinarily this had to be done once a year, or the claim lapsed,[2] and a not inconsiderable amount of custom and law arose upon this point, prescribing the exact terms under which a mining property fell vacant. Thus if any persons specially appointed to review bounds did by covenant with the would-be rebounder misuse the trust reposed in them, or suffered old bounds to become void, this fact was to be made a remitter to all old bounders and owners.[3] Or, again, we read that "bounding upon bounds, and bounds in reversion are to be void if there be but one corner well pitched out of sound ground, the same bound to be withdrawn and the bounds to be adjudged good."[4] A tinner might not bound stream works with the bounds of mine works, or mine works with the bounds of stream works.[5]

The precise amount of work which a tinner must do upon his claim in order to hold it was never fully defined.[6] We learn from

pt. ii, art. 24), and in Derbyshire (*ibid.*, pt. iii, art. 4). In Russia, Peter the Great made it a freehold (Scrivenor, 163).

1 Parl. Devon, 2 Hen. VIII, c. 21; 25 Hen. VIII, c. 1.

2 Add. MS. 6713, fol. 116; Harl. 6380, fol. 27, 30. Strangely enough, the tin shaft itself was not forfeit because the bounds were suffered to lapse. The pit remained but might not be extended laterally (Harl. 6380, fol. 30).

3 Convoc. Cornw., 16 Hen. VIII, c. 36.

4 *Ibid.*, c. 13.

5 Add. MS. 6713, fol. 128.

6 This was not the case in Germany, at least in early times, a fact which emphasizes the difference in origin of the English and the German mineral law. The German mine originally was a precarious holding allowed by the territorial lord on condition of continuous work, day and night, by the mine associates (Schmoller, Jb. xv, 672; Smirke, 83). Originally three days of absence from work brought with it the confiscation of the mine. Afterwards, when the miners had a greater stake in their holdings, the term was extended to three weeks or even to a year and a day, and a formal summons required of the magnate (Schmoller, Jb., xv, 672). The conditions under which the measured allotments were held, however, were much stricter, since every allotted "lane" had to have a special shaft of its own, and as many working places as could be operated without hindrance to the works (cf. Schmoller, Jb., xv, 695).

Roman mining law, as shown by a recently discovered inscription, insisted also upon continuous work. The mine owner must commence work within 25 days after taking possession, and cessation of work for ten days, or in special cases for six months, brought forfeiture of the mine (Neuburg, Zeitschr. f. d. ges. Staatsw., lxiii, 380).

Beare's account of the stannaries in 1586 that "the first year a man keeps his work by his bounds; the second by the charges he puts into it; and the third year by toll tin."[1] By a Penwith and Kerrier custom earlier in that century, the amount of annual cost which must be incurred by the tinner, as declared by twelve jurors at stannary court leet, was three months' cost for one man or one month's cost for three; otherwise, at the end of the fourth year, the work was void.[2] A later law gave the landowner the right to sue a tinner who did not "make sufficient" of his bounds.[3] It is undeniable, however, that the laxity of stannary law in this respect gave many opportunities for the abuse of the right of bounding, which became during the Middle Ages one of the great grievances harbored against the tinners by the people of Devon and Cornwall. Neither did the laws of the stannary contain any provision regulating the amount of land which might be included in a pair of bounds, and a possible outcome of this omission is seen in the fact that in 1786 all Dartmoor, comprising fifty thousand acres, was claimed by a single prospector.[4]

Included in the right of bounding were certain supplementary privileges, among them the right of free access to running water,[5] essential for cleansing the ores. In the tinners' charters this right is merely confirmed as having already existed time out of mind. It appears as the right "of diverting streams,"[6] and served not only to permit the washing of ore but also to lay bare the river beds in the search for stream tin. Connected with the above privilege was that of buying brushwood and faggots for the purpose of smelting tin ore,[7] amounting, doubtless, to a right of seizure in case the owner

[1] Harl. 6380, fol. 30. [2] Add. MS. 6713, fol. 116. [3] *Ibid.*, fol. 248, c. 61.

[4] Eng. Min. Almanack, i, 155–156. Of German mining law we may say, in general, that the original custom was one by which the allotments were small but unmeasured plots of ground. The only regulation was one which forbade too close an approach to a neighboring mine (Schmoller, Jb. xv, 693; Achenbach, 1871, p. 361). As time went on the bounds became certain definite measured surfaces, larger in extent than the old ones. In Saxony the typical allotment was seven "lanen," each of seven fathoms square (Schmoller, Jb., xv, 664, 667, 694; Smirke, 105).

[5] Convoc. Cornw., 30 Eliz., c. 11; Harl. 6380, fol. 26, 27. Cf. for Derbyshire, *Compl. Min. Laws Derb.*, pt. i, art. 9; and for Germany, Schmoller, Jb., xv, 678.

[6] "Et divertere aquas ad operationem eorum" (App. B.).

[7] App. B and D; Harl. 6380, fol. 26, 27. The necessity for a supply of fuel for mining operations is seen in the liberal privileges for purchase of brushwood and

refused to sell. In an age when coal was rarely used, save occasionally for domestic purposes, this was a concession of some importance, especially in view of the fact that the barrenness of the Cornish moors made it difficult at the best of times to obtain a sufficient stock of fuel. It was partly in consequence of this, and partly as a result of the approaching exhaustion of the Cornish and Devon peat beds, that the tinners were permitted to cut turf in the Forest of Dartmoor. For Devon this practice possibly derives its sanction from custom as ancient as that of bounding,[1] but there is no evidence that it was enjoyed by Cornwall until 1465, when we first find it conceded in a confirmation of Edward I's charter.[2]

The right of access to the highway is so obviously a concomitant of bounding that it may seem to be hardly worth mentioning. In fact it is nowhere expressly stated as a part of stannary law, although both the Dean[3] and the Derbyshire[4] miners claimed it among their privileges.

Exemption from ordinary taxation has already been noted.[5]

faggots which the King allowed the wardens of his own silver mines (Cal. of Pat., 1283, p. 69; 1339, p. 286; Cal. of Close, 1333, p. 152; 1337, p. 190; Pat., 15 Edw. IV, pt. i, m. 22). Needless to say, similar privileges of fuel were allowed the miners of the Forest of Dean (Houghton, pt. ii, arts. 26, 28, 29, 34), Alston Moor (Hunt, 148), and Derbyshire, where in addition the miners had pasturage and sufficient space to build their cottages (Add. MS. 6682, fol. 68; *Compl. Min. Laws Derb.*, pt. i, art. 11; Houghton, 7, art. xiii).

For this privilege as exercised by the German miners, see Reyer, 79; Smirke, 108; Schmoller, Jb. xv, 677, 678.

For France, see Louvrex, vii, 386–390. For Scotland, Patrick, 16. For Russia, Scrivenor, 163.

[1] In 1226 it is mentioned as an already existing right (Pat., 1 Hen. III, m. 5; Close, 3 Hen. III, i, m. 9 d., 23; 6 Hen. III, m. 6).

[2] Pat., 5 Edw. IV, pt. ii, m. 4, 7.

[3] Houghton, pt. ii, art. 13.

[4] *Compl. Min. Laws Derb.*, pt. i, art. 2.

[5] Such exemptions were not unusual favors. Freedom from tolls in the realm was a usual gild merchant privilege (Gross, i, 71). The workmen in the royal silver mines were almost invariably exempt from tolls and taxes (Pat., 27 Edw. I, m. 35; Cal. of Pat., 1299, p. 398; 1307, p. 14; 1313, p. 526; 1331, p. 74; Cal. of Close, 1315, p. 244). No trace of the privilege is to be found in the laws of the Forest of Dean, but in Derbyshire both miners and merchants were toll free (Add. MS. 6682, fol. 68).

By the edict of Charles VI of France in 1413, "merchants, masters, founders, refiners, and workmen" were exempted from aids and tallages (Louvrex, vii, 386–

This right, first granted in the charters of 1305,[1] seems, with the exception of the case of ship-money in the year of the Armada,[2] to have been thereafter recognized as absolute.[3] It included not only freedom from tolls and market dues at fairs and ports in the two counties on the tinners' own goods, but also relief from ordinary taxation, such as the levy of the tenth and fifteenth. Occasionally this privilege was attacked and attempts were made by royal officers to tax the tinners illegally,[4] but this was usually disclaimed by the Crown and the liberty upheld.[5] A case in point occurred in 1338, when an attempted levy of the tenth and fifteenth was answered by the miners' refusal to operate their works until their grievance had been redressed.[6] But legal protection against the tax collectors was also sought by the tinners in their own courts, and the bailiff, customer, or sheriff who included a tinner in his lists, came under the penal statute of the tinners' parliaments.[7] Freedom from ordinary taxation now no longer forms a part of

390; Smirke, 104; Trans. Roy. Geol. Soc. Cornw., vi, 239). This was reflected in 1471 by the edict of Louis XI (Louvrex, x, 623; Smirke, 112).

In Germany, in general, the same practice was followed with regard to miners (Smirke, 84), and the native tin workers of Banca, even to-day, enjoy this mediæval privilege (Diest, 31).

[1] Nominally only tinners on the demesne manors were so favored, but in practice the exemption seems to have applied to all.

[2] From the correspondence on the subject it would seem that the Queen at first endeavored through the warden to persuade the tinners to a voluntary contribution, and then at last ordered the amount to be levied (Acts of P. C., 1588, p. 148; S. P. Dom. Eliz., ccxii, 53; cclxii, 73).

[3] Pat., 12 Edw. III, pt. i, m. 23 d; 1 Edw. IV, pt. iii, m. 13; Close, 11 Edw. III, pt. ii, m. 20; 12 Edw. III, pt. iii, m. 13 d; Convoc. Cornw., 30 Eliz., c. 58; 22 Jas. I, c. 16; Parl. Devon, 2 Hen. VIII, c. 34; 16 Eliz., c. 32; Add. MS. 24746, fol. 92; Trans. Devon Assoc. viii, 381.

[4] Close, 12 Edw. III, pt. iii, m. 13; Anc. Pet., bdle. 108, no. 535.

[5] Cf. Parl. Devon, 2 Hen. VIII, c. 34. At the stannary court of Chagford in 1528, the mayor of Exeter and several others were convicted for having attached a lighter of salt for customs, being tinners' goods (D. O. MS. Vol., fol. 315).

[6] Close, 12 Edw. III, pt. iii, m. 13.

[7] This exemption from ordinary taxation gave rise to continual frauds on the part of men who attempted to become tinners merely to avoid the payment of rates (Pat., 12 Edw. III, pt. i, m. 23 d; 16 Edw. III, pt. iii, m. 2 d; 17 Edw. III, pt. i, m. 40 d; pt. ii, m. 5 d., 32 d; Close, 11 Edw. III, pt. ii, m. 20; Lay Subs. R., bdle. 95, nos. 12, 22, 31; Parl. Devon, 25 Hen. VIII, c. 1, 2; 16 Eliz., c. 32). No sufficient remedy for this abuse seems ever to have been found (S. P. Dom. Eliz., ccxxvii, 28; Stat. 16 Chas. I, c. 15).

a miner's privileges in the stannaries,[1] although it is difficult to determine the exact period at which the exemption vanished. In all likelihood it passed away during the Commonwealth,[2] along with several of the stannaries' own peculiar forms of taxation.

Doubt exists as to whether tinners were liable to impressment and forced labor. The charters of 1201 and 1305 state clearly that they were not to be called from their work save by the warden, and this fact seems to preclude any possibility of their being drawn upon by the lords of the manors. It does not, however, signify that they could not be made use of by the King, if so required, and although two royal patents are extant which give the tinners exemption from impressment for work in the royal mines,[3] the exemption seems to have been but temporary. Certain it is that on several occasions during the fourteenth century miners in Devon and Cornwall were impressed for the royal mines,[4] while in an early Receiver's Roll for the stannaries we find an entry of twenty-five tinners for six days at 2d. per day each.[5] It would seem, therefore, that when the King desired the services of miners from the stannaries he might take them through the medium of the warden, and this conclusion is supported by the fact that James I, on at least one occasion, had tinners impressed for service in Scotland.[6] In any case the tinners were not alone in submitting to this burden. Many other classes of artisans were similarly bound, and miners from Dean,[7] Derbyshire,[8] and the Mendip Hills[9] were frequently dragged from one part of England to another by the steward of a royal silver mine, or, it may be, by the lessees of royal mines, to whom among other powers had been given that to impress labor.[10]

[1] *Tin Duties*, 6.

[2] In 1697 the tinners were no longer free of fairs and markets (*Tinners' Grievance*).

[3] Cal of Pat., 1305, p. 331; 1308, p. 61.

[4] *Ibid.*, 1320, p. 537; 1328, p. 318; Cal. of Close, 1319, p. 134.

[5] Receiver, 31 Edw. III.

[6] Add. MS. 24746, fol. 120; S. P. Dom. Jas. I, Addenda, dxxi, 10, 101.

[7] Cal. of Close, 1319, p. 127. They were frequently used in the army (Nicholls, 17; Rymer, iii, pt. i, 78, 417; pt. ii, 762, 1021).

[8] Cal. of Close, 1288, p. 499; 1309 (Jan. 26); 1319, p. 212; 1326, p. 478; 1333, p. 52; Cal. of Pat., 1295, p. 179; 1380, p. 527.

[9] Cal. of Close, 1319, p. 127; S. P. Dom. Jas. I, clii, 9.

[10] Pat., 15 Edw. IV, pt. i, m. 22; Cal. of Pat., 1462, p. 194; 1468, p. 132; 1475 (Mar. 23). In Durham the bishop exercised a like privilege (Durh. Curs. Rec., 23

The tinners' privilege to be exempt from any summons save that of their warden found practical expression in their custom of mustering for military service under command of the lord warden apart from the remainder of the militia. This held true only of working tinners who were consequently relieved from all attendance upon the gentry of the county.[1] At the musters the force from each of the stannary districts was led by a colonel, each having under him four captains,[2] and the separate mustering of the tinners for militia service has been preserved in later statutes of the realm.[3]

The enumeration of the tinners' privileges, however, must be supplemented by an account of their trade rules before a judgment may be formed as to their proper place in the mediæval industrial order of England. Do these privileges and trade rules, to put the question plainly, justify an analogy between the tinners' organization and the mediæval craft gild?

The dominant spirit pervading the gild regulations[4] of the Middle Ages may be summed up in the phrase, "mutual protection by the exclusion or limitation of competition." Between different crafts this was effected chiefly by the operation of the "Zunftzwang," or pressure on the part of the gild to force all men of its own trade to join it. Between fellow gildsmen competition was limited by a host of petty rules, the object of which was to equalize opportunities and prevent the aggrandizement of a few masters at the expense of the many. Raw materials, in some instances, might be had only at prices fixed by town or gild authorities, and in reasonable amount. The number of apprentices and journeymen that a master might keep was often strictly limited, while labor conditions and wages

Hatfield, no. 31, m. 4 d; 29 Hatfield, no. 31, m. 5 d). The German miners, in most cases, seem to have been exempted from feudal or military service (Smirke, 84; Reyer, 35; Trans. Roy. Geol. Soc. Cornw., vi, 164; Schmoller, Jb., xv, 677), but it does not appear whether this exemption was extended to forced labor.

[1] Convoc. Cornw., 16 Hen. VIII, c. 27; Parl. Devon, 16 Eliz., c. 23, 24; Add. MS. 6713, fol. 113; S. P. Dom. Eliz., cxii, 23; ccix, 22; ccxvi, 49; cclxii, 73; cclxiii, 66; Jas. I, lxxviii, 36.

[2] S. P. Dom. Eliz., ccxvi, 48, 49; L. and P., Hen. VIII, xiv, pt. i, no. 1652.

[3] Stat. 38 Geo. III, c. 74; 42 Geo. III, c. 72, 90, sec. 154; 43 Geo. III, c. 55, sec. 12; 46 Geo. III, c. 90, sec. 47; c. 91; 51 Geo. III, c. 114; 56 Geo. III, c. 20; 55 Geo. III, c. 65[1]; 15 and 16 Vict., c. 50, secs. 26, 38.

[4] Cf. Schönberg's account of the gilds in Conrad's Jahrbücher für Nationaloekonomie, 1867, pp. 1–72, 97–169; Sombart, i, 127–131.

were frequently regulated by the central officials. Sometimes the gild went so far as to set an absolute maximum of production for each master. Even when the goods were put upon the market, we find the same careful supervision with regard to their sale, ensuring publicity and fairness, or in the case of custom-work providing that craftsmen might not entice away the customers of gild brothers or continue work which others of their trade had begun.

The aim of the gild system was, therefore, the solving of the problem of distribution by the strict regulation of production, and it is just this very factor which is most noticeably lacking in the laws of the stannaries.[1] The latter were designed to assist in the production, not in the distribution of wealth. The individual, in this respect, was everything and the organization nothing. Gild coöperation of mine owners in any of the processes of digging, smelting, and selling of tin did not exist. There is no trace of any system of coöperative purchases of iron, rope, or timber, for the use of the mines. There were no apprenticeship regulations, and in all probability no apprentices properly so called. The use of the "Zunftzwang" was precluded by the very constitution of the stannaries, inasmuch as any man engaged in the production of tin in any way became entitled *ipso facto* to all the privileges of the trade. The purchase of mining materials and the possibilities of enlarged production were in every way unregulated and fully as free as they are at present. The most ardent individualist could have found little that was amiss in the regulation of the tin production of the Middle Ages, if one excepts — and here is the exception which proves the rule — those inspired by the Crown with a view to the guarding of its fiscal perquisites.

It is obvious that as long as the King or the Prince of Wales maintained in any way a hold upon the miners, — and such a hold, of course, was furnished by the support given the stannaries in the

[1] Here obviously lies a temptation for theorizing as to the early status of the gilds. It has been supposed that they were at the start vested with some of the attributes which are described above, and traces of exclusive privilege may be found at an early period among some of the first craft-gild regulations. But is it not possible that they started as "regulated companies," like the miners, and later, not being subject to such rigid scrutiny on the part of the Crown, developed the "Zunftzwang" and other analogous practices in obedience to the mediæval spirit of close combination?

form of rights of free mining, freedom from taxation, and courts and parliaments of their own, as well as by the constant aid from the Crown in order to maintain these privileges against the lords,—no trade rules would be tolerated which threatened in the slightest degree to reduce production and so diminish the Crown revenues. Opportunities enough for combined effort were supplied the tinners, but their common activities were directed into quasi-political lines for the maintenance of their peculiar liberties, while the craft itself partook rather of the nature of the "regulated companies" of the seventeenth century than of the closely organized gilds of the mediæval towns.

Excluding such as were designed merely for the maintenance and regulation of the stannary courts, most of the stannary laws had to do with ordinary relations between one mine owner and another. Such, for instance, were laws guarding against thefts of ore,[1] or the forcible entry into another's tin work;[2] against the dumping of rubbish upon another miner's claim,[3] instead of in old shafts and pits;[4] and against the disloyalty of spaliards or hired workmen toward their employers.[5] Others concerned the relations existing between partners.[6] Another custom forbade the wearing of arms in the mine or at the washes,[7] a commentary on the semi-lawless character of the mining people, paralleled by a similar custom among the Mendip lead miners.[8] The giving of shares in a tin work to stannary officials or to powerful gentry was forbidden both in the stannaries[9] and among the miners of Derbyshire,[10] and was supplemented, probably in the later fifteenth century, by the prohibition of liveries upon other than manual servants,[11] and the forbidding of tinners to retain of any lord save the King, the Prince, the lord warden, or the vice-warden.[12]

Quite naturally, however, much attention was given in the regula-

[1] Convoc. Cornw., 16 Hen. VIII, c. 35; Parl. Devon, 25 Hen. VIII, c. 3, 4.

[2] Convoc. Cornw., 16 Hen. VIII, c. 12.

[3] Parl. Devon, 25 Hen. VIII, c. 6.

[4] Add. MS. 6713, fol. 191.

[5] Parl. Devon, 2 Hen. VIII, c. 17.

[6] *Ibid.*, 25 Hen. VIII, c. 9.

[7] Convoc. Cornw., 12 Chas. I, c. 31.

[8] Trans. Roy. Geol. Soc. Cornw., vi, 333.

[9] Convoc. Cornw., 16 Hen. VIII, c. 11. Cf. Star Chamber Proc., Henry VIII, bdle. 32, no. 90.

[10] Houghton, 22, art. 18; *Compl. Min. Laws Derb.*, pt. i, art. 24.

[11] Parl. Devon, 2 Hen. VIII, c. 14; Add. MS. 6713, fol. 103.

[12] *Ibid.*, fol. 280, c. 8.

tions to the production of tin. The washing of the ore, as it came from the mine, was subject to certain general restrictions, the object of which was to ensure publicity of work, to guard the interests of all partners in a mining enterprise, and in some degree to maintain, with gild-like firmness, the purity of the metal. The working of "private buddles or dishing places" in any secret place was prohibited.[1] Warning of the wash must be given the landlord [2] and all partners in the mine,[3] and the operation, together with the division of ore, which seems to have been a function of some consequence, was carried out under characteristically peculiar conditions. From amongst their number the partners of a mine chose a "meeter," or measurer, who took his seat, measure in hand, before the heap of dressed and pulverized ore, the black tin. This he scooped up in his hands and let trickle slowly into the measure from between his fingers, filling it as lightly as possible, until the "striker," a flat piece of heart of oak, bound with iron, might, when placed across the top of the vessel, rest upon the ore within.[4] The tin stuff was thus divided into perhaps fifteen shares, one of which was marked by the toller as belonging to the proprietor of the soil,[5] and the rest divided and subdivided by rule of thumb, until every miner received his portion.[6]

It was now ready for sale. "No man," so the law reads, "ought to buy or receive any black tin otherwise than openly at a wash, from the sheet, nor buy or receive any tin shift or the leavings of any blowing-houses from any suspicious person that is not known to be an adventurer for tin, or a worker of white tin, or owner in a blowing-house." [7] Similarly the owners of blowing-houses, crazing mills, or stamp mills, were forbidden "to suffer any disher of tin, laborer, or workers of tin or tin works, or miners for tin, to knock braws, rocks, or any other shift whereof tin may be made, black or white, without the said owner or owners will answer for the same

[1] Add. MS. 6713, fol. 237, c. 8. [2] See above, p. 160.

[3] Harl. 6380, fol. 48; Parl. Devon, 2 Hen. VIII, c. 24.

[4] Harl. 6380, fol. 35.

[5] Pryce, 187.

[6] *Ibid.* Cf. Trans. Roy. Geol. Soc. Cornw., vi, 332, on conditions in the Mendip Hills.

[7] Convoc. Cornw., 22 Jas. I, c. 26; 12 Chas. I, c. 21 (Add. MS. 6713, fol. 229); 2 Jas. II, c. 16.

tin upon pain of forfeiting a fine."[1] Tinners were forbidden to sell their ashes or leavings to pewterers or plumbers.[2] The measures by which alone black tin was sold were peculiar to the stannaries, and differed somewhat as between districts. The usual standard, the footfate, consisted in Blackmore of four quart-fates, each equivalent to two quarts.[3] In Foweymore, one footfate equalled two gallons and one bottle (wine measure), while a quartfate represented one bottle and one pint.[4] One footfate of good moor tin, according to Carew, ought to weigh about eighty pounds; mine tin, fifty-two pounds; and tin of the worst quality, fifty pounds.[5] All measures used in tin dealings must be taken to the leet twice a year, viewed by the inspectors, and sealed.[6]

Upon taking his ore to the blowing-house, the owner of black tin was obliged to make entry upon the blowing-house books of the amount of ore presented, together with the names of the persons, if any, from whom he had purchased it, and the blowing-house book must be kept open for public inspection.[7] The blower was subject to even more stringent laws. He might not smelt his own ore.[8] If proprietor of the establishment, he was obliged to appear at the Lostwithiel Exchequer[9] or at the stannary leet twice a year with a statement of the number of pieces of tin smelted by him, together with the names of the owners.[10] Twice a year also, he appeared with his workmen at the leet, and there they were made to swear to deal justly in their vocation, according to the laws of the stannaries.[11]

[1] Add. MS. 6713, fol. 294, c. 85.
[2] Convoc. Cornw., 22 Jas. I, c. 5.
[3] Harl. 6380, fol. 36.
[4] *Ibid.*, fol. 95.
[5] Carew, ed. 1811, p. 49, 50. Cf. S. P. Dom. Eliz., ccliii, 46.
[6] Convoc. Cornw., 2 Jas. II, c. 10; Add. MS. 6713, fol. 243, c. 32, 33; Parl. Devon, 16 Eliz., c. 8–11. The regulations of the stannaries extended also to the coal packs used by the colliers, each of which must contain three bushels, or sixty gallons (Convoc. Cornw., 2 Jas. II, c. 14). In Derbyshire, no ore might be sold other than by the "King's Dish" (*Compl. Min. Laws Derb.*, pt. i, art. 14, 15), and in the Forest of Dean by the vessel known as the "bellis" (Houghton, pt. ii, art. 31).
[7] Convoc. Cornw., 2 Jas. II, c. 16.
[8] Add. MS. 6713, fol. 242, c. 26; fol. 102–103. Cf. similar rules in the German mines (Schmoller, Jb. xv, 1001).
[9] Add. MS. 6713, fol. 103.
[10] Convoc. Cornw., 22 Jas. I, c. 3. Similar regulations were usual in the German mine codes (Schmoller, Jb. xv, 1000).
[11] Convoc. Cornw., 16 Hen. VIII, c. 17; 22 Jas. I, c. 2, 3; Add. MS. 6713, fol. 243, c.

This oath was aimed not only against the inclusion of foreign substances within tin blocks but against the misuse of the system of tin marking laid down by law. Several grades of refined tin were possible. The standard was "soft, merchantable tin." Inferior grades were "hard tin," "pilion tin," "cinder tin," and "relistian tin," each much poorer than the standard. By the laws of the stannaries the tin blocks must be stamped by the blower with the initial letter of their quality,[1] and to this must be added three other marks, that of the establishment at which the smelting took place, the "hot mark" of the blower's workman who smelted it,[2] and also the private mark of the owner of the tin.[3] These marks, which must be registered at the exchequer at Lostwithiel,[4] or perhaps later upon the steward's book at the stannary court leet,[5] and which might not be altered without previous notice,[6] served not only to trace smuggled tin,[7] but also to protect the consumer against fraud in manufacture.

The merchant exporter of tin, upon discovering that his blocks were of lower quality than represented, cut out the pieces containing the marks of identification and returned them to the stannary court, where the blowing-house proprietor and the workman who had smelted the tin, as well as the original owner of the tin, could be traced by the aid of the steward's register. The parties concerned were haled by the bailiffs before the steward, the marks shown them, and the tin officially melted down and tested by a select jury of three pewterers, three blowers, three merchants, and three tinners.[8] If the tin were found to be corrupt, the merchant was recompensed, the blower pilloried or fined,[9] the owner of the

35. Cf. Pat., 7 Hen. VII, pt. i; *Cornish Mining*, 15. Similarly we read that in Austria, in the Middle Ages, the smelters could hire no workmen without the knowledge and consent of the mine judge (Schmoller, Jb. xv, 1000).

[1] Convoc. Cornw., 16 Hen. VIII, c. 20; Add. MS. 6713, fol. 186; Parl. Devon. 2 Hen. VIII, c. 13.

[2] Parl. Devon, 2 Hen. VIII, c. 13.

[3] Harl. 6380, fol. 33.

[4] Add. MS. 6713, fol. 102, 103.

[5] Convoc. Cornw., 16 Hen. VIII, c. 18; Add. MS. 6713, fol. 241, c. 22; Parl. Devon, 2 Hen. VIII, c. 13.

[6] Add. MS. 6713, fol. 102.

[7] See above, p. 154.

[8] Convoc. Cornw., 16 Hen. VIII, c. 19; Add. MS. 6713, fol. 242, 243.

[9] Parl. Devon, 6 Edw. VI, c. 2.

block made fine to the Duke, besides forfeiting the block in question,[1] while apparently the owner of the blowing-house was not held responsible unless it could be shown that he had been privy to the fraud.[2] If, on the other hand, the tin proved to be as represented by the seller, the merchant was fined for having instituted the suit.[3] If, by any chance, it were found that a merchant and a tin owner were conniving in the distribution of corrupt tin, both parties might be tried and punished.[4]

The above laws, together with those of the coinage which we have already noted as prompted by the excessive smuggling out of uncoined tin, constituted the principal trade rules of the tinners. It may well be asked if this state of affairs was typical of free mining elsewhere in England. The question, upon the whole, can be answered in the affirmative. The laws of the Derbyshire miners, so far as they go, tally in the main with those of the stannaries, while in the Mendip Hills and Alston Moor the records, although scanty, reveal nothing analogous to the gild principle. But in the Forest of Dean a somewhat different system prevailed.

Few craft gilds, even in the fifteenth century, could have been more stiffly reactionary and exclusive than were the Dean miners, both of iron and of coal. They formed a close corporation which refused to admit as miner any whose father had not been a free miner[5] or who had not served a long apprenticeship, and which even prosecuted for trespass any who entered the bounds of the Forest without paying custom upon all his goods.[6] Within the Forest we find most of the ordinary gild rules grimly enforced. Strict custom required that the mines be worked by companies of four persons called "verns," or partners, who of necessity were free miners of the Forest, must have rented land and kept a house,[7] and who must proceed in the driving and working of levels or the sinking and working of a water pit by their own labor, with possibly the assistance of sons or apprentices.[8] These last were few in

[1] Add. MS. 6713, fol. 101, 102.
[2] Harl. 6380, fol. 33.
[3] Parl. Devon, 2 Hen. VIII, c. 15.
[4] *Ibid.*, 6 Edw. VI, c. 3.
[5] Cf. Houghton, pt. ii, arts. 30, 36, 37.
[6] *Ibid.*, art. 1–3.
[7] Award of Dean Forest Commissioners, 17.
[8] Fourth Rept. Dean Forest Commissioners, 6, 8.

number, and served anywhere from a year and a day to seven years before being admitted to full rights as miners.[1]

As regards production, the inference from prevailing conditions is that machinery was discouraged; under the old laws it could be erected only by express permission of the owner of the soil.[2] Free miners were forbidden to use more than four horses for carrying coal and ore,[3] a business, by the way, in which they were interested almost as much as in the digging itself, and of which their laws gave them a monopoly.[4] The prices at which they might act as carriers were fixed by a committee of six "bargainers" appointed by the mine law court, and underselling in any way was punished by expulsion from the fellowship.[5]

Under these restrictions, the rich coal and iron deposits of the Forest of Dean for centuries lagged in their development far behind those of the rest of England. The mines themselves were tiny grubbings, worked almost entirely by manual labor; while for the carriage of the minerals, as late as 1668, only pack-horses seem to have been used.[6] The time came inevitably when the entire Dean system fell to the ground. The field was too tempting to remain free from interlopers. Probably at an early date there had been outside capitalists who made it a practice to purchase claims or "gales" from free miners of the Forest, and to work them in defiance of Forest law, possibly with the connivance of the miners, by machinery and hired labor, instead of by apprentices.[7] These aggressions appear to have excited no marked opposition until the seventeenth century, when the Crown frankly ignored the rights of the miners and granted large sections of the Forest to various courtiers, together with monopoly privileges of mining. The consequence of this was that for

[1] *Ibid.*, 13. Occasionally but perhaps very rarely, at least in the Middle Ages, the title of "free miner" was conferred by the miners' parliament upon gentlemen who had never worked in the mines at all (*ibid.;* Nicholls, chap. iv).

[2] Award of Dean Forest Commissioners, 24.

[3] *Ibid.*, 14.

[4] *Ibid.*, 4. In 1676 this privilege was abandoned, but free miners were to have a preference in being loaded at the pit.

[5] *Ibid.*, 13, 14.

[6] Nicholls (p. 45) quotes an order of the mine court which forbade the use of carriages for this purpose. Cf. Galloway, 209.

[7] Award of Dean Forest Commissioners, 21; Fourth Report of Dean Forest Commissioners, 8, 9.

the Forest of Dean the century became a succession of uprisings and riots, interspersed with frantic appeals to the Crown on the part both of the miners and of the patentees.[1]

The miners, however, had committed the fatal error in 1613 of accepting a compromise with the Crown and its monopolists, by which they agreed that they were to hold all their privileges for the future merely out of grace and charity, and not as a right.[2] Their resistance to interlopers in time gradually slackened; the more burdensome of the miners' regulations were tacitly allowed to be set at naught,[3] and the miners' assemblies or law courts fell completely into disuse.[4] When in 1835 there came the inevitable reformation of the Forest of Dean mining law and the abolition of the liberties of the free miners,[5] it was found that few of the old miners owned mines of any value, but that nearly all of the mineral properties of the Forest were owned and operated by outside capitalists.[6]

What befell the Forest of Dean miners might well have been the fate of the tinners, had they endeavored or had they been able to make their organization conform to gild principles. That with the exception of a law which prohibited aliens from becoming tinners,[7] and another which prevented ore buyers from being their own smelters, the constitution of the stannaries gave practically a free rein to capitalistic development, was the result in the main of the fiscal interest taken in the industry by the King. The following chapter will show how, with this opportunity for expansion, the tinners' industrial organization developed.

[1] Cf. Rep. Hist. MSS. Com., xii, App., pt. i, 430; S. P. Dom. Chas. I, ccclxxv, 34; *Narrative concerning the Forest of Dean*, 2–4; Award of Dean Forest Commissioners, 14, 17; Nicholls, 24, 27, 28, 38, 39, 43, 47.

[2] Fourth Report of Dean Forest Commissioners, 5.

[3] Award of Dean Forest Commissioners, 17.

[4] Cf. Fourth Report of Dean Forest Commissioners, 67; Award of Dean Forest Commissioners, 19.

[5] Cf. Stat. 1 & 2 Vict., c. 43; Nicholls, chaps. v-vii.

[6] Fourth Report of Dean Forest Commissioners, 8–10. Cf. Lambert, 216. This was quite the reverse of the German practice, by which foreigners were encouraged to become miners (Schmoller, Jb. xv, 679; Achenbach, 1871, pp. 74, 313).

[7] Add. MS. 6713, fol. 304.

# CHAPTER VII

## INDUSTRIAL ORGANIZATION IN THE STANNARIES

It is one of the interesting and instructive features of industrial development in the mines, that numerous diverse types of organization, representing early and later stages of development, may be observed coexisting and actively functioning at the same period. More particularly is this true in the case of mines of the precious metals, where operations even on a small scale may yield a fortune. Thus in the gold mining camps, shortly after the first rush has subsided, one finds within a radius of a few miles the solitary placer or pocket miner working with shovel and pan, and companies of a half-dozen prospectors working a single claim in common, while larger syndicates financed by distant shareholders set to work upon the native lode with all the modern equipment of steam drills and derricks, operated by hired labor. This diversity of type seems to have characterized the mining industry from an early period, for instead of a slow and comparatively simple evolution from the solitary craftsman to the capitalistic organization we may discern the emergence of more complex forms almost as soon as mining records appear. For the stannaries, however, these early records are so scanty that it seems advisable to preface the account with a brief sketch of the progress of mining organization in Germany, already sedulously studied by numerous German scholars.[1] This may then be used as a background upon which to throw whatever facts can be brought forward with regard to the English mines, with due attention to the essential points of difference between industrial conditions of the two countries.

[1] Here, as in the chapter on early mining law, I have not hesitated to make large use of Schmoller's account of mining organization in Germany. I am aware that on some important points full agreement has not been reached among German students of mining history and that much remains for further exploration, but Schmoller's excellent summary represents, sufficiently for my purpose, the *communis opinio* at the present stage of investigation.

It appears, as has been already mentioned, that the earliest German mines were probably feudal enterprises of the territorial magnates.[1] To this period succeeded one in which the lords found it to their interests to turn over the mines to the workmen, with the result that there arose everywhere little autonomous associations of co-laborers, each with a mine of its own, which paid tribute to the lord and divided profits among the members.[2] The genesis of these associations[3] is to be found in the physical facts of the industry. The smallest silver mine would require not less than three men, especially if it must be worked day and night according to contract with the lord.[4] With the division of the miners' day into shifts of six or more hours,[5] more associates would be admitted and the division of labor further extended. Thus we find that by the thirteenth and fourteenth centuries mine companies usually consisted of sixteen able-bodied workmen, besides their children.[6] At their head would stand a foreman; next would come the men in charge of the surface works, the windlass and pumps, the cartage and preparation of the ore; then men in charge of the underground operations,[7] and finally bellows-makers, carpenters, carters, smiths, and hewers.[8]

Hardly, however, had these embryo companies sprung into being[9] when there began the inevitable trend towards capitalism. The deepening of the mines, the consequent complications in engineering, which required capital investments on a scale beyond the means of the manual laborer,[10] the increasing size of the lord's allotments,[11] which tended to enlarge to unwieldy proportions the number of partners, — these factors combined with others to reduce the originally free workmen to the position of wage-earners. The transition to this stage, however, would usually be marked by three other forms of industrial organization, the cost agreement, the tribute system, and the lease.

By virtue of the cost agreement one or more of the associates might

1 Inama-Sternegg, iii, bk. iv, 152.

2 Schmoller, Jb., xv, 686, 687.

3 Cf. Gierke, i, 455, n.

4 Schmoller, Jb., xv, 684.

5 *Ibid.*; Achenbach, 1871, p. 292.

6 Schmoller, Jb., xv, 685, 975.

7 *Ibid.*, xv, 708.

8 *Ibid.*, xv, 667, 686.

9 Gierke compares them to small industrial mark communities (i, 442, 443).

10 Inama-Sternegg, iii, bk. iv, 160–163.

11 Cf. *ibid.*, 165; Schmoller, Jb., xv, 693–695.

be excused from the common labor, if an equivalent money subsidy were given, or, as was sometimes the case, a substitute were hired by the inactive partner.[1] The convenience of this arrangement is seen in the fact that it made it possible for women and children, the wives or daughters of deceased associates, to retain membership in the association,[2] and furthermore that it was at that time the only means through which the necessary amounts of capital could be raised for the construction of the adits and engines which the enlargement of the works rendered essential. The sources state that the mine associates habitually granted to outside investors participation in the dividends of the mine in return for cash investments,[3] these foreign associates maintaining near the works special agents to safeguard their interests.[4] The wide diffusion of the cost agreement — especially in the thirteenth and fourteenth centuries — is shown by the special legal forms then instituted in the mine courts to facilitate the settlement of disputes arising under its operation.[5]

The tribute system, which first became prominent toward the end of the thirteenth century,[6] seems to have arisen primarily from a desire on the part of the seignorial owners to extend the building of adits and other costly devices for mine drainage.[7] To induce their mine associations to undertake these works, they had offered them holdings larger than could be readily worked by the partners. Their own numbers the latter did not wish to increase, as that would entail a further division of the profits. On the other hand, although, with the natural increase of population, numbers of workingmen were available, willing and eager for employment, the employing entrepreneur seems to have been lacking. From this dilemma the tribute system offered a means of escape. The associates gave over a portion of their mine to a company of laborers in return for a stipulated share in the profits,[8] the enfeoffed miners, or tributers, working the share assigned them according to their discretion and in their own interests.[9]

1 Inama-Sternegg, iii, bk. iv, 159, 162; Schmoller, Jb., xv, 699, 700, 702.

2 Bernhard, 9.

3 *Ibid.*, 10; Schmoller, Jb., xv, 706, 707, 986.

4 Bernhard, 11; Schmoller, Jb., xv, 987, 995.

5 Bernhard, 12, 13.

6 Schmoller, Jb., xv, 697.

7 Bernhard, 23.

8 Schmoller, Jb., xv, 697, 698; Bernhard, 24, 28, 29.

9 *Ibid.*, 21.

The tribute system, in its typical form, was operated, in substance, as follows. As soon as the mine associates had given over a portion of their holdings to tributers they were bound to put the latter into a position in which to begin operations. To this end they must furnish a supply of rope and leather, in order to free the mine from water.[1] According to decisions of the mine court of Iglau, timber and carpenters must also be supplied, together with horses and men to raise the ore. On the other hand the associates possessed rights correspondingly extensive, the general presumption always being that the tributers were their subordinates. A case in point is to be seen in the so-called "Durchschlagen," or breaking through. Frequently it happened that disputes would arise between laborers of different sections who had broken in upon each other's workings, and in that case when one part consisted of associates and the other of tributers, the former took precedence unless by special arrangement the tributers had insured themselves for such an event in advance. The amount of the compensation due from the tributers varied from one half the ore to one seventh or even less, and a special body of mine law arose to govern the division of the ore, one of the lord's officers, the "Rechenmeister," acting as a mediator between parties.[2]

The extension of the tribute system received a great impulse through the direct encouragement of the seignorial lords. The time was passing when these had received their profits in the form of shares in the mines. As they came to receive, instead, a fixed proportion, usually one tenth, of the gross product,[3] it became their interest, not that the associates themselves turn out as much as possible, but that the mine as a whole should be more productive. The magnates, therefore, extended the tribute system by all means in their power,[4] and their mine masters were in certain contingencies given the right to grant tribute pitches in a mine without consulting the partners themselves. The climax of these efforts is seen in the so-called Constitutions of Wenzel, in which the king of Bohemia made the presence of tributers obligatory in every mine and even provided for the subletting of tribute holdings.[5]

[1] Bernhard, 30; Schmoller, Jb., xv, 696, 1003.
[2] Bernhard, 31, 32, 33.
[3] *Ibid.*, 25, 27.
[4] Cf. *ibid.*, 24.
[5] *Ibid.*, 26, 27, 35, 36; Schmoller, Jb., xv, 696.

As in the case of the cost agreement, it is impossible to date the disappearance of the tribute system; but although it persisted in outward form down to the sixteenth century and even later, its real character was slowly changing to that of a mere wage contract. This transition was inevitable owing to the increasing disparity in bargaining power between the two parties concerned. The mine associates were, as time went on, for the most part no longer laborers; the tributers, by the provisions of the law itself, must be dependent solely upon the work of their hands. Even peasant folk were not as a rule admitted to tribute, at least under any but the shortest of contracts,[1] the rule being that the tributers must be those who besides their personal labor could contribute little or nothing.[2] It will be equally clear that only those tributers who made little or no profit would be content to remain in *bona fide* workingmen's organizations, the more prosperous of their number inevitably forming mine partnerships with tributers of their own beneath them.

But the immediate occasion of the passing of the system arose in connection with the disposal of the ore. The ordinary mine associate, even when he labored with his own hands, was as a rule economically self-sufficing, since besides his income from the mine he had his fields and cattle. It was thus comparatively easy for him to await the purchase of his ore at the lord's mint, although if necessary he could obtain an advance from the mint master. For the tributer such an advance was not available save when the ore was uncommonly rich. Freiburg records of the fifteenth century give abundant evidence of the increasing difficulties in selling, and the complaints of the tributers rehearse in no uncertain terms the straits to which they were reduced by the oppressions of the ore purchasers and smelters. Efforts to alter this condition often took the form of smelting houses established by the territorial lord himself to take over the tributers' ore. In 1512 the Emperor Maximilian erected an establishment of this sort, while Ferdinand attempted remedies of a like nature in the Black Forest; but neither of these experiments proved successful, and throughout the Empire the tributers slowly retrograded in their economic status. Achenbach believes that the system at about the end of the fifteenth century had taken on the character of a piece-work bargain, and that

[1] Schmoller, Jb., xv, 1002, 1003.

[2] Bernhard, 31.

it then declined, was discouraged by the laws, and so disappeared. At all events the later Saxon mine codes seem to refer to piece-work jobs exclusively, with weekly payments and with allowances for the hardness of the rock and other natural obstacles, in other words just such a change as is at present turning tribute work in the Cornish mines into tut and time work.[1] Similar conclusions may be drawn from an inspection of the Saxon codes of 1479–1509. These show the complete disappearance of the old tribute contract, the substitution to a certain extent of the piece-work system, and, what is still more significant, the appearance of time wage pure and simple.[2]

This drift towards wage work in the German mines is indicated even more strongly by the lease system, under which an outsider might lease for a fixed annual sum part of the territory of a mine association. This, it will be noticed, apparently differs little from the tribute system, but whereas the tributers were poor men who did their own work and paid a contingent rent only, the lessee was a capitalist. If the associates lacked the means to develop their holding, they surrendered a part to a man of wealth and enterprise, who in return for a fixed yearly rent was given the right to develop the area with the labor of his employees. Paying as he did a fixed rental, he incurred a risk beyond that of the tributers, and therefore received a correspondingly higher profit. This continued until in the course of time we find the lessee taking on more and more the character of a captain of industry, relieving the associates of not merely a part but of the whole of their claim.[3]

Although the fifteenth century has usually been fixed upon as the date of the disintegration of the primitive miners' associations,[4] one is justified in saying that even at that time probably every form of mining association here described might have been seen in full operation. In many districts, especially the newer and richer ones, companies of associates still labored together in the old way; elsewhere work was carried on by organizations of various sorts, — groups of tributers, fragments of primitive miners' associations working side by side with the substitutes hired by the more well-to-do partners,[5] and lessees with their hired labor. But in this mul-

[1] See p. 204.

[2] Schmoller, Jb., xv, 1004–1007.

[3] Bernhard, 38.

[4] Cf. Schmoller, Jb., xv, 705, 709, 710, 984.

[5] Schmoller, Jb., xv, 701, 702.

tiplicity of forms, differences between the new conditions and the old may be observed, and are perhaps nowhere so well exemplified as in the changes which had taken place in the composition of the mining partnerships. By the year 1400 comparatively few members did personal labor, and the mine shares, no longer connoting the necessity of mutual toil, took on a more modern character, became transferable with facility and carried with them merely the obligation of paying periodic assessments and the privilege of receiving dividends. The shareholders, who, for a time, had been composed of inhabitants of mine cities, gildsmen, merchants, craftsmen, and the landowners and priests of the neighborhood,[1] now included distant capitalists, merchants from the great trading centres, and financiers such as the Fuggers. Mine companies were organized and the sale of stock promoted with a recklessness and unscrupulousness worthy of modern high finance; while the spirit of friendly and intimate coöperation which had animated the older companies found little response in the new. The company meetings and settlements of accounts formerly held weekly became monthly or even quarterly. Payments of wages to the laborers grew to be wholly the affair, not of the individual members, but of the mine officers.[2] In short, we find in these late mediæval mining companies phenomena very similar to those of the corporations of to-day.

With the final cleavage between capital and labor and the depression of the poorer mine partners, or tributers, to the position of laborers,[3] the German miners had begun the formation of journeymen's gilds.[4] Originally started about the year 1400, as altar brotherhoods and charitable organizations for the insurance of members against accidents and illness, they speedily grew to be powerful fighting bodies, with common chests, closely organized for mutual offence and defence. The hewers, the smelters, the slagmakers, and other special crafts formed separate unions with from two to four aldermen at the head of each. The government of the united gilds was in the hands of a small council of the oldest alder-

[1] Inama-Sternegg, iii, bk. iv, 159; Schmoller, Jb., xv, 985–987.

[2] Schmoller, Jb., xv, 972, 987, 990–992, 702.

[3] Cf. Bernhard, 42, 65, 68, 69; Inama-Sternegg, iii, bk. iv, 160, 161; Schmoller, Jb., xv, 701, 702, 705, 709, 710, 1002–1007, 1013, 1014.

[4] Cf. Schmoller, Jb., xv, 1015–1018, Reyer, 79, 80.

men, under the ultimate supervision of the lord's officials. The aldermen were accustomed to go amongst the miners periodically to listen to complaints and to exercise a certain jurisdiction over members for matters of minor importance. We find the journeymen's unions early resorting to the strike as a frequent weapon in trade disputes and as a protest against unsatisfactory conditions of employment.[1] These revolts of labor occurred among the miners more frequently than among the other crafts and resulted in great losses to the mine owners and to the territorial lords. They came to a head during the period of the Peasants' War and ended in the humiliation and subjugation of the unions at the hands of the lords and capitalists.[2]

Despite the comparative paucity of information as to mining organization in England, there are sufficient indications that the development of mining in Germany is in many respects analogous to that of the tin mines of Cornwall. But two disturbing factors break the completeness of the analogy.

In the first place we must bear in mind the fact that what has been stated in the preceding paragraphs has had to do with silver mining in Germany, and silver is notoriously of all the chief metals the one which requires for its production the most extensive, skilful, and costly equipment for mining and smelting. It never occurs in "washes," or placer deposits, and even in the lode it is found in considerable masses only on rare occasions. To be successful, a silver mine must be driven deep into the ledge, and therefore requires preliminary expenditure of capital and intelligent coöperation among workmen to a far greater extent than in the case of tin. In an earlier chapter it has been seen how, until at least the sixteenth century, most of the tin of Cornwall was shovelled from alluvial deposits, from shallow pits drained by trenches or by the rudest of water-wheels. Operations of this nature necessitated no miners' associations or minute division of labor, such as Schmoller assumes to have taken place in the early silver mines. Often a single Cornishman, aided possibly by his son, could manage a stream work, and there may still be seen in Cornwall small affairs of this sort which yield a fair profit.

In the second place, when we compare conditions in the two

[1] Cf. Bernhard, 54.

[2] Schmoller, Jb., xv, 1008, 1015.

countries we must make allowance for the influence of the German territorial princes. What this meant for the regulation of mining has been already briefly indicated, in speaking of the enfeoffment of the miners by the seignorial officer, the necessity for continuous labor, and the interference of the mine master to promote or to regulate the granting of tribute holdings and similar contracts. The lords maintained at or near the mines, as agents for oversight and direction, a small army of officials,[1] mine masters, mine judges, mine headmen, "Hutreiters," tithers, mint masters, and small fry such as mine clerks, smelting-house clerks, check clerks, and assistants, the total force amounting in some cases to no less than sixty persons. In Kuttenburg in 1551 there were almost as many officials as laborers. The chief mine officials in the later Middle Ages were not infrequently the lord's creditors, and received from him tacit permission to wring from the associates the uttermost farthing, an abuse which, in the fifteenth century, called for remedial legislation. The under officers were chosen from amongst the skilled miners.

Thus, although nominally in private hands, the German mines remained in reality quasi-seignorial undertakings. The companies were checked and guided on every hand by the lords' officers. This fact becomes more noticeable during the revival of mining, actively beginning with the fifteenth century, when the lords began either to take over the mines or to subsidize, directly or otherwise, the less prosperous.[2] Once this change made and the laborers hired, not by the company but by the prince's servants,[3] the transfer of the mines to the state was wellnigh complete.[4]

All these details have an immediate bearing upon the subject in hand. Under the bureaucratic system prevailing in Germany, where every detail of mine administration was prescribed and supervised, where even the methods of accounting were laid down by rigid rules and the contracts between mine associates and tributers, lessees, and hired labor filled out according to prescribed form, there would

[1] Schmoller, Jb., xv, 673, 692, 973, 978, 997, 1004, 1009–1014, 1017, 1021–1026; Inama-Sternegg, ii, bk. iii, 334; Bernhard, 15–17, 55–57, 60–65, 68, 69.

[2] Cf. Schmoller, Jb., xv, 966, 967, 970, 971, 1018; Bernhard, 53; Reyer, 85.

[3] Schmoller, Jb., xv, 1010.

[4] Eng. Quart. Min. Rev., iv, 264; Eng. Min. Almanack, 1852, p. 193; Schmoller, Jb., xv, 970, 1027, 1028.

certainly prevail a much more fully organized and systematized administration than in Cornwall. There, as indeed in all England, existed a different conception of mining. It was not an affair of the state, but primarily of individuals. The state, indeed, opened the field to all comers, and endowed the mining classes with certain privileges for which through special taxation it made them pay roundly. But apart from the necessary safeguarding of its fiscal interests, it did not interfere with the private management of the tin mines, nor did it attempt to work them itself. From the comparative absence of state control and from the fact already mentioned that the tin deposits were largely alluvial, we may infer that the Cornish tin mines lacked the uniformity and the systematic administration of the German mines. It is doubtful, for example, if the free miners in any part of England ever kept written accounts of any sort, as the total lack of any mine document of this nature makes it seem probable that with the characteristic happy-go-lucky methods prevalent even in modern Cornish mines the mediæval free miners squared their accounts by the use of pebbles, just as Pryce represents them to have done as late as the latter half of the eighteenth century.[1]

The first observation which we make on examining the existing evidence is that the old stannary laws, like those of the Mendip Hills,[2] Derbyshire,[3] and the Forest of Dean,[4] are evidently based upon the assumption that a large part of the mining was to be carried on by groups of working shareholders. An instance of this may be seen in the law that in the case of several "partners working to-

[1] See p. 201.

[2] The old mining rules of Mendip furnish several illustrations of primitive partnerships. The status of the typical Mendip miner is shown by the law that "when any workman has landed ore, he may carry it for cleansing and blowing to what minery he wishes provided he pay one tenth to the lord of the soil where landed" (Trans. Roy. Geol. Soc. Cornw., vi, 329). Again we read that "if any pitch, gribb, or groof lie unlawful for the space of twenty-eight days, and the lead reeve cause proclamation upon the weigh day, in the hearing of twelve men who are miners, then it shall be lawful for the lead reeve after twenty-four days, if the old partners shall not work the same in mean season, according to custom to give the said gribb, pitch, etc. to any workman that will work the same" (*ibid.*, 330, 331). "Whoever throws the axe (the length of the pitch being determined by an axe throw), in any groof or gribb, shall be one of the eldest partners of the same" (*ibid.*, 331).

[3] Cf. the term "a miner and his grove-fellows" in the old mineral laws of Derbyshire.

[4] See p. 173.

gether in a tin work," if one of them endeavor to trick his fellows by allowing the work to be forfeited through lack of renewal of bounds, he is to be accounted as not to have done so.[1] It does not, indeed, require much investigation to discover these partnerships of small working entrepreneurs from the earliest times to the present. They are vouched for by Pryce in 1778, Jars in 1769, Carew in 1602, and Beare in 1586. Still farther back, in 1510, we meet a reference to them in the code of law confirmed in that year for the Devon stannaries, to the effect that each partner is to work in his own portion of the works without hindrance from his fellows.[2]

To this conclusion point numerous scattered references in the stannary records of the Middle Ages. Thus Henry Nanfan and his associates complained to the Black Prince that they were molested in their tin work in the moor of Lamorna.[3] In one of the old coinage rolls we find tin bills accounted for by Ben Rynwald and his fellows.[4]

In these rolls, especially those for the stannary of Devon, we meet with entries which could not possibly have been handed in by any but tinners on a small scale. An analysis of a typical coinage account gives, in this connection, some interesting results. Take, for example, that of Cornwall for the year 1300, the items of which are here tabulated.

| AMOUNT OF TIN PRESENTED FOR COINAGE | NUMBER OF TINNERS PRESENTING TIN | AMOUNT OF TIN PRESENTED FOR COINAGE | NUMBER OF TINNERS PRESENTING TIN |
|---|---|---|---|
| *Thousandweight* | | *Thousandweight* | |
| Under 1 | 19 | 10–11 | 2 |
| 1–2 | 32 | 11–12 | 2 |
| 2–3 | 23 | 12–13 | 2 |
| 3–4 | 10 | 13–14 | 3 |
| 5–6 | 4 | 14–15 | 3 |
| 6–7 | 9 | 15–16 | 1 |
| 7–8 | 6 | 16–17 | 2 |
| 8–9 | 1 | over 17 | 7 |
| 9–10 | 2 | | |

[1] Harl. 6380, fol. 43.
[2] Parl. Devon, 2 Hen. VIII, c. 31.
[3] Smirke, 26.
[4] Accts. Excheq. K. R., bdle. 261, no. 1. According to Smirke (Eng. Mining Almanack, 1849, p. 155), so little did the customs of the stannaries, as late as 1495, con-

Here of a total of one hundred and twenty-eight men, two thirds presented four thousand-weight or under. The small tinner preponderates, but the list shows the existence of a class of large producers; the highest class contains seven men who present amounts varying from thirty-three to two hundred and ninety-four thousand-weight. The figures for Devon, two years later, are even more striking. Out of one hundred and thirty-four men, one hundred and nine produced less than a thousand-weight each; sixteen from one to two thousand; seven from two to three; one from four to five; and one from nine to ten.[1] In 1394, for Devon, out of one hundred and nine producers, all but six had less than a thousand each.[2] In 1398, one hundred and one were less than a thousand-weight each; seventeen were from one to two thousand-weight; and five were higher.[3] The same story might be repeated for almost any year in either county, the significant feature being the great number of men whose income from tin must have been exceedingly small. This is strikingly exemplified in 1524, when of tinners presenting less than a thousand-weight each, Cornwall showed four hundred and thirty-two and Devon seven hundred and thirty-seven.[4] The stannary tax lists offer further information to the same effect, the assessments in the tinners' villages coming down to 2d. per man, as contrasted with 6s. or 8s. for many of the wealthier tinners in the boroughs.[5]

Such an analysis indicates the further fact that in all probability some of these tinners were not wholly dependent upon the mines for their living. Many perhaps were artisans or small farmers. We find casual mention of this double occupation in documents of the fifteenth century. "John Aunger, the blower," was also a husbandman,[6] and one John Lawe a husbandman and tinner.[7] In later centuries, when the mines were said to be decaying, a constant subject

template the possession of tin bounds by any but working tinners, that an order of Prince Arthur provided that "no persone, neyther persones having possession of lands and tenements above the yearly value of £10 or noone other to theyr use be owners of eny tyn work or parcell of eny tyn work with the exception of persons claiming by inheritance or possessed of tyn works in their own freeholds."

[1] Accts. Excheq. K. R., bdle. 260, no. 24.

[2] *Ibid.*, bdle. 263, no. 25.

[3] *Ibid.*, bdle. 264, no. 3.

[4] Accts. Excheq. K. R., bdle. 271, nos. 9, 12.

[5] Lay Subs. R., bdle. 95, nos. 12, 33.

[6] Cal. of Pat., 1426, p. 108.

[7] Receiver, 11 Hen. VII. Cf. for Germany, Schmoller, Jb., xv, 687.

for complaint was that the tinners were forsaking the stannaries and turning to husbandry.[1] At all events the quantities of small tin bills presented, especially in Devon, where most of the amounts seem to have been under two hundred-weight, justify the inference that, with prices as they were, the tinners must have resorted to subsidiary employment.

There are to be found in the stannaries abundant traces of the cost agreement, tribute system, and lease. The following law against defaulting partners, for example, throws light on forms of organization. "Where there are many co-partners in a tin work, such owners as upon warning given shall not set their parts to farm, nor bring in their men and money account to their parts within one month and pay the cost and spale for that month, shall be excluded from entering the said work or adventure therein during that adventure, and shall be contented with such farm as the work is set for by the rest, and in case where the rest work their own right, the farm shall be rated and assessed by three indifferent tinners to be chosen, one by the workers, one by the owners not adventuring, and the third by the steward of the stannary court."[2] The period of mining development to which this provision refers cannot be accurately ascertained, since, like nearly all the stannary laws, the written enactment represents the formal registration of old custom rather than new ordinance. But whatever the precise date, we may here perceive clearly the simultaneous existence of several forms of organization: first, the old association of miners, many of whom "work their own rights;" second, the cost agreement, in which the cost-giving partner has begun not only to pay a money contribution, "cost and spale," but also to hire a substitute to do his labor;[3] third, the tribute, since an owner "may set his part to farm," the "farm," to all intents and purposes being the tribute as we have seen it in Germany and as it persists in Cornwall to-day.

[1] Lans. 19, fol. 99; 86, fol. 67.

[2] Convoc. Cornw., 22 Jas. I, c. 21; 12 Chas. I, c. 16 (Add. MS. 6713, fol. 228). Cf. also Convoc. Cornw., 22 Jas. I, c. 23, where the tinners are said to work their tin works by themselves, their wage-men, or their farmers.

[3] Another illustration of the survival in Cornwall of the cost principle is the fact that under the stannary laws, valid even to-day, a member of a cost book company may bring in his own goods, if he choose, and not pay to have them brought in one lot with those of his associates (Convoc. Cornw., 2 Jas. II, c. 5, 6).

Just when and how the wage work system first appeared in the stannaries is a matter of some doubt. Certainly nothing analogous to the German journeymen miners' unions is found in Cornwall, where the journeymen miners from the earliest times down to the present have had no organization of their own. This fact is doubtless due in part to the continued prevalence of the tribute system, which, in itself, was a virtual partnership with the owners. In Germany we find it rapidly giving place to piece work or simple time wage; in Cornwall as late as the seventeenth century the tribute system, so far from dying out, was stronger than any other and completely dominated the mines.

There is, however, strong evidence of the fact that capitalism, in one form or another, had invaded the Cornish mines at an early date. The cost-giving system was probably fairly well developed by the end of the fourteenth century, since we find in contemporaneous sources references to the holding of mine shares by others than manual laborers. Among other documents preserved in the White Book of Cornwall is a mandate of the Black Prince in which, in consequence of the general abandonment of mining after the Black Death, he warns the tinners that their shares in the mines will be forfeit unless they continue to expend the same costs and the same labor as in times past.[1] Less ambiguous is the record[2] of "doles," or shares in tin works, owned by Cornish tin merchants, and bequeathed to the warden of a parish church in order that masses be said for the donor's soul.

Nor is direct evidence upon this subject wholly lacking. In the Chancery of Henry V appears the case of John Thomas, a Cornish miner, whose tin work in Crukbargis Moor had been taken from him and tin to the value of £40 seized, "for which he is in the debt to divers laborers their working, twenty marks."[3] "Abraham the Tinner" in 1357 is said to have owned two mine works and four stream works,[4] in which he employed over three hundred men,

[1] White Bk. of Cornw., i, 25 Edw. III (Feb.).

[2] Early Chanc. Proc., bdle. 17, no. 232. The tendency towards capitalism is possibly further exemplified by the repeated laws against the giving or selling of tin doles to powerful persons for maintenance (Convoc. Cornw., 30 Eliz., c. 16; Parl. Devon, 2 Hen. VIII, c. 11).

[3] Chanc. Proc. Eliz., i, p. xiii.

[4] White Bk. of Cornw., i, c. 15, 19, 20, 22.

women, and children. In 1342 occurs the case of "certain of the wealthier of the tinners of Cornwall," Michael de Trenewyth, Michael his son, John Billyon, Hervey his son, Ralf Reslack, Walter le Beare, John Carnignon, and William Scarlet, who are said "to have usurped divers stannaries by force and duress, and to have compelled the stannary men to work in these contrary to their will, for a penny for every other day, or a bit more, whereas before they worked 20d. or more worth of tin per day, and for a long time have prevented tinners from whitening and selling their tin worked by them; wherefore the stannary men have ceased working, and some of them are impoverished."[1] Whatever may have been the nature of the coercion here exercised, whether "force and duress" or economic pressure, the incident shows a bold endeavor on the part of certain mine owners to procure abundant hired labor to work their holdings. In this connection also it must not be forgotten that among the complaints raised against the tinners, — the earliest dates back to the year 1237,[2] — we find the grievance that not only the tinners but their servants who labored for them were admitted to the franchises of the mines.[3]

Similar inferences can be drawn from other sources. We find the tithingman of Tregarrek presenting a just hue and cry against certain men for having invaded a tinner's work and driven out his servants;[4] while at another time an inquest taken for the purpose of finding who were true and who were false or feigned tinners, selected certain men for the former category "who work the stannaries by their servants or by themselves."[5] In the coinage rolls we may also note the ownership of tin by persons who evidently could not have worked the mines with their own hands and among whom some at any rate were probably owners of mines or shares rather than mere purchasers of ore. For example, we find that John the mercer presents five hundred-weight;[6] Henry, Earl of

[1] Pat., 16 Edw. III, pt. ii, m. 15 d. They are again complained of in the following year (Pat., 17 Edw. III, pt. i, m. 38 d.).

[2] *Annales de Burton* in *Annales Monastici*, ed. Luard, p. 256.

[3] Parl. R., ii, 343, 344.

[4] Ct. R., bdle. 156, no. 27. Cf. *ibid.*, bdle. 161, no. 81 (Penwith and Kerrier, 30 Edw. III): "Ricardus la Wynne arrestus quia recessit a servitio Martini Smyth, contra formam statuti Domini Regis."

[5] Lay Subs. R., bdle. 95, no. 12.

[6] Accts. Excheq. K. R., bdle. 265, no. 25.

Devon, one hundred-weight;[1] John, Earl of Cornwall, ninety-four thousand-weight;[2] Thomas the goldsmith, four thousand-weight;[3] while Richard the smith, Thomas the pewterer,[4] John Trenagoff the clerk,[4] Michael the skinner,[5] John vicar of Bodmin,[6] Ralph the rector of the church of St. Ladoce, Ralph Doly the clerk of Lostwithiel,[7] Joanna the widow of Ralph Barson,[8] Ralph the chapman,[9] John the merchant, Philip the prior of Tywardratch, and Alfred the prior of Mt. St. Michael,[10] all figure in the lists. Some of these "tinners," therefore, were women,[11] others churchmen, others small tradesmen and artisans. Certainly the evidence, scattered and inferential though much of it is, warrants the conclusion that in the stannaries in the Middle Ages as at a later date we are confronted with a variety of classes and conditions. We have the small prospector working his own claim, perhaps usually in conjunction with partners, at the same time that elsewhere is visible the gradual rise of a class of adventurers who merely hold shares in mines, the actual work in which is carried on by hired labor.[12]

[1] Accts. Excheq. K. R., bdle. 271, no. 13.
[2] *Ibid.*, bdle. 262, no. 26.
[3] *Ibid.*, bdle. 265, no. 20.
[4] *Ibid.*, bdle. 265, no. 2.
[5] *Ibid.*, bdle. 261, no. 6.
[6] *Ibid.*, bdle. 261, no. 7.
[7] *Ibid.*, bdle. 262, no. 29.
[8] *Ibid.*, bdle. 263, no. 1.
[9] *Ibid.*, bdle. 262, no. 21.
[10] *Ibid.*, bdle. 261, no. 1.

[11] Living, many of them, in the towns rather than at the mines (Lay Subs., R., bdle. 95, nos. 12, 29).

[12] Cf. Add. MS. 24746, fol. 61; Parl. Devon, 2 Hen. VIII, c. 5, 17. The status of other English miners in the Middle Ages is not quite clear. On the one hand, in the northern coal fields most of the mines were probably worked by capitalists and hired labor (see p. 216). The condition of this laboring class, so far as it is possible to judge from the scanty evidence available, seems to have been inferior to that in the stannaries (cf. Galloway, 48, 102). As the mines were worked or owned by the lords, the miners were perhaps under the same disabilities as the ordinary villein (cf. Galloway, 118; Trans. Fed. Inst., vii, 613). Galloway believes that the colliers were in a state of serfdom (49, 75, 76, 269).

In the Mendip Hills the wage work system, although probably only on a small scale, existed early in the Middle Ages, as may be seen by a perusal of the old mining laws (Trans. Roy. Geol. Soc. Cornw., vi, 330–332).

In Derbyshire we likewise find traces of a class of "servientes" (Add. MS. 6681, fol. 122), working along with the "grove fellows," or mine partners (Eng. Min. Almanack, 1850, p. 221; Houghton, 28, art. 56: remedy against the withholding of a miner's wages).

Opinions have differed with regard to the Scottish mediæval miner. It seems likely that the so-called "slavery" of the Scottish colliers, which lasted until 1798

In the absence of account rolls for the tin mines, nothing can be learned of the internal management of the larger works. But an interesting sidelight is thrown upon the genesis of capitalistic mining concerns by the account rolls which survive of the royal silver mines at various points in Devon, with reference to which it must be borne in mind that the labor employed there was largely, if not wholly, recruited from the free mining communities of Derbyshire, Cornwall, and Devon, bordering on the silver mine district.[1] For the sake of clearness we may recapitulate what has already been said concerning the external side of their administration. The mines of Beer Alston, Beer Ferris, Combe Martin, and Birlond seem to have been worked for the King's profit as mines royal under the guidance of a warden appointed by the King, with whom were usually associated a controller and a steward, the latter to preside at the miners' court. Workmen were obtained by impressment. They received the King's wage, were free from ordinary taxes and tolls while in service, were under none but the mine jurisdiction save for serious offences, and in various other ways enjoyed the King's special patronage.

It was especially these mines which the King was continually pledging to his creditors, and on more than one occasion the mines were in this way actually turned over to private management.[2] One of the contracts or leases is still extant, by virtue of which Edward I in 1299 made over a Devon mine — which of the half-dozen we do not know — to Coppus Josephus and various other financiers of the Florentine company of the Frescobaldi.[3] The

or thereabouts, was not a relic of mediæval serfdom, but was brought about by the wide and harsh application of vagrancy laws at the beginning of the seventeenth century, and that the mediæval Scottish miner was free to choose his master. The whole subject is exhaustively discussed in the Edinburgh Review, clxxxix, 120–148. Cf. also Patrick, introd. lxv–lxvii.

[1] According to Atkinson (51), a thousand men were employed in the mines of Beer Ferris and Combe Martin after their reopening in 1485. They yielded the King £44,000 per annum in silver and lead, and were worked for five years, when they were abandoned because of their increasing depth.

[2] Cal. of Pat., 1299 (Apr. 27); 1301, p. 625; 1308 (Aug. 16); Cal. of Close, 1309, p. 95.

[3] L. T. R. Mem., 69 c. m. 34. The Florentine companies will be remembered in connection with the pledging of the stannary revenues, and the lease of the preëmption to Antonio Pessaigne (Cal. of Close, 1335, p. 458; Cal. of Pat., 1336 (May 7);

lease contains eleven clauses, the purport of which is as follows. First, the King makes over to the lessees all the ore, and the miners must deliver it pure and clean and well washed according to the agreement between the King and the miners. Second, the lessees, having accepted the ore, are to write off a certain sum per load from the amount which the King owes them. Third, the King agrees to supply the lessees with all the men necessary for the mine, said laborers to receive the same wages as those given before the contract, or less if the merchants can beat them down. The merchants are to allow the miners for each load five shillings sterling, or less if they can so agree, and to pay all other costs, besides rendering to the King twenty shillings per load. Fourth, the King promises to supply buildings for the men and horses of the lessees. Fifth, he agrees to deliver to them all the machinery, tools, and property appurtenant to the mine. Sixth, he agrees to allow the lessees to cut wood for the charcoal needed in smelting and for machinery, tools, and houses, if required for the mine operations, the lessees to pay the price the King used to give. Seventh, the King agrees to reimburse the lessees for whatever new adits or machinery they may have constructed, the terms to be discussed with the controller whom the King will keep at the mine. Eighth, the lessees may employ as many kinds of craftsmen as the King has been wont to use. Ninth, the King agrees to take the lessees and their company under his protection, just as he does his own miners. They are to be free of all tallages and taxes and are not to be haled before any court save that of the Treasurer and Barons of the Exchequer for anything touching the said mine. Tenth, the lessees and their attorneys shall have safe conduct by the King's sheriffs whenever they carry money to the works or silver from them. Finally, a clerk is to be appointed by the King to reside at the mine to watch over his interests and to act as controller, as against the lessees.

These miners, therefore, whom the King or the Florentine capitalists hired, were free workmen. They contracted for a money wage, a good one as we shall see presently, and when it was not paid promptly, as on one occasion at least, they struck work.[1] They

Close, 11 Edw. III, pt. i, m. 12; 13 Edw. III, pt. ii, m. 33; 15 Edw. III, pt. ii, m. 18; Pat., 5 Edw. III, pt. i, m. 16).

[1] Cf. Cal. of Close, 1330, p. 13.

worked by the piece for a stipulated price per load of ore; some of them were housed at the mine, at their employers' expense; they paid no taxes or tolls and for most offences were responsible only to a court at the mine, composed of a steward and a jury of their fellow workmen.

The working force included miners, pumpmen, carpenters, smiths, chandlers, sawyers, washers, roasters, refiners, charcoal-burners, and carters.[1] Most interest, of course, centres about the miners. In one account roll their number is put at seventy-nine, but many were evidently only casuals not regularly in the employ of the mine, as we learn that the number of days' work in the year from each miner varied from one to two hundred and sixty-one. Their wage of 4d. per working day compares favorably with that of the skilled artisan of the same period,[2] but they were paid extra wages for special jobs, such as cleaning the shaft or piercing a level.[3] This particular account roll for Beer Ferris,[4] late in the fifteenth century, contains no trace of anything corresponding to associations on the part of the miners. In a roll for Birlond, one hundred and fifty years earlier,[5] the miners were paid, according to agreement with the warden, from 3s. to 5s. per load of ore,[6] and many of them worked together in companies, under the direction and leadership of one of their number.[7] Both for ordinary ore hewing and for special work, paid at contract rates per fathom,[8] the men were paid in cash.[9]

[1] The following information is taken from various account rolls in the years when the King ran the mines. It is not likely that the lessees made any important alterations in the internal economy.

[2] 1480–1481. In 1308 we have the case of twenty-five miners whose work in building an adit was rewarded by 9d. per week each and expenses, while the carpenter received only 10d. without his living, and the overseer of the surface works of the mine 10d. (Accts. Excheq. K. R., bdle. 261, no. 9).

[3] Cf. Accts. Excheq. K. R., bdle. 263, no. 10.

[4] *Ibid.*, bdle. 266, no. 25.

[5] Accts. Excheq. K. R., bdle. 262, no. 11.

[6] Again, in 1343, we find the miners paid so much per load of a certain number of dishes (Accts. Excheq. K. R., bdle. 263, no. 17). In an account roll of 1292 (*ibid.*, bdle. 260, no. 3), they receive 3d. per day each.

[7] Cf. Accts. Excheq. K. R., bdle. 263, no. 16.

[8] *Ibid.*, bdle. 263, no. 17.

[9] Sometimes also in ore, which they sold (cf. Accts. Excheq. K. R., bdle. 261, no. 9).

Beer Ferris in 1480[1] maintained two smiths in constant attendance at the mine forge at 22d. and 20d. a week respectively,[2] a figure somewhat under the regular miner's wage, four carpenters at from 4d. to 5d. per day each,[3] for varying lengths of time, a chandler at fourpence for forty-nine days, and three sawyers whose wage was dependent upon the amount of lumber they handled.[4] In earlier account rolls we find mention made of other craftsmen, masons, and masons' laborers,[5] as well as of foresters.[6] The ordinary method of draining the Beer Ferris mine was by windlass and bucket, several laborers drawing pay in this way as "water winders." A small water wheel was used as well and also suction pumps,[7] which seem to have required the undivided attention of three somewhat highly skilled mechanics.[8] The argentiferous lead ore which the mine produced was broken and washed,[9] probably by hand, and then carried to the furnace, where another relay of workmen, furnace-men, and bellows-blowers[10] drew the mixture of metals from the rock. To separate the silver from the lead, however, a special refining process was necessary and a test of consolidated wood ashes. The "fining-mill" employed as many as ten hands, including a chief refiner whose wage in 1480 was 5d. per day,[11] the rest receiving 4d. each. Other expenses were incurred for the wages of forty-seven charcoal-burners, each receiving twopence per quarter of coal as well as a regular wage of 4d. per day,[12] woodcutters,[13] the salaries

[1] Accts. Excheq. K. R., bdle. 266, no. 25.

[2] In the Birlond counter-roll of 19 Edward III (*ibid.*, bdle. 262, no. 11) the smith received 10d. per week, while in 1292 (*ibid.*, bdle. 260, no. 3) we find the "workmen" getting 3d. per day.

[3] The carpenters were employed in building refining mills and huts. Their wage in 1292 seems to have been about the same as that of the miners, 2½d. or 3d. per day (*ibid.*, bdle. 260, no. 3).

[4] In 1292 they received a fixed wage of 3d. per day each (*ibid.*, bdle. 260, no. 3).

[5] *Ibid.*, bdle. 262, no. 11. They were paid 2½d. and 1½d per day respectively.

[6] *Ibid.*, bdle. 261, no. 25. They received 10d. per week each.

[7] "Ordenaunce."

[8] *Ibid.*, bdle. 266, no. 25. They received 2s. 8d., 2s. 4d., and 5d. per day respectively, besides extra pay for constructing pipes.

[9] *Ibid.*, bdle. 263, no. 17.

[10] *Ibid.*; cf. also bdle. 261, nos. 9, 24, 25.

[11] Early in the fourteenth century we find this highly skilled workman in receipt of 18d. per week (*ibid.*, bdle. 261, nos. 9, 24).

[12] *Ibid.*, bdle. 266, no. 25; bdle. 263, no. 17.

[13] *Ibid.*, bdle. 260, no. 3.

of the warden and clerks,[1] the overseer or foreman,[2] and the steward who kept the court. There were also payments for the cartage of ore from the mines to the washing places and from there to the smelting, and for the cartage of the metal from the smelting furnace to the refining mill; for brushwood and timber conveyed from the forest to the mine; and for periodic trips to Tavistock or Plymouth, where purchases were made of iron, hides for bellows,[3] and pump suckers, "winding hooks," sea coal for the forges, canvas for bags for carrying the ore, "Normandy ropes,"[4] shovels, and other utensils.

What most impresses one in reading these accounts is the relatively high status of the workmen and the highly organized form in which these mining operations took place. What the real position of the workman was may be seen, not only from the fact that as a rule the wages he received were as high as or higher than the average wage paid to skilled artisans throughout the land, but also from the varied nature of his duties. The rolls on which the foregoing account has been based mention several different classes of laborers, but as a matter of fact often the same set of men performed several functions. To quote from the roll of 1480 alone, we find that one of the three sawyers was also a carpenter and a pumpman; several of the carpenters acted as carters; while the miners themselves, aside from their regular employment, took part in nearly every other occupation, some acting as ore washers, others as smelters, others as refiners, carpenters, water winders, and carters; while one man, Thomas Robyn, who on other occasions figures as the captain of a gang of miners, was sent to London as a messenger to the barons of the Exchequer.[5]

The aggregate wage which many of the men received for their different tasks constituted oftentimes a very fair income. Thomas Dery worked in the mine for eighty-three days at 4d. per day, and made £2 by his work as ore washer, in all £3 7s. 8d., which, presumably, was earned in considerably less time than a working

[1] Accts. Excheq. K. R., bdle. 266, no. 25. Eight clerks were employed in 1480.

[2] One man superintended the entire course of operations both above and below ground (*ibid.*, bdle. 261, no. 24; bdle. 263, no. 17).

[3] Bellows and ladles were made on the premises (*ibid.*, bdle. 263, no. 17).

[4] *Ibid.*, bdle. 263, no. 17; bdle. 266, no. 25.

[5] *Ibid.*, bdle. 262, no. 11.

year. Roger Payne worked one week in the mine at 4d. per day, but made £4 3s. by ore washing. Simon Russell was a miner for thirty-four days at 4d. per day, but his ore-washing brought him £7 12s. 11d. and his work as water winder 7s., in all £8 11s. 3d. John Staunton worked as a miner eleven days for 4d. a day, and thirteen days as a smelter for the same wage. Robert Martin earned six shillings for thirty days' work as a miner and 1s. 4d. for four days' work as a smelter, and also received 14s. 2d. for smelting one hundred and seventy bowls of ore, at a penny a bowl. William Martin, by working two hundred and twenty-one days as a miner and fifty-three as a water winder, made £4 11s. 6d. It is safe to say that, if the mining classes fared equally well in other parts of the country, their economic position was relatively a high one. Since the labor to which we have just referred was free but impressed, it is possible to assume that the free miners of Derbyshire, of the Forest of Dean, and of the stannaries were as well if not even more favorably situated.

It is probable that the capitalistic state of mining organization indicated in the above account of the royal mines in Devon is not equally characteristic of English mining in general, since the royal silver mines doubtless showed in this respect the more advanced development. That mining, however, was one of the first industries to generate capitalistic organization, there can be little question. The scattered hints in the existing records, and the almost stationary nature of English mining technique until the sixteenth or seventeenth centuries, lead one to the conclusion that, did we possess more account rolls of mines, we should find that little or no advance over mediæval forms of organization took place until comparatively recent times,[1] and that in the main these forms, thus early developed, were in advance of the growth of capitalistic enterprise in other branches of industry.

Up to this point, therefore, we have found that the industrial history of the stannaries presents points of similarity to that of the German silver mines, the mines in both cases, although not necessarily in the same order, having passed through the hands of organizations

[1] The account of the iron works of Sir Charles Coote in Ireland about the year 1640 does not differ materially from those of the mines royal in Devon, centuries earlier (Scrivenor, 64).

which we have styled respectively the mine partnership, the cost agreement, the tribute system, and the lease. Capitalistic enterprise may be traced to an early date in Cornwall, and from a digression to the silver mines of Devon we have seen something of the manner in which such adventures were conducted. It remains now very briefly to trace the development of mining organization in the centuries which follow the Middle Ages.

First of all let us take the account of the stannaries as described in the year 1586 by Thomas Beare, the bailiff of Blackmore. He begins by enumerating the various classes of stannary workers. There were the charcoal peddlers, who went from blowing-house to blowing-house with their packs.[1] There were the blowers and the owners of the blowing-houses. These smelters, presumably, began by owning their own shops and ovens, but by this time many instances existed where the blowing-house owner had become a small master,[2] as indeed he had been a century before, although he let out part of his house to independent workmen.[3] Then there were smiths, carpenters, and various other classes employed about the mines,[4] and finally, the miners themselves.

"The most part of the workers of the black tin and spaliers are very poor men,—and no doubt that occupation can never make them rich,—and chiefly such tin workers as have no bargains but only trust to their wages, although they have never so rich a tin work, for they have no profit of their tin if they be hired men, saving only the wages, for their masters have the tin. Now, if they should chance to be farmers themselves and their work fall bad, then run they most chiefly in their masters' debt and likely to incur more and more rather than to requite any part thereof, for of these two choices, to be a hired man or a farmer, the one is a certainty and the other an uncertainty. The farmer knoweth not how the work will do until time that he have proved it and must needs live in hope all the year, which for the most part deceiveth him." [5]

Here in the sixteenth century we have the same gradations, according to Beare, as in the Middle Ages. Many of the mines were worked by their original companies of miners adventuring in partner-

[1] Harl. 6380, fol. 37.
[2] *Ibid.*, fol. 32, 39.
[3] Cf. Add. MS. 6713, fol. 101, 104.
[4] Harl. 6380, fol. 34, 37, 38.
[5] *Ibid.*, fol. 56, 57.

ship[1] under an elected captain,[2] and it is doubtless men of this stamp who are meant when reference is made to a wealthier sort of tinner working side by side with the poor spaliard, the latter in this case probably hired by the cost-giving associates. "The tinner," says the bailiff of Blackmore, "in my opinion is he that giveth wages by the year to another to work his right in a tin work for him as a dole or half dole more or less, or else works his right himself as many do." [3] These laborers were paid by the amount of ore excavated and received part at least of their wages in tin.[4] Other mines were leased on shares to other spaliards, or laborers, who to all intents and purposes worked them by the tribute system, leasing portions of the mine for various percentages of the product; but as Beare states that most of the spaliards worked for wages, the implication is that the tribute system at that time was comparatively limited, and, on the other hand, that cost-giving or perhaps a system under which the adventurer worked his claim as a small entrepreneur, with hired labor, was still in the foreground.

Carew, who wrote only a few years later, tells us that small undertakings were worked single-handed but that usually the discoverer of a lode took others as associates, because the charge "amounteth mostly very high for any one man's purse, except lined beyond ordinary." [5] The adventurers were either working miners or capitalists who put in hired labor. The larger works were under direction of a captain, and toll was paid to the lord of the soil, or to the lord and the bounder. The produce of the mine was shared out in doles and a proportionate division made of the charges.[6] Each adventurer had his share of the black tin, after the payment of the toll, and each man carried his portion to the blowing-house and after the coinage sold the white tin either to the London merchants or to the wealthier tinners.

For stannary conditions in the later seventeenth century our sole source of information is the anonymous article in the Philosophical Transactions for 1671. This tells us little beyond the fact that the proportion of workmen was three hewers, or beelmen, to two barrowmen, which was as much as a drift would hold.[7]

[1] Harl. 6380, fol. 6. [2] *Ibid.*, fol. 58. [3] *Ibid.*, fol. 6.
[4] Lans. 76, fol. 34; Doddridge, 94, 95. [5] Carew, 10; Worth, 53.
[6] Carew, 14. [7] Phil. Trans., 1671, p. 2104.

It is not until 1765 that another cross-section of the industry is obtainable. This is given by a Frenchman, Gabriel Jars, who visited Cornwall in 1758, 1765, and 1769. He says that "the usage established in all the mines is to give out the extraction of the metal to entrepreneurs; the latter employ workmen at wages who work according to their orders; some are workmen themselves."[1] On a day previously set, the account continues, those who are interested in a mine, the adventurers or their agents, assemble, and the contractors make offers, bidding downwards. The extent of each bargain is called a "pitch," and is usually of fifteen, twenty, or twenty-four fathoms of level, the distance from one "winze" or air-shaft to another, and seven fathoms in depth. The workmen find themselves the necessary tools, light, and powder. The adventurers provide merely for the maintenance of machinery and ropes. The number of workmen who do the work is usually seven, eight, or nine, and the time of contract six months. The consideration for the contract is a portion of the mineral extracted, that is, the contractors receive a third, fourth, or fifth of the value raised.

According to this statement, therefore, the actual workmen now appear for the most part to have been hired for wages. The original mine partnership has now definitely developed into a company of non-working shareholders, or gentlemen adventurers. The workmen were under contractors who agreed to excavate the ore for a certain proportion of the selling price. Nevertheless Jars states also that in some cases the entrepreneurs are actual workmen, and in other passages says that single workmen often commence the exploitation of a mine at their own risk.

In 1778 Dr. Pryce gave a descriptive account in his *Mineralogia Cornubiensis*. "Mining," he says, "is so expensive and uncertain, that few Cornish mines are carried on at the risk of one or two persons. Many partners are united, four, ten, sixteen, twenty-four, or thirty-two in number. The shares in these adventures are often so fractional and intricate that a stranger, although a tolerable mathematician, would be greatly at a loss to decide and apportion the doles, or shares, with that precision which is familiar to many illiterate miners, who can cast a piece of ground and assign the proportions of a parcel of copper or tin ore with the utmost accuracy

[1] Jars, iii, 202.

by means of twenty shillings, pebbles, or buttons." [1] "Deep and chargeable mines," he continues, after having described the meetings of these mining companies, "are carried on by persons of fortune or great skill, but shallow mines are occupied indifferently by such or by the laboring miners, and frequently by both. In the large mines there is a superintendent or captain who has direction of the works above and below ground." [2] He then proceeds to speak of the wage systems.[3] "It is a good and customary way for the owners to set their dead ground either in or out of the lode, to be sunk, driven, stoped, or cut down by the fathom, but if there is no choice in respect to saving the ore drawn or the like, they set it to be sunk . . . upon tut bargain, that is, a piece or part of unmeasured ground by the lump, for such price as can be agreed upon;" and from the same passage we learn that the work was done by a small gang of laboring tinners, who supplied their own tools and materials.

When the lode had been tapped, two methods, again, might be employed. The ore might be broken either by the fathom, that is to say, for a stated price per unit of length or volume, or by the tribute system; only instead of being merely a matter of bargain between "entrepreneurs" and adventurers, it was more complex. First to be noted is the fact that, as in Jars' account, tin works were often given over to one tributer.[4] "Adventurers very often lease a mine on tribute. Some miner takes the mine of the adventurers for a determined time, that is, for half a year, a whole year, or seven years. If it is a tin mine, he articles first to pay the lord, or the lord and bounder, if any, their shares or doles free of all cost. . . . Of the remainder, he pays the adventurers one moiety or one fourth part according to the agreement, it being more or less in proportion to the richness of the mine." Often, also, the tributer was associated with several others who clubbed together to provide the necessary capital for machinery or the payment of wages.[5] Much more commonly, especially in the larger works, the mine was divided into pitches and auctioned off to small associations of laborers,[6] who, it deserves to be noted, seem to have ousted the single entrepreneur tributers mentioned by Jars a few years earlier.

Upon the whole the process of transformation in the stannaries

[1] Pryce, 173. [2] *Ibid.*, 174. [3] *Ibid.*, 180.
[4] *Ibid.*, 187. [5] *Ibid.*, 188. [6] *Ibid.*, 189.

from the Middle Ages until 1778 is fairly clear. The mines started, it will be recalled, under the entire charge of several or even of one working adventurer, the small entrepreneur who plays so prominent a part in the history of mining in Germany. This primitive type of organization always maintained a certain standing. The bailiff of Blackmore refers to it.[1] Carew mentions it in 1602;[2] Jars in 1765[3] and Pryce in 1778[4] distinctly state that this type existed in their times, although becoming somewhat rare. We may go a step farther and say that the working adventurer still survives, here and there, although by the nineteenth century, if not earlier, he was confined to stream works.[5]

Nevertheless, as we know, there had early arisen in the associations of working miners cost-giving members whose substitutes, or spaliards, we have seen at work in 1586. Of these last we hear nothing further until 1765, and then only indirectly. For it will be noted that, although in Jars' account the "entrepreneur" is said to be compelled to have workmen of all kinds, yet no express mention is made by the French writer of the preliminary work to be done at the mine, the sinking of shafts or driving of levels in preparation for the extraction of the ore itself. Now in Pryce's account, a little later, this work of preparation was known as "tut work" and was done upon contract made with companies of laborers. It is a fair assumption that these tut workers of 1778, who evidently approach more nearly the status of the ordinary wage earner than do the tributers, passed to tut from the piece work system mentioned by Pryce as concomitant with it, and are the same men who were employed by Jars' "entrepreneurs" in 1765 to open the ground, and who in 1586, as spaliards, accepted a wage from cost-giving mine associates or from small masters, the latter conducting small mines or shares in mines upon a capitalistic basis.

The tribute system, as we have had occasion to see from an examination of ancient stannary law, was probably an early development in the stannaries, and had become extensively employed by the time that Beare was writing his account. But just as might have been expected, these tributers had become by 1765, how and when we do not know, what some of the non-working

[1] Harl. 6380, fol. 6. [2] Carew, ed. 1811, p. 30–34. [3] Jars, iii, sec. 10.
[4] Pryce, 178. [5] Lit. Pan., iii, 1238–1241 (1808).

mine partners were already, small entrepreneurs with hired labor. Even then, however, the common miner, possibly under the steadying influence of the Wesleyan movement, had begun to improve his position, and we find him, probably in the latter part of the eighteenth century, superseding his erstwhile employer and taking the tribute system for his own use.

Out of this conglomeration of wage systems, the two of the greatest importance and interest in the nineteenth century have been those of tut and tribute. The Cornish miners' year is divided into various periods, and at the end of each occurs what is known as a "setting day." Some time before that day the agents or captains go through the mine and decide what work is to be done in the succeeding period, what shafts are to be sunk, and what levels driven. They determine also how many pitches or divisions of the lode are to be worked. They estimate the quantity and quality of the ore which these pitches are likely to yield and the amount of labor that must be devoted to the work. In a similar manner they decide upon the amount of labor which can be done in a given time in preparing the way for the extraction of the ore, determining, for example, the quality of the ground through which the shafts, or winzes, are to be sunk or the levels driven.

On the setting day the men employed in the mine, together with those who have come from elsewhere desirous of work, assemble around the account houses or a platform where the chief agent or captain takes his stand. He reads the rules under which the mine is to be worked, and then auctions off the different pitches or pieces of work in the mine to the lowest bidders, who in this case represent small groups or companies of from two to eight men or boys. When these groups of men go to work together, they are charged for the material they use, the tools, candles, powder, and other necessaries, as well as for the cost of hauling the rubbish to the surface. At the end of the period for which the contract is let a balance sheet is prepared; they are credited with the amount of work they have done and debited with its cost, and frequently also with a subscription for medical attendance and the maintenance of a club, which supplies them or their families with aid in case of accident. Sometimes these tut workers are in addition credited with the small percentage of ore that may be extracted in the course of their operations, in order to induce them to keep it as separate as possible from the rubbish, or "deads," and during the progress of the work they frequently receive payments on account.

In the case of the actual excavation of ore, the work to be done is divided into small compartments, or pitches, and in like manner put up to auction, except that in this case the tributers offer to do the work in consideration of receiving a percentage of the price realized from the ore, after paying all or nearly all the cost of reducing it to a salable state. As was the case with tut work, the miners pay their own costs in tools and materials, while, in order to provide for their support until the balance is handed over to them, they obtain a certain customary advance from the adventurers, known as "subsist." [1]

It should be added in closing that both the tut and the tribute systems, particularly the latter, are and have been for perhaps a century giving way to the ordinary wage system common elsewhere. The details of this movement we need not stop to analyze. Suffice it to say that its mainspring is probably to be found in the increase of engineering skill and geological knowledge among mine captains and agents of mining companies, which tends to increase their caution in the allotment of work to tributers and lessens the latter's chances of making lucky strikes. This fact has brought into unpleasant relief the casual and fluctuating nature of the tributer's compensation and the fact that he may be forced to work for weeks and sometimes for months without pay and dependent for support upon the advances of his employers. All these considerations, present also to a somewhat less degree in tut work, have brought about a steady drift on the part of ore excavators from the tribute to the tut form of payment, and on the part of the original tut workers, in an even more strongly marked degree, to ordinary piece or time work.[2]

The employers of these tut and tribute workmen up to fifty years ago, and in large measure even to the present day, had in essentials maintained a close connection with that nucleus of the present day mining company, the original association of working adventurers. Cornish mining companies to-day work for the most part under the

[1] Journ. Roy. Stat. Soc., li, 494–566; Rep. Stan. Act. Amend. Bill (1887), Q. 36.

[2] Journ. Roy. Stat. Soc., li, 519–528, 560–562; Rep. Stan. Act Amend. Bill (1887), Q. 1475–1490, 2616–2621.

so-called "cost book" system. Under the cost book system two or more men secure the lease of a property [1] and induce some others to join them, and if the property seems promising these few will include a banker, a smelter, merchants of iron, timber, candle, and cordage, and possibly a dealer in new and second-hand machinery. This loose association then registers under the cost book a company of perhaps five hundred and twelve shares and is ready to begin business. A meeting is called, a purser elected to take charge of the accounting, and a call made on each shareholder for perhaps £1 per share. The following month a second meeting is called and perhaps £1 more levied, and so matters continue until one of three things happens: the mine becomes self-supporting; it earns a profit, in which case a dividend is declared; or there comes a call to which one of the shareholders refuses to respond. Then if the others refuse or are unable to take up the shares of the defaulting associate, the mine is wound up and the lease sold, together with the machinery. If the proceeds from the sale are not sufficient to liquidate the liabilities of the concern, the adventurers are called upon to contribute *pro rata;* and as long as a single moneyed man remains among them, the creditors are sure of recovering.

The main features of the cost book company which distinguish it from an ordinary corporation or partnership may be summarized as follows: first, the absence of any fixed capital; second, the right to transfer shares without the consent of one's associates, simply by giving written notice to that effect to the purser; [2] third, the right of any adventurer to relinquish his interest [3] upon written notice; fourth, the right of joint management; fifth, the continuance of the undertaking as long as any remain in the association; sixth, the frequent and periodic settlement of accounts in a cost book, and the enforcement of contributions by the purser in the stannary court; seventh, the perfect register of adventurers for the time being; eighth, the right of excluding defaulters from participation in the profits; ninth, the liability of mine, machinery, and produce to the claims of creditors; and finally the fact that the adventurer's interest

[1] *Cornish Mining,* l; Bartlett, 24.

[2] Pike, 52.

[3] The retiring shareholder is entitled to withdraw in cash his share of the machinery and materials, the value of which is usually left to arbitration (Rep. Stan. Act Amend. Bill, 1887, Q. 14).

lies, not in the capital which he may have advanced, but solely in the mine itself.[1]

It should be noted that the liability of the adventurer, though restricted in kind,[2] is unlimited in amount, so that during his adventureship he is liable for the whole of the engagements incurred in the ordinary course of mine management.[3] Should a mining company attempt to default its dues, its creditors have action solely against the purser [4] by petition in the vice-warden's court, and upon non-payment the court would order a sale of the mine materials and produce. The purser, in turn, might obtain his remedy in the stannary courts against adventurers who refused to meet their obligations.[5]

Under this system the control of operations rests with the purser. He keeps the accounts, authorizes the purchase of materials and stores, hires the laborers, recommends the assessments, summons the shareholders to monthly or bi-monthly meetings, and at each reads his report and presents his estimates.[6] Next in importance to the purser stands the chief captain, or manager, who superintends the entire mine [7] and the general routine of surface work, and under him are usually selected from the most intelligent workmen subcaptains, who act as foremen or inspectors of the work underground, are held strictly responsible for the work of their section, and are stimulated by prospects of advancement.[8]

To point out the easy process by which the old Cornish mine partnership developed into this simple yet effective type of industrial organism, peculiar to the stannary districts,[9] would be needless. It is enough to repeat once more what has already been said, namely, that nowhere more clearly than in the mining industries is it possible to see ancient types of economic organization in full vigor at the present day. Like the tut and tribute systems, the cost book

[1] James, 22, citing Dickinson *vs.* Valpy, 10 B. & C. 128; Hawlayne *vs.* Bourne, F. M. & W. 595.

[2] Tredwen *vs.* Bourne, 6 M. & W. 461; James, 22.

[3] Convoc. Cornw., 2 Jas. II, c. 6.

[4] James, 15.

[5] Cf. Pryce, 173.

[6] Bartlett, 24; Watson, *Compendium of British Mining*, 11.

[7] Pryce, 174.

[8] Eng. Quart. Min. Rev., iv, 265.

[9] The cost book is valid solely by virtue of stannary law and custom (James, 22).

company is a survival of ages long since past, and unlike them it still retains most of its pristine vigor.[1]

[1] All the Cornish mines were cost book concerns until the Companies Act of 1862, which introduced the principle of limited liability into the stannaries (Rep. Stan. Act Amend. Bill, 1887, Q. 19). Even to-day nearly all the mines are worked on the former basis.

# CHAPTER VIII

## CAPITAL AND LABOR IN THE STANNARIES

THE freedom of early mining law from anything approaching the restraints of the gild system, while it led, as we have seen, to the early introduction of various forms of capitalistic enterprise, also gave rise to another phenomenon not so early observable to any notable extent in other mediæval industries. This was the entrance of the middleman, the tin dealer.

Here again there seems to be a difference in commercial organization between Germany and England. In the former country the rôle of middleman was apparently played by the smelter, who, at the beginning of the Middle Ages, had become entirely dissociated from mining properly so called.[1] The smelting huts were small affairs quite distinct from the mines, and in the hands of a somewhat well-to-do class of capitalists.[2] Depending as they did for their existence upon concessions granted by the lords, they were under official supervision and had no free position as regards purchase and sale, but they possessed what amounted practically to a monopoly right, since as a rule the precious metals might not be carried from the mining district and the refined silver must be sold at the lord's mint.[3] The almost inevitable result was that the smelters arranged among themselves as to the price which the miners should receive for ore, and would purchase it only if they felt that they could make a large profit from the transaction. If not, the owners of the ore were obliged to have the smelter reduce it for them at their own expense, which was usually as great as the smelter could make it.

It was these obstacles to the disposal of their ore at a fair price which brought about the decline of the tributers, and it has been seen how the lords endeavored to remedy their difficulties by set-

[1] Schmoller, Jb., xv, 667, 687–690, 692.

[2] As a general rule mining law forbade the ownership of a smelting establishment by miners and *vice versa*.

[3] For copper also there existed no freedom of sale (Schmoller, Jb., xv, 692).

ting up smelting establishments of their own and purchasing the ore at a just price, as well as by other similar means. These projects failed to remove the evil. The restrictions placed upon the removal of ore weighed heavily upon the poorer miners, whose richer associates were enabled to purchase the ore at low prices.[1] In this connection we notice also the rise of a merchant class acquiring ore of the poorer miners, so that in the fourteenth century among the regulations of the mines in the Black Forest is one which orders that all ore be brought to a shed each Saturday and sold under fixed rules through the agency of the mine master.[2]

Similar conditions, so far as we have information, seem to have been common throughout mining centres everywhere in Europe. In France it was the merchants who, in the often quoted edict of Charles VI, are said "to open mines at their own cost," [3] probably from the fact that they advanced capital to the miners. In England, in the Forest of Dean, although on this point our information is slight, a similar relationship may well have existed between the miners and the owners of the neighboring forges.[4] In the free lead mines of Derbyshire the situation is shown by the following laws, which date back certainly to 1288 and are probably very much older.[5]

"And if any miner or other take costage of a merchant and may not find mine [6] to that merchant, the said merchant will make no more cost to find ore in the same place; and after he laboreth and findeth the mine to the profit of another merchant; and if there be two or three, the same first merchant that first made his costage shall have the third stone till he have recovered his costage, and the other merchant shall have the other two parts, for that the mine was found at his costages. . . .

"And if the miners have received any money of any other man beforehand for their ore, then the miners shall pay their debts with-

[1] Schmoller, Jb., xv, 978, 1004, 1005.

[2] *Ibid.*, xv, 691. Cf. also *ibid.*, xv, 1005. A usurious class of middlemen also sought to obtain control of the tribute contracts and depress the tributers.

[3] Smirke, 104.

[4] Cf. Cal. of Pat., 1385, p. 118; Cal. of Close, 1229, p. 166; Smiles, 29. A man becoming a smithy holder lost his privileges as a free miner (Houghton, pt. ii, art. 33).

[5] Add. MS. 6682, fol. 65 *et seq.* (printed from Escheat Rolls, 16 Edw. I, No. 34).

[6] Ore.

out neglect of the Burghmaster so that this be without fraud or deceit." [1]

Of these laws, the first may refer simply to the cost agreement. A merchant has advanced sums of money to a miner or miners, as his part of the undertaking. If the miners endeavor to defraud him of his share of the ore by declaring the mine to be worthless and afterwards find ore in sufficient amount to induce others to contribute cost, the first cost-giver may not be passed by. The second law, however, brief as it is, refers clearly to advances made to the miners by ore dealers on security of ore not yet raised.

These tendencies show themselves still more strongly in the stannaries, and for two reasons. In the first place, the miners there, as far back as authentic records exist, were forbidden to sell their tin save at two periods of the year, at the Midsummer and Michaelmas coinages. This course would tend to play at once into the hands of the middleman. He alone had ready money and could afford to wait, while the unfortunate tinner, who had piled up ore for six months without having been permitted to realize upon it, would be obliged when coinage day came to sell at once and at whatever price was offered.

In the second place, although to a large extent this lack of the power freely to dispose of ore was characteristic also of the German silver miners, yet the manner in which the lords regulated all mining in their territories made conditions considerably easier for the German. To name only a few of the laws which characterize this mediæval labor legislation, we note in various German territories laws restricting the hours of labor in the mines, laws giving the hewer who had been promised cost by a mine partner a lien upon his employer's share in the mine or upon his ore,[2] laws prescribing weekly settlements of wages by tallies,[3] laws forbidding the mine partners to compel the acceptance of truck payments instead of cash,[4] or to dismiss a man for refusing to accept truck, laws forbidding the landlord, butcher, or baker, to whom a miner might be in

[1] Cf. Smirke, 58: "Item, they present that the first dealer in tin delivering money to a tinner upon his having possession thereof delivered to him for his security shall enjoy the same."

[2] Schmoller, Jb., xv, 702, 705.

[3] Bernhard, 55, 56.

[4] Schmoller, Jb., xv, 1010, 1011, 1013. More and more it came to be the rule that wages should be paid only in lawful coin (*ibid.*, xv, 1012).

debt, from claiming a lien on his wages, or even being present at their payment, laws regulating the place of the payment of wages, payment to be made in the presence of seignorial officers, not at places so far distant from the mine that the miners would be inconvenienced in attendance, and the "Schichtmeister" and "Steiger," officers representing the company for the payment of wages, forbidden to sell beer at the mine, or to keep inns, or to force the miners to receive their pay at their (the officers') houses, and laws, finally, which limited the amount and kind of property which the creditors of a miner might seize, and protected, through the medium of the mine judge, the orphans of miners in their inheritance.[1]

In Cornwall, on the other hand, with true *laissez-faire* spirit the English mineral law left the unorganized [2] tinners to a much greater extent unprotected, and handed them over to the tender mercies of the middleman and regrator with the damaging restriction on time of sale above mentioned.

Of the beginnings of the relations between the miners and the dealers we know little or nothing. When De Wrotham wrote his much quoted letter in 1198, he divided the stannary people into four classes, — diggers, smelters, ore buyers, and tin dealers, — each of which had its well known customs. His letter also informs us that the "mark of the second smelting" must be paid by the merchants before the tin might be removed from the custody of the keepers. Later, in 1304, we find the merchant buyers of tin in Cornwall petitioning the King in Parliament that they may have two days in which to pay the coinage duties, namely at the Feast of All Saints and at the Feast of St. John the Baptist,[3] a fact which would seem to imply that at this time as in the seventeenth century[4] the tinners pledged their tin in advance to the dealers, and upon getting their vouchers from the receiver of the coinage delivered them to their

[1] Schmoller, Jb., xv, 1012–1014.

[2] A factor which may have been unfavorable to tinners of the lower ranks was their lack of anything corresponding to journeymen's unions. The fact that Cornish miners as a rule preferred the illusory independence of the tribute system rather than frank acceptance of the wage system tended to make them an easier prey for the dealer.

[3] Parl. R., i, 163 a. Cf. also Pat., 9 Edw. III, pt. i, m. 8; 16 Rich. II, pt. iii, m. 14; Close, 7 Edw. II, m. 10.

[4] Cf. S. P. Dom. Eliz., civ, 54.

merchant creditors, who then discharged the dues and claimed the tin.

Further light upon this subject is shed by a petition to the King in Council, probably of the early fourteenth century. The tinners of Cornwall complain that the fixing of the staple for tin at Lostwithiel is detrimental to their interests, inasmuch as the town is far from the centre of mining, and because — in consequence of some reason illegible in the manuscript — the merchants of Cornwall who have been accustomed to advance them cash on the purchase of tin now refuse to do so.[1] These merchants were probably those mentioned by the tinners in their petition against the lease of the preëmption to Pessaigne, as "coming to Cornwall to buy tin, and trading for it wines, cloths, silks, and iron wares to the Cornish."[2] Among the middlemen were Cornish traders,[3] Londoners, and aliens. Of the dealings of the Jews in tin we have specific record[4] before their expulsion from England in 1290. We are told in 1455 that Italian merchants with ready money go about the county, and seeing the needs of the poor tinners buy the tin cheap, as well as wool from the wool-growers.[5] Still later we find, in 1492, a proclamation from Henry VII, appointing Southampton a staple for tin and decreeing two extra coinages "because the poor tinners have not been able to keep their tin for a good price when there are only two."[6]

The tinners' grievances were evidently smouldering through a long period, until under the Tudors matters reached an acute stage. The reason for this is to be found in the price revolution of the sixteenth and seventeenth centuries. To the enhanced cost, not only of food,[7] clothing, and other necessities of life, but also of

[1] Anc. Pet., 315 E, 197. The petition is undated but the handwriting is evidently of the period stated in the text. Much of its contents is illegible but enough remains to indicate its purport.

[2] App. E.

[3] Close, 11 Edw. II, pt. i, m. 28, 33. William de Carlisle of Bodmin, Michael Piers of Lostwithiel, William de Pencors of Cornwall, and William Collan of Truro, all merchants, are mentioned as owing stampage to the King.

[4] App. A. The question as to the presence of the Jews in the stannaries has never been clearly settled. Cf. Journ. Roy. Inst. Cornw., 1867, p. 324; Gent. Mag., i, 483; xiii, 701; Penzance Nat. Hist. and Ant. Soc., i–ii, 350; Trans. Roy. Geol. Soc. Cornw., iv, 75; White Bk. Cornw., i, c. 15, 19, 20, 22.

[5] Parl. R., v. 334 b.

[6] Pat., 7 Hen. VII, pt. i.

[7] Cf. Journ. Roy. Stat. Soc., i, 72.

timber, rope, iron, candles, and other mining requisites[1] daily becoming more and more important with the increasing depths of the mines and the change from stream to lode mining, corresponded not, as one might expect, a threefold or fourfold rise in the price of tin, but an increase in price of only about one hundred per cent.[2] The earliest quotation of the price of tin,[3] that of 1199, was £3 per thousand-weight, or about 6s. per hundred-weight. After this, in spite of the increasing output, a steady increase in price ensued, owing to the spread of the use of the metal. In 1294, according to the assessments of white rent, it was 8s., in 1297, 10s., a figure which it held for some time. According to Pegolotti's account, the usual price quoted for tin in the fourteenth century was one mark, or 13s. 4d. per hundred-weight, from which figure it rose somewhat later to about one pound sterling. This seems to have been the usual rate until the general rise in prices in the sixteenth century. Thereafter, with less trustworthy statistics, it becomes difficult to account for the rapid oscillations in prices. In 1559, for example, tin is quoted at a price of 65s. 4d. per hundred-weight. Two years later, however, we find it rated at but 33s. 4d. at the Southampton customs. In 1571 the Queen purchased tin at the rate of 54s. 7d. per hundred-weight. In 1588 it was back at 40s. or 46s., and at the close of the century does not seem to have advanced far beyond that figure. Nominally tin had risen in price; actually, in comparison with the prices of necessities, it fell.[4]

The first adequate exposition of the situation among the miners, thus become acute, is to be found in the complaint against Brokehouse under date of 1554. "The merchant tinners and the poor laborers," so we are told, "use always to receive and take money beforehand to set themselves at work and so bind themselves in obligation in great sums to deliver tin for such money as they receive at the day of deliverance. By this grant they will be forced to forfeit their bonds, for the whole tin will be delivered to Brokehouse. If the poor tinner offer Brokehouse the halfpenny per pound during the three days, and pay deliverance of his own tin for the saving of

[1] Rogers, iii, 287–296, 346–356, 396–400; iv, 328–331; 335–337, 449–452, 570.

[2] Harl. 6380, fol. 6. The sellers of black tin at the wash received payment in 1586 at the rate of £20 per thousand-weight of white tin. This price is less than twice as great as the price a century earlier. See App. U.

[3] App. U.

[4] Reyer, 85.

his bond, Brokehouse may say that he has sold his tin to another, and thereby defraud the poor man so that his bond be forfeited. Thus a number of merchant tinners which keep a number of poor laborers at work and use to deliver unto some £5, unto some £10, and unto others £20 beforehand, after the rate of such quantity of tin as they yearly use to make, whereby the said merchant tinner shall disburse among the poor men £1000 before he receive one pennyworth of tin, he cannot disburse such a sum beforehand without taking money beforehand of the merchant buyer. Now the merchant buyer will disburse no money to the merchant tinners for so much as he can make him no surety to the deliverance of any tin for the same. Thus the merchant tinner will lack funds to keep his tinners working, so in time many will be idle, and the whole tin business undone." [1]

Among the Lansdowne Manuscripts in the British Museum appears another document, twenty-one years later, which illustrates still further, and in slightly different fashion, the practice here described. According to the author, the distressed state of the tinners is due to the usurious contracts of the merchants. "There are three kinds of usury. In the first, the tin masters and owners of the tin [2] take money of the merchant buyers, and enter into bond to deliver tin in certain quantities at the next coinage, be the price high or low." By this contract the merchant usually gains sixty per cent. The second usury is where the tin master uses the same and harder practice toward the laborer or poor miner, "for he delivers to him so much money as he needs, on bond to deliver him a certain quantity of tin within a few days, a condition well known to be hard to perform by reason of the poor man's necessity. When this bond is broken the laborer is at his master's mercy for his whole wage, which wage is usually paid the poor man in tin, so that either he redeems the bond with double as he should have paid, or enters into a new usurious contract on further interest. The third usury occurs when the tin master does not deliver the laborer money, but some other commodity, such as linen cloth, woolens, etc., and so contracts with him in manner as aforesaid, by which the master, by these

[1] S. P. Dom. Mary, iv, 5.

[2] This apparently refers to the middlemen dealing in ore, on the one hand, and, on the other, to the owner of the ore who has not sold to the middleman.

wares delivered, gets ordinarily twopence in the shilling in the price agreed upon, and if he gets money, then upon such money for every hundred-weight of tin they have many times used to take five shillings for the loan of their money." [1]

Beare, to a certain extent, corroborates these accounts. Writing in 1586, he describes the "master," by whom he evidently means the ore buyer, as delivering the money to the tin workers before the tin was dug and thus becoming its owner long before it was produced. This brings the miners into the master's debt, and he in turn is in debt to the merchant buyer. The situation is aggravated by the practice of holding post coinages, for just as prices have been arranged between the tin merchant and his customers a rival seller may appear with some of the post coinage tin and spoil the market. The merchant recuperates himself for this uncertainty by paying less to the master, who, in turn, squeezes the miner.[2]

But the most concrete, if not the clearest, picture of these conditions is that given by Carew in 1601. "When a western gentleman," he writes, "wants money to defray his expense at London, he goes to a tin merchant for a loan. Uusually he has to give bond for a thousand-weight of tin for every £20 he borrows, the said tin to be delivered at the next coinage. But the business goes still farther. The merchant, that he may be sure to have tin for his money at the time of coinage, lays out great sums beforehand unto owners of tin works, who are bound to deliver for the same so many thousands of tin as the money shall amount to after the price agreed upon at the coinage. To them resorts the poor laborer desiring some money before the time of his payment at the coinage time. The other at first says he has none to spare, and in the end when the laborer is hardest pushed, he delivers to him wares instead of money, and the laborer is under bond to deliver tin at the coinage. And this ex-

[1] Lans. MS. 76, fol. 34. The truck system is mentioned as prevalent among the English iron artificers in 1604 (Third Rep. Hist. MSS. Com., p. 11). The German miners were protected against such abuses by stringent legislation (Schmoller, Jb., xv, 1012–1014; Jars, iii, 495).

[2] Harl. 6380, fol. 109. He mentions the existence also (*ibid.*, fol. 59) of a class of dealers in refuse tin. In 1595 we have also Thomas Middleton's report to Elizabeth to much the same effect (S. P. Dom., Eliz. ccliii, 46). The ultimate profits of the tin dealers must have been large. It is stated on two separate occasions that whereas "the poor tinners" sell to the usurers for £15 or £16 per thousand-weight, the latter resell to the pewterers for £28 or £30 (S. P. Dom. Eliz., ccxliii, 13; Jas. I, vi, 138).

treme dealing of the London merchants is imitated by the wealthier sort of dealers in black tin.[1] The wealthier tinners,[2] laying out part of their money beforehand, buy black tin from the poor laborers at so much per mark, *i. e.* look how many marks there are in the price made at the coinage for the thousand-weight, so many twopence halfpenny, threepence, or fourpence, partly after the goodness, and partly according to the hard conscience of the one and the necessity of the other shall he have for the foot, as if the price be £26 13s. 4d. per thousand-weight, therein are forty marks, then shall the poor tinner get of him who deals most friendly per foot of black tin forty times fourpence, or £20 per thousand, and less for the worst." [3]

This species of "usury," therefore, played directly into the hands of the merchant buyers at the top, apparently at this time confined mainly to a few London haberdashers,[4] who controlled the second group, consisting of non-laboring shareholders, small, independent miners, and, in some cases, ore dealers.[5] This latter group, in turn, recouped itself by pressure upon the dependent classes, the tributers, and the wage workers.

For the conditions prevailing among the workers at this time of depression we must rely in the main upon the somewhat general statements of local historians. Unquestionably it had fallen considerably below mediæval standards. Twenty pence per day is the profit which Trenewyth's impressed workmen in 1342 claimed that they could make before their seizure.[6] In Beare's account of the stannaries in 1586, the laborer worked by the dole or share and re-

[1] Carew, ed. 1811, p. 49.

[2] It is difficult to say whether by this is meant the gentlemen-adventurers or the buyers of black tin. Probably, however, the same practice was common to both.

[3] Carew, ed. 1811, p. 49.

[4] Welch, i, 268; ii, 10; S. P. Dom. Eliz., ccliii, 46. A few French and Breton merchants are said to have resorted to Cornwall to barter wine and cloth for tin (Lans. 24, fol. 50).

[5] A strikingly parallel situation existed at about the same time in the Newcastle coal fields, where the "hostmen," a select company of dealers holding a monopoly of the trade, oppressed the colliery lessees, and through them the miners (cf. Cal. S. P. Dom., 1637–1638, p. 387; Galloway, citing Grey, 24, 25).

In Yorkshire the lead miners were, as in Germany, under the thumb of the smelters (Malynes, 269).

[6] See p. 190.

ceived usually £3 per dole, working for the year. Out of this he was obliged to find himself, which usually cost about 2d. per day, and to support his wife and family, to say nothing of payment for rent and clothes.[1] This wage, taken by itself, can hardly have been sufficient to sustain human life,[2] being far below that paid elsewhere to ordinary unskilled labor.[3] Raleigh, in 1601, claimed that his exercise of the preëmption had increased the tinner's wage from 2s. to 4s. per week, and that work was no longer scarce.[4] According to Carew in the same year, the laborers, or "hirelings," received 8d. per day, or from £4 to £6 per year.[5]

On these starvation wages, the tinners of the lowest class, in number "ten thousand or twelve thousand of the roughest and most mutinous men in England,"[6] lived a life such as might have been expected of men in a like situation. The wretchedness of their existence became proverbial.[7] They lived in hovels and bred like rabbits. Working as they did in short four-hour shifts, that being as much as a miner could endure in the ill-ventilated shafts and levels,[8] their life was irregular and broken. Tippling and ale-houses abounded most in the parishes richest in tin.[9]

The sixteenth century ended in an extraordinarily severe depression throughout the stannaries. The mines had been deepened to a point where the drainage engines of the time failed to keep out the water;[10] capital and labor turned from mining to husbandry;[11] and the annual production of tin declined. It was agreed, therefore, by the government, that measures must be taken to set the tinners on their feet, and, as might have been expected, the remedy took the form of the usual Tudor-Stuart panacea for social and industrial ills, a monopoly. In an earlier chapter the policy of leasing the preëmption of tin has been mentioned in the discussion of its utility as a source of revenue. One of the motives leading to its

[1] Harl. 6380, fol. 57.

[2] Cf. S. P. Dom. Eliz., ccxliii, 13.

[3] Cf. Rogers, vi, 614.

[4] D'Ewes, 299; Edwards, i, 273.

[5] Carew, ed. 1811, p. 34. On Crawford Moor in Scotland in 1619 a large body of men were employed at washing gold at the rate of 4d. per day (Atkinson, 22).

[6] S. P. Dom. Eliz., Addenda, xxix, 126. Cf. the account given by Beare (Harl. 6380).

[7] Cotton, Titus, B. v, fol. 402; S. P. Dom. Eliz., ccliii, 46.

[8] Carew, ed. 1811, p. 35.

[9] *Ibid.*, 53.

[10] Lans. 19, fol. 99.

[11] *Ibid.*; cf. Lans. 86, fol. 67.

employment, with scarcely a break from 1599 until the time of the Commonwealth, was the idea of relieving the tinners from the necessity of dependence upon the tin dealers by a guarantee of a fixed price for their tin for a term of years in advance. The success or failure of such a scheme depended primarily upon two factors. First of all, the price offered the tinners must be high enough to ensure them a reasonable profit and at the same time must be automatically raised to correspond with the rise of general prices. In the second place, granted the continuance of the tin coinage, — and no one till Cromwell's time seems to have thought of abolishing it, — a substitute must be found for the advances of money which the dealers had been accustomed to make to the stannary people.[1] The Crown in its leases attempted, although with indifferent success, to cope with both these problems.

The price at which, down to 1645, the various patentees of the Crown agreed to purchase the tin stood as a rule little higher than the ordinary cut-throat figure at which the tinners had been accustomed to supply the tin dealers. In 1599 Elizabeth is said to have taken the preëmption at the rate of £28 per thousand-weight.[2] In 1601 it was given to Brigham and Wemmes for the same price.[3] In 1603 James raised it by ten shillings.[4] In 1607 Bludder succeeded in having it reduced again to £28;[4] and this seems to have lasted through the patent of December, 1608.[5] In 1613 the price rose to £30,[6] and the same terms were continued in the double patent of 1615,[7] the patent of 1621, and that of 1628.[8] A slight increase was made during the continuance of the preëmption of Job Harby and Robert Charlton, in the form of £2000 distributed annually among the tinners. So far, then, as concerns the price of tin the stannaries were probably no better off than they had been before.[9]

The other point in connection with the tin monopoly was the occasional establishment of "banks," or loan funds for the use of the tinners between coinages. From the late sixteenth century the idea

[1] Cf. Norden, 16.
[2] S. P. Dom. Eliz., cclxxiii, 74.
[3] *Ibid.*, cclxxxvi, 26.
[4] Lans. 1215, fol. 226–230.
[5] S. P. Dom. Jas. I, Oct., 1614.
[6] Receiver, 11 Jas. I.
[7] *Ibid.*, 13 Jas. I.
[8] *Ibid.*, 9 Chas. I.
[9] Cf. S. P. Dom. Chas. I, cccxxii, 1; cccxxvi, 2, 8, 60; *The Case of the Tinners Plainly Stated* (1636).

of a loan fund seems to have been prominent in the minds of most would-be preëmptors.[1] Sir George Carey in 1577, in his proposal to take over the monopoly, agreed upon reasonable security to allow the tinners the use of £4000 at low interest.[2] Buckhurst, a few years later, offered to loan them money at eight per cent; they were then paying forty.[3] Bevis Bulmer would loan them £10,000 without interest.[4] Elizabeth during her preëmption disbursed £8000 annually among the miners without interest,[5] and Brigham and Wemmes £1000.[6] Later preëmptors neglected to provide loan funds, and we find the Cornish parliament of tinners in 1624 asking the King that a £4000 fund be granted to the tinners upon reasonable security.[7]

In spite of this relief, however, it is not likely that the tinners were benefited by the preëmption. General prices continued to rise, while that given them for their tin by the monopolists was kept stationary. Westcote, writing in 1630, describes the condition of the day laborers as wretched in the extreme.[8] In 1636 the Cornish tinners complained to the King that the mines were decaying on account of the daily increase of charges upon the work, while the price received for their tin had not advanced.[9] The justices of Cornwall stated that the losses on tin mines often wiped out all profit.[10]

What the tinners themselves at this time thought of the policy of preëmption there is some difficulty in ascertaining, and it is doubtful whether, if we knew, we would be better able to judge of the merits of that system. Drowning men proverbially catch at straws, and if the tinners welcomed the establishment of these monopolies, it signified merely that, half starved and ruined as they were, they were glad of anything which promised a change. As a matter of fact, however, accounts of contemporaries differ as to the popularity of the preëmption in the stannaries. A document previously cited pictures the distress which would arise among the tinners if

[1] A similar practice obtained in some German mining districts (Jars, iii, 494, art. xxviii).

[2] Lans. 24, fol. 44, 47, 48, 50.

[3] S. P. Dom. Eliz., ccli, 71, 121.

[4] *Ibid.*, cclxxiii, 9.

[5] *Ibid.*, cclxxvi, 18.

[6] *Ibid.*, cclxxxvi, 26; Jas. I, xxiii, 57.

[7] Convoc. Cornw., 22 Jas. I, c. 10.

[8] Westcote, 52, 53.

[9] S. P. Dom. Chas. I, cccxxii, 1. Cf. *ibid.*, cccxxvi, 2.

[10] S. P. Dom. Jas. I, cxxxi, 66.

they were forced by the lease of the preëmption to forfeit their tin bonds to the dealers.[1] On several occasions, tinners either of Cornwall or of Devon put themselves on record as demanding the free sale of their commodity;[2] yet, on the other hand, a writer of the later Stuart period claimed that the working tinners desired the lease of the preëmption.[3] The rich tinners opposed it, "since they make money out of the necessity of the poor men, who cannot wait for a better market, but are forced to borrow at twenty per cent and to sell their tin cheaply." Certain it is that about the same time we find the tinners much concerned at the surrender of the Ford preemption,[4] and the two contracts which Anne made with the tinners seem to have been highly popular.[5]

To this period of monopoly alternating with usury, followed in the years 1650 to 1660 a policy on the part of the Commonwealth of complete *laissez-faire* as regards the stannaries, and certainly it must be admitted that in this respect, where the Stuart nostrums had failed, Cromwellian non-interference was accompanied by a return in the stannaries to a condition of abounding prosperity. With the abolition of the coinage duty and of the preëmption,[6] there began a new era. "The preëmption being resigned by the farmers, the price of tin rose. Multitudes of tradesmen left their callings for that of mining. Still the prices rose. Old, abandoned works were filled again and new ones taken."[7] And all this, says the chronicler, was owing to the tinners' freedom to sell at all times and at the best price. The price of tin had actually risen from £3 to £6 per hundred-weight when, with the Restoration and consequent reaction, the coinage rules were reimposed,[8] and coincidently the wave of prosperity subsided.[9]

With the exception of two short contracts for the preëmption

[1] S. P. Dom. Eliz., ccxliii, 113.

[2] Cotton, Titus, B. v, fol. 402; S. P. Dom. Jas. I, viii, 136; Cal. MSS., House of Lords Journ., viii, 439. See also S. P. Dom. Eliz., cclv, 46; Add. MS. 6713, fol. 331.

[3] D. O. MS. Vol.

[4] S. P. Dom. Chas. II, clxxv, 44.

[5] Cf. Treas. Papers, xc, 22.

[6] Cal. Treas. Papers, i, 211; ii, 44. An excise tax instead of coinage was collected at the blowing-house.

[7] *Tinners' Grievance.*

[8] Cal. Treas. Papers, i, 211.

[9] The market price of tin did not fall, but the tinners were no longer able to take advantage of it (*Tinners' Grievance*).

which Charles II made in favor of Bellot and Ennis, and of Ford, and the period 1703 to 1717, when Anne monopolized the tin output,[2] the way was again open for the tin dealers. The coinage regulations as usual assisted their old tactics,[3] and the tinners suffered deplorably in the prolonged depression in the tin trade which marked the closing decades of the seventeenth century.[4] Wages, which under the Commonwealth are said to have been 30s. per month, had by 1667 fallen to about 28s. for pickmen, 20s. for common tinners, and 16s. for other laborers;[5] and then they dropped still lower, until the ordinary tinner received only 14s. or 15s. per month with which to support himself and his family.[6] No clothing trade existed to serve as a by-employment. In the weeding season or in harvest time he obtained a little work with which to eke out his miserable wage;[7] but when the mines shut down, as they frequently did, "some became beggars and others resorted to pillage." We read of a horde of tinners who in 1690 appeared at Falmouth and plundered a ship laden with salt.[8] Many families never saw meat save on rare occasions when they could carry off diseased sheep, or cattle that had died in the fields. In winter their ordinary food was potatoes and barley bread, with gruel thickened with barley meal; in summer, barley bread and milk, and little even of that. The series of lean years in the stannaries had reduced their strength, yet when they did find work they were forced by their employers to labor night and day.[9] It is scarcely to be wondered at, therefore, that in 1703 a huge mob of laborers surrounded the convocation

[1] Lans. 1215, fol. 226–230; S. P. Dom. Chas. II, xxix, 93; xcix, 12; Welch, ii, 131. In the case of both these farms the price given, namely 90s. per hundred-weight, shows to what extent the market price had risen in the years 1650 to 1660. The farmers were ordered to take the tin at four periods in the year instead of two, but this was but a feeble palliative.

[2] Anne provided an ample loan fund for the tinners (Add. MS. 6713, fol. 437–442).

[3] *Tinners' Grievance.*

[4] Hunt, 53. Cf. Yarranton's description of the tinners (*England's Improvement*, pt. ii, 169). He suggested as an amelioration the establishment of a "lumber (Lombard) house" where tin (and iron from the Forest of Dean) might be deposited and money advanced on it as security (*ibid.*, 170). A like proposal was made by a Mr. Vane in 1680 (*Plea for the Bringing in of Irish Cattle*, 14).

[5] D. O. MS. Vol.

[6] Cf. colliers' wages (Galloway, 217, 218).

[7] *Tinners' Grievance.*

[8] News Letter Entry Bk., iii, 15; H. O. Letter Bk., ii, 9.

[9] *Tinners' Grievance.*

house at Truro and threatened the parliament with violence unless it accepted the Queen's offer of preëmption.

The eighteenth century and the early nineteenth, while not witnessing such scenes as have just been described,[1] nevertheless cannot be said to have brought the laborer any great change for the better.[2] Norden, writing in 1728, assigns him a wage of 8d. a day, or £6 per year.[3] Carew's editor, Tonkin, at about the same time states that the best men received from 20s. to 27s. per month, with extra pay for overtime, and all tools found. The captains received from 30s. to £4, and the carpenters from 40s. to 50s. Fifty years later, the men's pay had risen by only 3s. or 4s.[4] Another quarter-century sent wages up 10s. further, from 30s. to 40s. per month, the stamp mill employees, however, receiving only 24s. or 30s.[4] In 1836 the average of all classes, tut workers, tributers, and laborers, in the Consolidated Mines, was less than £3 10s.[5] per month; and to this day the Cornish miner, although comparatively speaking a highly skilled and intelligent laborer, receives an inadequate wage.[6] Although he is no longer subject to the oppression which he suffered in the sixteenth and seventeenth centuries, he is still kept more or less in a state of semi-bondage by the custom of obtaining money advances, or "subsist," from his employers.[7]

Toward the end of the seventeenth century, however, a change is noticeable in the composition of the class of tin merchants. Up to about 1650 these had consisted for the most part of Londoners,

[1] With the possible exception of the severe depression at the close of the eighteenth century (Unwin, *Letters and Remarks*, 14, 37).

[2] In the collieries of the north of England, the year 1763 is memorable for the first signing of the so-called yearly bond between the colliery owners and their workmen, stipulating that the latter were to work for a year without strikes, desertions, etc. Little is known of the yearly bond previous to this period, but it perhaps existed from the abolition of serfdom, whenever that may have occurred (Galloway, 269). For wages and conditions of labor in the mines in the middle of the eighteenth century, see Galloway, 283–284, 320, 352.

[3] Norden, 12.

[4] Worth, 58. Many, however, had by-employments, such as fishing (Pryce, 35). The captain, who was usually a shareholder in the mine, received five or six guineas per month (Jars, iii, 205). The duration of a shift meanwhile had been extended to eight hours (Pryce, 175; Trans. Roy. Geol. Soc. Cornw., iii, 64).

[5] Delabeche, 692.

[6] Hardly more than £1 per week (cf. Worth, 58).

[7] Journ. Roy. Stat. Soc., li, 532, 533.

but during the prosperous times of the Commonwealth the Cornish storekeepers, who then became more interested in the stannaries, had purchased shares in mines, or had established blowing-houses of their own;[1] and after the Restoration we find them dealing with the tinners not merely as the factors of the London merchants but as middlemen and tin exporters on their own account. It is, however, their connection with the stannaries as smelters which arouses most interest.

Presumably, at the dawn of mining, the two occupations, mining and smelting, were merged in the same person; but when the recorded history of the stannaries begins, the occupations were separate. De Wrotham included in his enumeration of the classes engaged in tinning not only diggers (*foditores*), but smelters (*funditores*) as well, having their own laws and customs.[2] These smelters doubtless began as small craftsmen, but when Beare in 1586 wrote his account of the stannaries a change had taken place in their position. The smelter had become to some extent a capitalist, employing a number of blowers, or renting sections of his blowing-house to small smelters working on their own account.[3] The owners of the blowing-houses smelted the miners' tin for a certain percentage of the product,[4] the smelter having bargained for the parcel of ore brought him and given his note to deliver at the ensuing coinage the quantity of white tin agreed upon.[5] These notes, which were transferable by endorsement,[6] the always needy tinners sold, as we have already seen, to the merchant dealers. But the transformation of the smelter into a capitalist, and the identification of the smelters with the tin dealers, which began in the seventeenth century, resulted during the eighteenth in the retirement of the merchants of London, and from that day to this the smelters have acted as the financiers of the stannaries.[7]

[1] *Tinners' Grievance.* [2] App. A. [3] Harl. 6380, fol. 34.

[4] Cf. Add. MS. 6682, fol. 297; Carew, ed. 1811, p. 41; Pryce, 291; Salmon's Min. and Smelt. Mag., v, 7.

[5] According to Borlase (*Nat. Hist. Cornw.*, 81), the owner would receive about twelve twentieths of the product, and sometimes thirteen, or, if the ore were very poor, not more than eight. Cf. Add. MS. 6682, fol. 297.

[6] Pryce, 292.

[7] Parallel to this may be cited the relations between the lead miners and the smelters in Derbyshire in the eighteenth century (Farey, 378, 379).

At first they simply bought back at a discount their notes from the tin owners,[1] and as long as the old coinage system lasted this system of indirect purchase was bound to continue, for to buy an article of fluctuating value which could not be sold save at two periodic coinages would have been too speculative a business for the smelter to undertake. To the tinner it obviously made no difference where he took his ore, inasmuch as all the smelters made similar assays and charges; while his bill, of course, he was careful to sell to the highest bidder.[2] Since the abolition of the coinage, the smelters have laid aside their former methods and have bought the tin from the owner direct.

To understand how heavily this system bears upon the tin mining industry at present, it will be necessary for us to examine at some length the conditions under which the ore is sold. The chief of these is known as the tin standard. This is an amount paid by the smelter per hundred-weight of metal contained in the ore, as calculated from the results of a dry assay after the deduction of one and one fourth for returning charges from the produce of every twenty.[3] By an old custom there is also a deduction on the weight of the parcel of tin ore of three pounds per hundred-weight; and it is customary also to reckon the price by the nearest eighth of a pound sterling above or below the calculated price.

These provisions are anything but fair. The smelter buys, not on the actual contents of the ore, but on the contents which he assumes he will recover in the process of smelting. The difference between that and the wet assay, which gives the actual contents of the ore, is stated variously at from five to ten per cent in favor of the latter; but as we have to deal with commercial values, that consideration may for the present be dismissed by taking it for granted that the smelter loses such a percentage in the process of smelting, and is therefore entitled to an allowance from the seller.

But this is not all which the smelter claims, for since the standard is the price which he pays for the metallic contents of the ore, it follows that the difference between such a price and that at which he sells, *i. e.* the market price, represents gross profit. A second

[1] Pryce states unequivocally that the tinners were at their mercy (pp. 292, 293).

[2] Salmon's Min. and Smelt. Mag., v, 68.

[3] *Cornish Mining*, 16.

consideration is the returning charge, or the assumed cost of smelting deducted in mineral from each batch of black tin sold by the miner. To be fair, the charges for smelting should be upon a cash basis, varying solely with the rise or fall in the cost of labor, fuel, or fluxes. As it is in kind, the higher the price of black tin the higher the price which the miner must pay the smelter for his work.[1] Thus the miners, in the years 1883 to 1900, have paid a yearly average of forty per cent over even what the smelters claim as their actual cost, and the returning charge, which according to the standard the miner believes to be only six and one fourth per cent, is, on a 62.25 black tin, really ten per cent. As, moreover, most ore sold by the miners is higher than that, it follows that the returning charge increases proportionately with the quality of black tin, and the higher market price ruling for white tin.[2]

By draftage, an old trade custom, the smelter is allowed three pounds per hundred-weight on every purchase of black tin. At its inception this was for the "turn of the scale" on all the black tin purchased by the smelter, conditionally upon his allowing the same draftage upon all the white tin he delivered. In the days of barter pure and simple the arrangement was perfectly equitable, but with the passing of the coinage duties the smelter, who might reasonably have been expected either to abolish the three pounds draftage or retain it in its entirety, continued to enforce the clause as applied to the receipt of black tin from the miner, but waived it in his delivery of white tin. It can readily be figured that the miner, under the draftage allowance, has to turn over to the smelter the rough equivalent of five per cent in cash on his gross turnover, while the smelter will have the assurance that even at the worst of times the allowance is not likely to be worth less than forty-four per cent on his working costs (as it was in 1896), and with a good price for tin it may be worth ninety-three (as in 1900),[3] while there is no reason to fear that the average for the next twenty years will fall short of that from 1883 to 1900, viz., sixty-three and one fifth per cent. The smelter makes certain of a profit, therefore, even when the miner works at a loss.

[1] £4 per hundred-weight of metal would be a liberal estimate of the cost of smelting, yet since 1883 only once has the smelter received less, while on an average he has received fifty per cent more (*Cornish Mining*, 16).

[2] *Cornish Mining*, 17.

[3] *Ibid.*, 18.

The actual loss of metal in the smelting of the tin ores has never been accurately determined,[1] but in any case it is a question for which the smelter can hardly claim serious consideration, since he buys his black tin on a dry assay, not on the actual metallic contents, but on the assumed contents which he will recover by his method of smelting. As therefore, by his conditions of purchase, he has already safeguarded himself by securing an allowance equivalent to his probable loss in smelting, he can hardly ask for any further consideration. Moreover, as he sells his "ashes" at prices based on their tin contents, he is actually reimbursed in cash for at least a portion of the loss in smelting which the miners had already allowed him in full and in kind.

Smelters' profits, taken as a whole, are subject to wide fluctuations. Thus in 1900 they were but 2.91 per cent, but in 1899 they had been twenty-five. The number of years in which over ten per cent has been made in profits seems to predominate over the bad years in the proportion of ten to eight. For the period 1883 to 1900, an average of 12.48 per cent clear profit has been realized.

The present relations between tinner and smelter are the result largely of the apathy of the Cornish mining companies, their adherence to customary methods, and their slowness to grasp the principles of ordinary business management. At present, however, it would be difficult, in the face of organized opposition on the part of the smelters as well as with the existence of the present system of leases, under the terms of which the tinner is usually debarred from smelting his own tin, to bring about the much needed consolidation of the mining and the smelting businesses under a single management.

[1] *Cornish Mining*, 20.

# CONCLUSION

A GENERAL outline of the foregoing chapters, and the chief points which, in the present essay, I have attempted to present, can be given in a few sentences. After an introductory chapter which dealt with the history of technical conditions in the stannaries, an attempt was made in the following chapter to set before the reader the salient features of the development of one of the oldest, and, in former centuries, most important of English industries, namely that of tin mining, together with an outline of the history of the distribution and consumption of the metal.

It is clear that this industry could not be investigated as an isolated subject, detached from the background of the general history of mining in Europe. A digression into the history of mining law on the continent revealed the fact that the fundamental principles and development of mining law, as usually stated by continental writers, apply to England only with serious limitations. In five old mining districts of limited area, of which the most noteworthy were the tin regions of Cornwall and Devon, we found much to gainsay the prevalent assumption that the freedom of mining originated in the seizure of mineral rights from the landlords by the territorial lords; and it was suggested tentatively that the right of free appropriation of mineral properties was probably, in view of the great age of the workings, of very ancient origin.

Taking up from that point the study of the stannaries as a political unit, we examined first of all the relations, administrative and judicial, in which the tinners both of Cornwall and of Devon stood to the Crown. The growth of a stannary judiciary in the thirteenth century, coupled with the laxity of the laws restraining the tinners from encroachment on private lands in their search for ore, led to centuries of strife between the stannary people and the manorial lords, in the course of which the scope of the tinners' charter was defined and redefined, and the jurisdiction of the stannary courts subjected to a variety of interpretations. There followed a discussion of the origin and extent of the magisterial and judicial powers of the warden and vice-warden, of the composition of the stewards' courts and court leet, and of the tinners' parliaments in the two counties. In Chapter V, a study was made of the fiscal relations in which the stannaries stood to the Crown, or Duchy, the kinds of taxes paid by the tinners, their amount, and the manner in which they were collected.

In Chapter VI, we turned to the privileged status of the tinners, and found on examination that the various liberties which they enjoyed were typical in the main of the rights of free miners all over the island. An investigation of the trade rules under which the miners worked showed that in the stannaries a gild organization was not developed. The principle underlying all stannary law and custom was freedom of production, so far as was compatible with a regard to the interests, not, as in the case of the gilds, of brother workmen, but of the duchy authorities, as receivers of stannary taxes. This principle of stannary regulation, in marked contrast to that animating the exclusive rules of the Forest of Dean miners, was doubtless an important factor in preserving the stannary organization intact down to the nineteenth century.

The freedom of stannary industrial conditions, however, facilitated the early and rapid introduction of capitalism in various forms. In this respect the development of industrial organization in the German mines shows results analogous to those in Cornwall. In both regions we find an early resort to the cost agreement, under which members of a company of mine adventurers were enabled to commute their personal labor for a periodic money payment. A second stage is shown in the tribute system, where the original mine partners leased portions of their mine to groups of laborers for a percentage of the product. In the lease, a concomitant development, parts of the mine were let on fixed rents to small masters, who worked them with hired labor; and this form of industrial development, as well as the preceding, is to be found to a certain extent in the stannaries. The entrance of wage work probably occurred at an early date in the tin mines; but as we can find no trace of any definite class organization of wage earners, such as in Germany was evidenced by the formation of journeymen's gilds among the miners, it is more likely that in Cornwall wage work long played a part secondary to the tribute system.

One is more readily led to this conclusion upon a study of the development in the sixteenth and succeeding centuries. We find the continued survival of the primitive mine partners, although in greatly diminished numbers. The original substitutes, whose wages were paid by cost-giving members of the association, gradually rose to the position of tut or contract workers, similar to the tributers save that the gangs of tut workers were used for the preliminary work of mining, rather than for the actual excavation of ore. The original tributers of whom mention is made in 1586 became in later years small masters, having under them hired laborers for the working of their tribute pitches; but at some period in the eighteenth century the latter seem to have risen in turn and to have taken the tribute contracts for themselves, thus bringing about a reversion to the earlier state. The nineteenth century has witnessed a further development from the tribute and tut systems to simple wage work, a transformation induced for the most part by the increase of technical skill on the part of the mine agents and the consequent lessening both of the tributers' chances for making rich finds of ore, and of the tut workers' opportunities for easy contracts. Meanwhile the old mine partnership has developed by easy stages into the modern cost book company.

Another characteristic of the tin mining industry, brought early into prominence by its freedom from all gild-like regulations, was the rise to power of the middlemen or tin dealers. As early as 1198 these constituted a distinct class, and during the Middle Ages seem to have made a practice of advancing money to the tinners upon promise of future delivery of tin at prices stipulated in advance. Whatever may have been the hardships which this system of money advances imposed on the tinners, and especially upon those of the lowest economic status, no urgent protest was heard until the great rise in prices in the sixteenth and seventeenth centuries, when the condition of the laboring tinners became such as to call for the interference of the paternally inclined Tudor government. But Tudor and Stuart assistance was associated with grants of monopolies for the purchase of tin, and failed to attack the root of the tinners' hardships, namely their inability to sell their tin save at semiannual coinage periods. The result was that, except during the period of the Commonwealth, when the common-sense remedy was applied of abolishing both the preëmption and the coinage duties, the condition of the laboring classes in the stannaries was unfavorably affected and the influence of the limitation of time of sale has been felt long after the removal of the restriction.

In the meantime the position of the middleman, formerly assumed by the tin merchants, had been taken over by the Cornish tin smelters. The smelters are still at the present day the purchasers of the ore of the stannaries, and the antiquated and in some respects apparently unjust business relations between smelter and tinner are responsible for a share of the depression which has rested for a number of years on the tin mines of Cornwall.

# APPENDICES

# APPENDICES

## A

### *De Wrotham's Letter of* 1198.

HUBERTO Dei gratia Cantuariensi Archiepiscopo et Domino G. filio Petri et omnibus baronibus Domini Regis de Scaccario, vicecomes Devoniae et Cornubiae et Willielmus de Wrotham, Ricardus Flandrensis, Robertus de Champeaus, Ricardus Peverel de Hininton, Ricardus filius Walteri, Ricardus filius Gileberti, Lucas filius Bernardi, Henricus de Alneto, Odo filius Frawini, salutem.

Sciatis vicecomitem Devoniae et Cornubiae recepisse per manum Willielmi de Wrotham praeceptum Domini Cantuariensis Archiepiscopi in haec verba. Hubertus Dei gratia Cantuariensis Archiepiscopus totius Angliae Primas et Apostolicae Sedis legatus vicecomiti Devoniae et Cornubiae salutem. Praecipimus tibi ex parte domini Regis, quod loco G. filii Petri, committas Willielmo de Wrotham omnes stannarias domini Regis in balliva tua et omnia quae ipsas stannarias contingunt. Et stagnatores ei habere facias in ea libertate quam habere debent et solent, facias etiam ei habere omnes illos viros legales quos idem Willielmus tibi nominabit quos ad hoc expedire videritis, qui auxilium et consilium ei conferant ad custodiendum cuneos domini Regis, et omnem exitum ipsarum stannariarum et ad disponendum de profectu ipsarum. Prohibe etiam omnibus hominibus communiter in balliva tua ne sine licentia ejusdem Willielmi stagnum aliquod asportent per mare vel per terram. Tu ipse etiam ei tantum auxilium facias in praesenti negotio domini Regis expediendo quod bene procedat, nec pro defectu tui dominus Rex aliquod dampnum incurrat. Teste Stephano de Turneham apud Westmonasterium xx die Novembris.

Et sciatis vicecomitem Devoniae et Cornubiae recepisse aliud praeceptum domini Cantuariensis Archiepiscopi in haec verba. Hubertus Dei gratia Cantuariensis Archiepiscopus totius Angliae Primas et Apostolicae Sedis legatus vicecomiti Devoniae et Cornubiae salutem. Praecipimus tibi quod per sacramentum duodecim liberorum et legalium hominum de balliva tua qui melius sciant rei veritatem facias diligenter inquiri quae fuerint pondera primae funturae et quae pondera secundae funturae, et ea pondera facias de cetero observari sicut dicet tibi lator praesentium Willelmus de Wroteham. Teste G. de Bocland apud Salopesberi vii die Januarii.

Item sciatis nos recepisse literas domini G. filii Petri ut essemus cum Willelmo de Wrotham loco domini G. filii Petri tanquam justiciae ad faciendum domini Cantuariensis praeceptum de admensuratione ponderum primae et secundae funturae et de dispositione utilitatis profectus domini Regis de stagno, unde est quod vicecomes Devoniae et ego Willielmus de Wrotham, Ricardus Flandrensis, Robertus de Champeaus, Ricardus Peverel de Henninton, Ricardus filii Walteri, salutamus dominum Hubertum Cantuariensem Archiepiscopum et dominum G. filium Petri et omnes barones domini Regis de Scaccario.

Et mandamus quod in nono decimo die Januarii noni anni regni Regis Ricardi in Devonia in Comitatu apud Exoniam inquisivimus per sacramentum istorum scilicet Rogeri Rabi, Gervasii Mercer, Admeri de Movinton, Martini Prudhome, Hugonis de Movinton, Walter filii Turberni, Willielmi Hastement, Sansonis nepotis Radulphi, Galfridi Faierchild, Willielmi Daci, Johannis filii Radulphi, Johannis Caperun, Philippi de Sede, Aluredi de Brente, Walteri le Bon de Totenais, Walteri Bolt, Humfredi Faber, Ailwardi Burgeis, Oseberti Prigge de Asperton, Ailwardi Faermund, Johannis Prigge, Roberti de Prato, Willielmi de Esse de Plinton, Roberti de la Cnolle, Aluredi de Lega, Helye Mewi, sapientiorum et discretiorum de justis ponderibus stagni Devoniae quod justum et antiquum pondus civitatis Exoniae per quod antiquitus et nunc et semper solebat fieri secunda funtura stagni est de tali quantitate et semper debuit esse quod justum et antiquum pondus de prima funtura stagni antiquitus et nunc et semper octies ponderatum ex se faciebat nonum et facere debet per pondus civitatis Exoniae secundae funturae et hac de causa, scilicet, quia de quolibet miliari per majus pondus ponderato dantur domino Regi de antiqua consuetudine xxx denarii ad firmam de stannariis in Devonia, et pro custo vecturae ad villas marcandas, et pro eo quod stagnum decidit in secunda funtura, et secundum eorum sacramentum in tali quantitate coram nobis admensurantur et in stannariis Devoniae et in villis constituuntur. Item mandamus quod in suprascripto decimo nono die Januarii noni anni regni Regis Ricardi Willielmus de Wroteham primo recepit stannariam Devoniae et tunc primo cepit operari per pondera in justam et antiquam quantitatem per sacramentum praedictorum juratorum admensurata et constituta. Item vicecomes Cornubiae et ego Willelmus de Wroteham, Ricardus filius Gileberti, Lucus filius Bernardi, Henricus de Alneto, Odo filius Frewini inquisivimus in Cornubia in Comitatu apud Lanzauenton xxv die Januarii noni anni regni Regis Ricardi per sacramentum istorum scilicet Hereberti, praepositi de Bodmine, Nicolai filii Margareti Sawel de Bodmine, Johannis filii Aldredi, Ade Dai, Osberti mercatoris, Stephani filii Beli, Roberti Keustek de Triueriu, Rogeri mercatoris, Ailmari fratris Bernardi, Willielmi de Polstad de Liscaret, Willielmi de Lanteisin, Roberti Dreienos, Fulconis de Liscaret, Roberti Tropinel, Willielmi Cole, Ailsi praepositi de Lanzauinton, Dudemanni de Lanzauinton, sapientiorum et discretiorum de justis ponderibus stagni Cornubiae quod justum et antiquum pondus secundae funturae stagni in villis marcandis Cornubiae antiquitus et nunc et semper solebat esse et debet esse de tali quantitate quod justum et antiquum pondus primae funturae septies ponderatum faciebat et facere debet ex se octavum per pondus secundae funturae et quod justum et antiquum pondus primae funturae stagni Cornubiae antiquitus et nunc et semper solebat esse et debet esse majus de xxx numeratis stagni quam pondus primae funturae Devoniae et hac de causa, scilicet, quia de quolibet miliari per majus pondus ponderato dantur domini Regis de antiqua consuetudine v solidi ad firmam de stannariis in Cornubia et pro custo vecturae ad villas marcandas, et pro eo quod stagnum decidit in secunda funtura. Et super eorundem sacramentum in tali quantitate coram nobis admensurantur et in stannariis Cornubiae et in villis marcandis constituuntur.

Item mandamus quod in suprascripto xxv die Januarii noni anni regni Regis Ricardi Willelmus de Wroteham primo recepit stannariam Cornubiae et tunc primo inde cepit operari per pondera in justam et antiquam quantitatem per sacramentum praedictorum juratorum admensurata et constituta.

Mandamus item quod quia dominus Rex amodo postquam Willielmus de Wrotham suscepit stannariam habebit de novo redditu annuali de quolibet miliari ponderato per pondus secundae funturae unam marcam argenti quae pondera secundae funturae in Devonia et Cornubia per equalem ponderationem admensurantur et per equalem quantitatem constituuntur.

Item mandamus quod nec marca domini Regis novi redditus de stagno nec antiquae consuetudines de firma stagni nec pondera de stagno, sicut nobis visum est nullo modo potuerunt melius legalius subtilius commodius ad utilitatem domini Regis et populi provideri quam nunc provisum est per capitula quae de stagno et stannariis in Devonia et Cornubia statuta sunt per consilium nostrum et per providentiam Willielmi de Wroteham anno nono Regis Ricardi, scilicet ista:

Omnes foditores et nigri stagni emptores et de stagno primi fusores et de stagno primae funturae mercatores habent justas et antiquas consuetudines et libertates in Devonia et Cornubia constitutas.

Item justa et antiqua pondera primae et secundae funturae stagni per sacramentum suprascriptorum juratorum statuta et de cuneo domini Regis consignata teneantur.

Item omnes homines communem habent emptionem de stagno in justis et antiquis et liberis consuetudinibus sicut debent et solent per marcam de quolibet miliari secundae funturae.

Item in burgis et in villis mercandis in quibus capitalis custos stannariarum statuerit secundam funturam de qua dominus Rex de quolibet miliari de novo debet habere unam marcam, fiat ponderatio secundae funturae stagni per pondus civitatis Exoniae et illud pondus signetur cuneo domini Regis de marca.

Item pondus civitatis Exoniae constitutum semper sit custoditum in custodia duorum legalium hominum in villis mercandis et in custodia clerici ex parte domini Regis constituti.

Item cuneus de marca per quod pondus debet signari et omne stagnum secundae funturae semper sit custoditus sub sigillo custodis ponderis secundae funturae et clerici ex parte domini Regis constituti, nisi dum per eum signant.

Item custodes secundae funturae et clericus diligenter et memoriter sicut propria corpora sua inbreviare faciant omnia miliaria et centum et libras quae per pondus et cuneum custodiae suae ponderentur et signentur in toto anno.

Item in qualibet villa ubi fuerit secunda funtura constituantur duo legales et divites homines qui recipiant de mercatoribus marcam domini Regis coram custodibus et clerico secundae funturae et cunei ad ponderationem et signationem suam et clericus et custodes cunei non permittent stagnum asportari donec thesaurarii domini Regis marcam domini Regis et consuetudinem de stagno receperint.

Item thesaurarii domini Regis de marca faciant de averio domini Regis dicas et cyrographa contra custodes et clericum ponderis et cunei. Et in cyrographis continuantur dies receptionis et nummus recepte et numerus miliarium et centum librorum unde fuerit receptio et nomina mercatorum qui stagnum acquietaverint.

Item thesaurarii domini Regis simili modo per dicas et per hujusmodi cyrographa tradant capitali custodi de stannariis averium domini Regis.

Item capitalis custos vel aliquis minister suus nullo modo praesumat vexare thesaurarios domini Regis in vita eorum vel heredum eorum post obitum patrum dum rationabile acquietaverint se de receptionibus averii domini Regis secundum dicas et cyrographa facta contra custodes secundae funturae et clericum domini Regis.

Item custodes cunei et ponderis et clericus domini Regis salve custodiant semper in communi custodia sua sub sigillis suis dicas et cyrographa facta contra thesaurarios et in tali firmura quod quilibet eorum habeat clavem suam et diversam.

Item clerici notarii de marca domini Regis et de cyrographis marcarum retineantur ad stipendium et liberationem domini Regis.

Item in qualibet villa ubi fuerit secunda funtura extra muros civitatis Exoniae et extra villam de Bodmene constituatur una domus per conductionem domini Regis ubi fiat tota secunda funtura et ponderatio et signatio, et nullus praesumat alibi facere secundam funturam nec ponderationem nec signationem sicut amat se et sua.

Item in civitate Exoniae et in villa de Bodmene fundatur stagnum in secundam funturam in locis sicut semper solebat eo tamen servato, quod nullus praesumat sicut amat se et sua facere secundam nisi coram custodibus ponderum et funturae et clericis ex parte domini Regis constitutis.

Item non praesumat aliquis in villis mercandis habere aliqua pondera per quae ponderet stagnum nisi coram custodibus ponderis prius per pondus domini Regis admensurentur et per cuneum domini Regis de marca signentur.

Item ad custodiendum pondus primae funturae et cuneum de firma de quibus dominus Rex habebit antiquas consuetudines suas pertinentes ad firmam stannariae constituantur duo legales homines et clericus ex parte domini Regis coram quibus omne stagnum primae funturae ponderetur et signetur.

Item custodes et clericus primae funturae stagni sicut amant se et sua diligenter et memoriter inbreviare faciant omnia miliaria et centum et libras quae fuerint ponderata et signata per pondus et cuneum custodiae suae per totum annum.

Item non praesumat homo nec femina Christianus nec Judaeus vendere vel emere aliquid de stagno primae funturae vel donare vel asportare nec extra stannarias nec extra loca ad ponderationem et signationem primae funturae constituta donec ponderetur et signetur coram custodibus et clericis ponderis et cunei de firma.

Item non praesumat homo nec femina Christianus nec Judaeus in stannariis vel extra stannarias habere aliquid de stagno primae funturae ultra quindenam nisi sit ponderatum et signatum per custodes et clericum de pondere et cuneo firmae.

Item non praesumat homo nec femina Christianus nec Judaeus in villis et burgis mercandis in mari vel in terra ultra tresdecim septimanas habere stagnum de prima funtura ponderatum que signatum per pondus et cuneum primae funturae nisi sit positum in secundam funturam et acquitatum de marca.

Item non praesumat homo nec femina Christianus nec Judaeus aliquo modo asportare stagnum nec per terram nec per mare extra Devoniam et Cornubiam nisi prius habeat licentiam capitalis custodis stannariarum.

Item constituantur boni et legales homines in portibus circiter Devoniam et Cornubiam qui capiant sacramentum omnium sturemannorum et marinellorum navium ibi applicantium quod non asportabunt nec asportari permittent in navibus suis stagnum nisi sit per regias consuetudines ponderatum et signatum et nisi habeant breve capitalis custodis stannariarum.

Item cuneus de firma semper sit custoditus sub sigillo custodis et clerici nisi dum per eum signant in locis constitutis.

Item in omni ponderatione de stagno quod linguae staterae rectum judicet inter pondus stagnum ita quod pro voluntate emptoris non trahatur lingua staterae versus stagnum ultra judicium equitatis staterae.

Et ut dominus Rex et dominus Hubertus Cantuariensis Archiepiscopus, et dominus G. filius Petri, et omnes barones de scaccario sciant nos diligenter et memoriter imposuisse omnem curam et studium et sollicitudinem nostram cum Willielmo de Wroteham ad utilitatem domini Regis de stagno providendam, et quod dominus Rex de cetero non possit leviter per aliquem de mundo falli nec decipi; quatuor cyrographa inde fecimus unum mittentes domino Cantuariensi et baronibus de scaccario, alterum domino G. filio Petri, tertiam in communi custodia thesaurariorum domini Regis de marca et custodum cuneorum et ponderum Devoniae et

Cornubiae tradentes, quartum Willielmo de Wroteham in testimonium liberantes. In quolibet cyrographo sigilla nostra apponentes.

*Black Book of the Exchequer*, No. 10.

## B

## *Stannary Charter of* 1201.

Rex archiepiscopis etc. salutem. Inspeximus cartam quam dominus Johannes rex pater noster fecit stannatoribus nostris in Cornubia et Devonia in haec verba: — Johannes Dei gratia rex Angliae etc., archiepiscopis, episcopis, abbatibus, comitibus, baronibus, justiciariis, vicecomitibus, forestariis et omnibus ballivis et fidelibus suis salutem. Sciatis nos concessisse quod omnes stannatores in Cornubia et Devonia sint liberi de placitis nativorum, dum operantur ad commodum firmae nostrae vel commodum marcarum novi redditus nostri, quia stannariae sunt nostra dominica, et quod possint omni tempore libere et quiete absque alicujus hominis vexatione fodere stannum et turbas ad stannum fundendum ubique in moris et in feodis episcoporum et abbatum et comitum, sicut solebant et consueverunt, et emere buscam ad funturam stanni sine vasto in regardis forestae, et divertere aquas ad operationem eorum in stannariis, sicut de antiqua consuetudine consueverunt, et quod non recedant ab operationibus suis pro alicujus summonitione, nisi per summonitionem capitalis custodis stannariarum vel ballivorum ejus. Concessimus etiam quod capitalis custos stannariarum et ballivi ejus per eum habeant super praedictos stannatores plenariam potestatem ad eos justificandos et ad rectum producendos et quod ab eis in carceribus nostris recipiantur, si contigerit quod aliquis praedictorum stannatorum debeat capi vel incarcerari pro aliquo recto; et si contigerit quod aliquis eorum fuerit fugitivus vel utlagatus, quod catalla eorum nobis reddantur per manus custodis stannariarum nostrarum quia stannatores firmarii nostri sunt et semper in dominico nostro. Praeterea concessimus thesaurario et ponderatoribus nostris ut sint fideliores et intentiores ad utilitatem nostram in receptione et custodia thesauri nostri per villas marcandas, quod sint quieti in villis ubi manent de auxiliis et tallagiis dum fuerint in servitio nostro thesaurarii et ponderatores nostri, quia nichil aliud habent vel habere possunt per annum pro praedicto servitio nostro. Hiis testibus W. comite de Sarisberio, Petro de Stokes, Warino filio Geroldi, Roberti de Ros, Pedro de Rupibus, thesaurario Pictavensi. Data per manum S. Wellensis archidiaconi apud Bonam Villam super Tokam xxix die Octobri anno etc. tertio.

Nos autem praedictam concessionem ratam habentes et gratam, eam pro nobis et heredibus nostris concedimus et confirmamus sicut praedicta carta rationabiliter testatur. Hiis testibus . . . Data per manum nostram apud Merton, ii die Aprilis.

*Charter Roll*, 36 Henry III, m. 18. (Charter Roll, 3 John, is wanting.)

## C

### *Disforesting of Cornwall.*

Johannes Dei gratia Rex Angliae etc. Sciatis nos dedisse concessisse et hac carta nostra confirmavisse hominibus nostris Cornubiae quod morae de Fainmora, Gundiwragh et bruwii de Warham et Kellibullugh quae retinuimus in manu nostra in foresta quando deafforestavimus totum residuum Cornubiae et carta nostra confirmavimus deafforestentur, ita quod soluti et quieti sunt de nobis et heredibus nostris inperpetuum de omnibus quae ad forestam vel forestarium pertinent. Et quod omnimodam venacionem capiant sine impedimento in predictis moris et bruillis, quia praedictae morae non sunt de dominico nostro sed militum et vicinorum et libere tenencium. Praeterea concessimus eis quod habeant libertates et liberas consuetudines quas habere solebant tempore Henrici Regis patris nostri scilicet ne perdant aliquid de serviciis vel consuetudinibus quas habere solebant aut de hominibus et nativis suis occasione stagnariarum, sed eant homines sui ad stagnum qui voluerint salvis serviciis et consuetudinibus quas habent de aliis nativis suis non euntibus ad stagnum.

Concessimus etiam eis ne nativi sui recipiantur in libertates burgorum nostrorum vel aliorum sine assensu dominorum suorum aliter quam consueverunt tempore Henrici Regis patris nostri. Concessimus etiam eis quod vicecomes patriae habeat custodiam portuum et escaetarum et non alii quam vicecomes sicut tempore Henrici Regis patris nostri consuevit habere.

Concessimus etiam eis ne feoda de Moretoini graventur de scutagiis aliter quam fuerunt tempore Henrici Regis patris nostri quae dare solebant xii solidos et vi denarios quando viginti solidi dabantur de scuto. Et quod liberi sint de donis auxiliis et hospiciis vicecomitis inperpetuum. Quare volumus et firmiter etc., quod praedicti homines Cornubiae et heredes sui habeant et teneant inperpetuum de nobis et heredibus nostris haec omnia supradicta bene et in pace libere quiete et integre cum omnibus libertatibus et liberis consuetudinibus sicut praedictum est. Testibus [etc.]. . . . Data per manum magistri Ricardi de Mariscis, Cancellarii nostri apud Novum Templum, London xxii die Aprilis anno regni nostri xvi[mo].

*Charter Roll*, 16 John, m. 2.

## D

### *Stannary Charter of 1305, Cornwall and Devon.*

Rex, archiepiscopis, etc., salutem. Sciatis nos ad emendationem stannariarum nostrarum in comitatu Cornubiae et ad tranquillitatem et utilitatem stannatorum nostrorum earundem concessisse pro nobis et heredibus nostris quod omnes stannatores praedicti operantes in stannariis illis quae sunt dominica nostra dum operantur in iisdem stannariis sint liberi et quieti de placitis nativorum et de omnibus placitis et querelis curiam nostram et heredum nostrorum quoque modo tangentibus; ita quod non respondeant coram aliquibus justiciariis vel ministris nostris seu heredum nostrorum de aliquo placito seu querela infra praedictas stannarias emergenti, nisi coram custode nostro stannariarum nostrarum praedictarum qui pro tempore fuerit, exceptis placitis terrae et vitae et membrorum, nec recedant ab operationibus suis per summonitionem alicujus ministrorum nostrorum seu heredum nostrorum, nisi per summonitionem dicti custodis nostri. Et quod quieti sint de omnibus tallagiis, theoloniis, stallagiis, auxiliis et aliis custumis quibuscumque in villis, portubus, feriis, et mercatis infra comitatum praedictum de bonis suis propriis. Concessimus etiam eisdem stannatoribus quod fodere possint stannum et turbas ad stannum fundendum ubique in terris, moris et vastis nostris et aliorum quorumcumque in comitatu praedicto, et aquas et cursus aquarum ad operationes stannariarum praedictarum divertere, ubi et quotiens opus fuerit, et emere buscam ad funturam stanni, sicut antiquitus fieri consuevit, sine impedimento nostri vel heredum nostrorum, episcoporum, abbatum, priorum, comitum, baronum, seu aliorum quorumcumque, et quod custos noster praedictus vel ejus locumtenens teneat omnia placita inter stannatores praedictos emergentia, et etiam inter ipsos et alios forinsecos de omnibus transgressionibus, querelis et contractibus factis in locis in quibus operantur infra stannarias praedictas similiter emergentia. Et quod idem custos habeat plenam potestatem ad stannarios praedictos et alios forinsecos in hujusmodi placitis justiciandos et partibus justiciam faciendam prout justum et hactenus in stannariis illis fuerit usitatum. Et si qui stannariorum praedictorum in aliquo deliquerint per quod incarcerari debeant, per custodem praedictum arestentur et in prisona nostra de Lostwythiel et non alibi custodiantur et detineantur quousque secundum legem et consuetudinem regni nostri deliberentur. Etsi aliqui stannatorum praedictorum super aliquo facto infra comitatum praedictum non tangente stannarias praedictas se posuerint in inquisitione patriae, una medietas juratorum inquisitionis hujusmodi sit de stannatoribus praedictis, et alia medietas de forinsecis. Et de facto totaliter tangenti stannarias praedictas fiant inquisitiones sicut hactenus fieri consueverunt. Et si quis eorundem stannatorum fugitivus fuerit vel utlagatus, vel aliquod delictum fecerit pro quo catalla sua amittere debeat, catalla illa per custodem praedictum et coronatorem nostrum comitatus praedicti apprecientur, et per ipsos proximis villatis liberentur ad respondendum inde nobis et heredibus nostris, coram justiciariis itinerantibus in comitatu praedicto. Volumus insuper et firmiter praecipimus quod totum stannum tam album quam nigrum, ubique inventum et operatum fuerit in comitatu predicto, ponderetur apud Lostwithiel, Bodmynran, Liskiriet, Treueru, vel Helleston per pondera nostra ad hoc ordinata et signata, sub forisfactura totius stanni praedicti, et quod totum illud stannum coignietur in eisdem villis singulis annis coram custode praedicto ante diem Sancti Michaelis in

Septembri, sub forisfactura praedicta. Et concessimus pro nobis et heredibus nostris quod omnes stannatores nostri praedicti totum stannum suum sic ponderatum licite vendere possint cuicumque voluerint in villis praedictis, faciendo inde nobis et heredibus nostris coignagium et alias consuetudines debitas et usitatas, nisi nos vel heredes nostri stannum illud emere voluerimus. Quare volumus et firmiter praecipimus pro nobis et heredibus nostris quod stannatores nostri praedicti habeant omnes libertates, liberas consuetudines et quietancias supradictas, et quod eis sine occasione vel impedimento nostri vel heredum nostrorum justiciariorum, escaetorum, vicecomitum et aliorum ballivorum seu ministrorum nostrorum quorumcumque, rationabiliter gaudeant et utantur forma praedicta . . . Data per manum nostram apud Westmonasterium decimo die Aprilis anno regni nostri tricesimo tertio.

*Charter Roll*, 33 Edward I, m. 8, No. 40.

(On the same date a similar charter was issued to the Devon tinners, in which the only difference is that the places of coinage are at Tavistock, Ashburton, and Chagford, and the stannary prison at Lidford. See *Charter Roll*, 33 Edward I, m. 8, No. 41.)

## E

## *Tinners' Petition against Pessaigne.*

A nostre Seigneur le Roi et a son Conseil prient ses Gentz de Cornewaille remedie, de l'Esteym q'il ad grante a Antonyn de Pisane en le Conte de Cornewaille, en grant damage de luy et de son people; pur ceo que le dit Antonyn de Pisane par ses marchaunz fet destreindre les gentz du pais, qe le Esteym ount vendre, del dit Esteym carier a Lostuthiel et les facent peiser a sa volonte demeyne par nient leal peys, et ne donne pur le miller que xliis. ou il poient de altre marchant prendre vi marces. Dount, par encheson qe les dytz Estaymours ne pount aver la value de lur overaygne de l'Esteym, quore ou il y aveynt treys mil Estaymours overanz en Cornewaylle pur le profist le Roy a la venue del dit Antonyn, meyntenant il ny ount mye cynk cenz Esteymours overanz; dount le Roys perdera de son coignage en cest an vi cent livres a meyns, et plus de an en an si remedie ne soit mys, dunt la myne del tut se anentira. E del hour qe les mores seyent estopez, et les Estaymours seyent pris as altres overaynes come avant estoyent, et est a damage du Roi saunz estimation, et son poeple destruit de meisme le pais, que prest sont averrer que le dit Antonyn par ses marchantz lour ad tollet par estorsion vi M li. en ceux deux annz et demy, en apernont lour Esteym contre lour volente, et en peysant come desus est dit, en donant pur miller meyns quatre marces xii s. iiii d. come home lem pout as autres marchandz vendre. Dont par le dit esteyme les gentz del dit pais soleynt estre servi dest Drapes, Vines, Fere, Seel et autres marchandz dount ore rien ne vient a regard, totes queles chosez susditz prest sont d'averrer, pur le profit nostre Seigneur le Roi et du poeple, par inspection des Aceountes des Viscontes du pais ou par enquest, a la volente nostre Seigneur le Roy et de son Consail.

*Parl. R.*, i, 308.

## F

### *Petition of Devonshire against the Tinners.*

A tres-excellent et tres-redoute Seigneur, nostre Seigneur le Roi, supplie sa povre Commune del Counte de Devenshire, que luy plese, par l'avys des Prelatz, Countes, Barons, et autres Sages en cest present Parlement, ordeigner remede, de ceo que les Esteymours, et les Ministres de l'Esteymerye el dit Countee ont long temps ala dit Commune, si bien as Seigneurs come as autres, fait et font de jour en autre diverses extortions, oppressions, et grevaunces, par colour de les Fraunchises a eux grantez par les Chartres nostre Seigneur le Roi, et de ces Progenitours, encontre la Ley et le purport des ditz Chartres, et par lour malveys interpretation d'ycelles: Et que les dites Chartres et les Franchises comprises en ycell, puissent estre lieuz et declarez d'article en article, si que la Commune du dite Countee puissent estre apris droiturelment d'ycelles, et que cest declaration soit mys en record. Et si nul article y soit en les dites Chartres que touche Custumes ou Usages, que plese a nostre dit Seigneur le Roi d'ordeigner, et maunder en brief temps suffisantz Justices, Seigneurs et autres apris de la Ley, a celles parties, d'enquere des dites Custumes et Usages; et q'ils eyent poair d'oier et terminer touz les conspiracies, confederacies, alliaunces, champerties, extorcions, oppressions, grevaunces, fauxines et meyntenaunces, queux les ditz Esteymours et lour Ministres ont fait a la dite Commune, ou a nul de eux qi pleyndre se vorra; et ce auxi bien al seute le Roi come de la partie, entendantz, que le Roi nostre Seigneur ent gaygnera molt. Et d'autre part, si remedie ne lour y soit ore fait, ils serront en brief temps pur la greyndre partie desheritez et destruitz a touz jours, que Dieu ne voille.

Le tenour d'ascuns des Articles de les dites Chartres que lour busoignent de declaration s'ensuent cy en apres.

Primerement, c'est assavoir;

Sciatis nos ad emendationem Stannariarum nostrarum in Comitatu Devoniae, ad tranquillitatem et utilitatem Stannatorum nostrorum praedictorum earundem, concessisse pro nobis et heredibus nostris, quod omnes Stannatores praedicti operantes in Stannariis illis quae sunt Dominica nostra, dum operantur in eisdem Stannariis liberi sunt et quieti de Placitis Nativorum et de omnibus Placitis et Querelis Curiam nostram et heredum nostrorum qualitercumque tangentibus; Ita quod non respondeant coram aliquibus Justiciariis vel Ministris nostris seu heredum nostrorum, de aliquo Placito seu Querela infra praedictas Stannarias emergentibus, nisi coram Custode nostro Stannariarum nos rarum praedictarum qui pro tempore fuerit; exceptis Placitis terrae, vitae, et membrorum. Nec recedant ab operationibus suis per summonitionem alicujus Ministrorum nostrorum seu heredum nostrorum nisi per summonitionem communem dicti Custodis nostri. Et quod quieti sint de omnibus tallagiis, theoloniis, stallagiis, auxiliis, et aliis custumis quibuscumque, in Villis, Portubus, Feriis, et Mercatis, infra Comitatum predictum, de bonis suis Propriis etc.

Sur quoi plese declarer, si autres persones que les Esteymours overantz en les Esteymeryes averont et enjoyeront la Fraunchise grante par la dite Chartre du Roi desicome la dite Chartre voet, "Quod omnes Stannatores praedicti operantes in Stannariis illis sint liberi etc." Et autres persones que les Overours, C'est assavoir lour Maistres qi les louent, et lours Servantz et autres claymont mesme la Fraunchise. Et auxint plese declarrer, si les ditz Overours y averont les Fraunchises en autre temps que quant ils overont enmesme l'Esteymerye, desicome la Chartre voet "... dum operantur in eisdem Stannariis sint liberi etc."

Endroit de les ditz paroles, "Operantes in Stannariis illis, et dum operantur in eisdem Stannariis ..." soient clerement entenduz "... De Operariis laborantibus dumtaxat in Stannariis illis sine fraude et dolo, et non de aliis nec alibi laborantibus."

Item soit declarre, si mesmes les Overours averont mesmes les Fraunchises tant come ils overont aillours qu'en les Demesnes que feurent au Roi l'aiel nostre Seigneur le Roi q'ore est, luiquel Roi aiel lour grantast la dite Chartre, au temps del dit Grant des Fraunchises, desicome la Chartre voet "... Quod omnes Stannatores praedicti operantes in Stannariis illis quae sunt Dominica nostra, dum operantur in eisdem Stannariis sint liberi etc." Et ils les clayment d'avoir, tout soit il einsi q'ils overont aillours q'en les dites Demesnes le Roi l'aiel.

Endroite de ceste Article; pur ce q'il y a une autre Article en mesme la Chartre que lour donne congie et licence de fouer "... In terris, moris, et vastis ipsius Domini Regis et aliorum quorumcumque in Comitatu praedicto, et aquas et cursus aquarum ad operationes Stannariarum praedictarum divertere ubi et quotiens opus fuerit, et emere buscam ad functuram Stanni, sicut antiquitus fieri consuevit, sine impedimento Domini Regis, Heredum suorum, Episcoporum, Abbatum, Comitum, Baronum, seu aliorum quorumcumque etc." Il semble bien busoignable chose en ce cas, que lours Custumes et Usages soient diligeaument enquiz. Et que le Gardein de l'Esteymerye soit chargez, q'il ne soeffre nul Overour del dit Esteymerie fouer en prees ne autry boys, ne ne abate autry boys, ou autry mesons, ne bestonerer eawe ou cours de eawe par malice. Et si par cas le dit Gardain se y vorra excuser, que les ditz Esteymours n'y voillent obeire a ses mandementz, ne cesser lour malice pur luy, que tanttost il ce face monstrer al Grant Conseil le Roi et due et hastive remede ent serra ordeignez.

Item soit declarrez en especial, comen les Justices q'ore serront assignez d'aler celles Marchees pur ent faire la dite Enquerree, prendront l'issue du Pays si ascun y chiece entre parties; et coment ceste Article precedent touchant les Custumes et Usages estoit usez devaunt la fesaunce de la dite Chartre l'aiel; et par queux gentz tielle issue serra trie: C'est assavoir, lequiel par Foreins soulement, ou par Esteymours soulement, ou par ambedeux etc.

Endroit de ceste Article, en soit l'advis pris du Grant Conseil; et y soient les Recordz en Eyre, si nulles y soient, et autres Evidences et Remembrances deinz le Tresorie le Roi, et aillours, et auxint les Remembrances des Seigneurs queux y ont este pur le temps, serchez et duement examinez: Et auxint soient les livres et evidences quelles les ditz Esteymours ent ont envers eulx veuez et regardez, issint que l'en y purra le mieltz venir al droite veritee.

Item soit declarre, si le Gardein de l'Esteymerye puisse tenir Plee entre Esteymour et Foreyn, de querele sourdante aillours que en les lieux ou ils sont overantz, desicome la Chartre voet "... Quod Custos noster praedictus, vel ejus Locum tenens, teneat omnia Placita inter Stannatores praedictos emergentia, et etiam inter ipsos et alios forinsecos, de omnibus Transgressionibus, Querelis, et Contractibus factis in Locis in quibus operantur infra Stannarias praedictas similiter emergentia etc." Qar il tient Plee des tieux Quereles sourdantz chescune part deins le dit Countee.

Et endroit de cest article, se ent extende la Jurisdiction clerement solonc les paroles del dit Chartre; C'est assavoir "... In locis ubi iidem Operarii operantur . . . " et nemye aillours, ne en autre manere.

Item plese declarrer de ceo que la dite Chartre voet einsy, "... Et si qui Stannatorum praedictorum in aliquo deliquerint per quod incarcerari debeant, per Custodem praedictum arrestentur, et in prisona nostra de Lydeford et non alibi detineantur, quousque secundum legem et consuetudinem Regni nostri deliberentur..." Et en cas que Esteymour soit pris pour felonie, et liverez au Gardein, il est suffert sovent d'aler a large, de quoi grant peril avient moelt des foitz; et auxint de ce que la delivrance del dit gaiole n'est pas faite une foitz en dis ans. Et que pis est, par colour de mesme ceste Article, le dit Gardein prent hors d'autry prisone les emprisonez pur arrerages sur accomptez, et les mette a Lydeford, ou ils sont en tant favorez q'ils n'y font force de jamais faire gree a lour Seigneurs.

Endroit de ceste Article, en soit enquiz diligeaument, devant les Justices q'ore y serront proscheinement assignez d'enquere, par quele auctoritee ils y font einsi; depuis qu'en mesme la Chartre sont exceptez par especial touz Plees de terre, et de vie, et de membre; et celle Enqueste retournee, soit declaree en especial s'il busoigne.

*Rot. Parl.*, ii, 343, 344. (The duplicate petition from Cornwall substitutes Lostwithiel for Lidford.)

## G

### *Resolution of the Judges in 1608.*

The Resolution of all the Judges by force of his Majestyes letters concernyng the stannaries in Devon and Cornwall, upon the hearing of the Councell learned of both partyes at severall dayes and what could be alledged and shewed on either party and upon viewe and hearing of the former proceedings in the Courts of the Stannarie, both before and since a certaine Act of Parliament made concernyng the Stannaries in quinquagesimo Edwardi tercii vicesimo sexto Novembris, millesimo sexcentesimo octavo at Serjeants Inne.

First we are of opynion that as well Blowers as all other Laborers and workers without fraud or covyn in or about the Stannaries in Cornwall and Devon are to have the priviledge of the Stannaries duryng the tyme that they doe worke there. Secondly, that all matters and things concernyng the Stannaries or depending upon the same are to be heard and determyned in those Courts according to the custome of the same, tyme out of mynd of man used. Thirdly, that all transitory accions betweene Tynner and Tynner or worker and worker (though the case be collaterall and not perteyning to the Stannarie) maye be heard and determyned within the Courts of the Stannaries according to the Custome of the said Courts, albeit the cause of accion did rise in any place out of the Stannaries, if the defendant be found within the Stannarie, or may be sued at the Common Lawe at the election of the Plaintife, but if the one party only be a tynner or worker, and the cause of accion being transitorye and collateral to the stannarie doe rise out of the said stannaries, then the Defendant maye by the custome and usage of those courts plead to the jurisdiccion of the Court that the cause of accion did rise out of the Stannaries and the Jurisdiccion of those courts which by the custome of the Court he ought to plead in proper person upon oath. And if such plea to the Jurisdiccion be not allowed, then a prohibicion in that case is to be graunted, and if in that case the defendant doe come to pleade to the Jurisdiccion of the Court upon his oath, he ought not to be arrested eundo redeundo vel morando, at the suite of any subject, in anye Corporacion or other place where the said Courts of the Stannerie shalbe then holden.

Fourthly, if the Defendant maye plead to the Jurisdiccion of the Court in the case before mencioned and will not but plead and admitt the Jurisdiccion of the Court and Judgment is given and the body of the defendant taken in execucion, the Party cannot by Lawe have any accion of false imprisonment, but the execucion is good by the custome of that Court; but if in that case it doth appeare by the Plaintife's owne shewing that the contract or cause of accion was made or did rise out of the Stanneries and the Jurisdiccion of those Courts or if it appeare by the condicion of the bond whereupon the accion is grounded that the condicion was to be performed in any place out of the Jurisdiccion of those Courts then all the proceedings in such cases upon such matter apparant are coram non Judice.

Fifthly, We are of opynion that noe man ought to demurre in that Court for want of forme, but only for substance of matter, as if an accion be brought there for words which will beare no accion or an accion of debt upon a contract against executors or admynistrators or such like. In such cases a Demurrer maye be upon the matter and that the proceedings there must be according to the custome of those courtes used tyme out of mynd of man for that noe writt of error doth lye upon any Judgment given there but the remedy given to the party grieved is by appeale as hath byn tyme out of mynd of man accustomed.

Sixthly, that the Courts of the Stannary have not any Jurisdiccion for any cause of accion that is locall rising out of the stannary.

Seaventhly, that the Priviledge of the workers in the Stannaries do not extend to any cause of accion that is locall rising out of the Stannaries nor for any cause locall rising within the Stannaries whereby any Freehold shall be demaunded, for that makers of life, member, and Plea of Land are excepted by expresse words in their charters and noe man can be exempt from Justice.

Thos: Fleming. Edw: Coke.

Et memorandum quod vicesimo quarto die Januarii anno regni domini nostri Regis Jacobi Angliae Franciae et Hiberniae sexto et Scotiae quadragesimo secundo, Willielmus comes Pembrochiae venit in Cancellariam dicti domini Regis et protulit tunc et ibidem chartam praedictam sub manibus praedictorum Thomae Flemyng militis, capitalis Justiciarii dicti domini Regis ad placita assignati, et Edwardi Coke militis capitalis Justiciarii dicti domini Regis de Banco signatam, et petiit ut eadem in rotulis cancellarie dicti domini Regis irrotulari possit ad cujus quidem comitis instanciam de verbo in verbum ut praescriptum est. Irrotuletur vicesimo quarto die Januarii anni praedicti.

*Close Roll*, 6 James I, pt. v.

## H

### *Resolution of the Judges in 1627.*

Apud Sergeantes Inne in Fleet Streete decimo quarto Novembris anno regni regis Caroli terci› 1627.

Whereas sithince a Resolucion heretofore made by all the then Judges uppon the six and twentieth day of November in the sixth yeare of the raigne of our late gracious soveraigne lord Kinge James concerninge the Stannaryes of Devon and Cornwall, which resolution is inrolled in the high Courte of Chancery, some difference have growne and arisen about the true intente and meaninge of some of the articles in the said resolucion by the misinterpritacion whereof, and by the enlarginge the said resolucions in some parte, contrary to the expresse words thereof, and by some miscarriage of some of the under officers there, and likewise by the Bayliffs, great greivances and vexacions have of late happened to the inhabitants of the said counties.

And whereas the right honorable William, Earle of Pembroke, lord Steward of the King's houshold, Lord Warden of the Stannaries and one of his Majesty's most honorable privie Counsell hath referred the explanacion of the said resolucion and the orderinge of the differences to all his Majesty's judges — Wee haveinge perused the said resolucions and haveinge heard the Counsell learned on both sides and haveinge alsoe heard his Majesty's Attorney Generall, doe explaine the first article of the said resolucion in this manner followinge.

That is to saie that as well Blowers as all other laborers and workers without fraude or coven in or about the Stannaries in Cornewall and Devon are to be taken for such tynners as are to have the priviledges to sue and to be sued in the Courts of the Stannaryes, duringe the tyme they worke there and not longer, and noe other Tynners whatsoever, for although many persons may be styled Tynners, as the Jurates of the severall Stannary Courts, owners, adventurers, undertakers in Tynne Mynes and such like, yett nevertheless the Tynners which are to have the priviledge only to sue and to be sued in the Courts of the Stannaryes and not elswhere are such Tynners as are the blowers and all other labourers and workers of the said workes, whose personall attendance are necessary to be imployed in the said Tynne workes duringe the time they worke or attende there and not longer, and noe other Tynners whatsoever, which doth appeare soe to be by former resolucions in parliamente as well as the fiftieth and one and fiftieth yeare of Kinge Edward the third, when Richard then Prince and Duke of Cornewall interceded, as alsoe in the eight yeare of the raigne of the said Richard, when hee was Kinge which was finally resolved in these words followinge. Pro stannatoribus in comitatu Devoniae Rex vicecomiti Devoniae salutem. Cum inter ceteras libertates et quietantias stannatoribus nostris in comitatu praedicto per Cartam domini Edwardi quondam Regis Angliae progenitoris nostri filii regis Henrici quam dominus Edwardus quondam Rex Angliae proavus noster ac Dominus Edwardus nuper rex Angliae avus noster per Cartas suas confirmaverunt concessum sit iisdem quod omnes Stannatores praedicti operantes in Stannariis illis quae sunt dominica nostra dum operantur in eisdem Stannariis sint liberi etc. volumus ac dictus avus noster per literas suas patentes post modum declaravit quod quaedam verba superius expressa sub modo et forma sequenti capiantur et intelligantur videlicet quod operantes in stannariis illis et dum operantur in eisdem stannariis intelligantur clare de operariis laborantibus duntaxat et Stannariis illis sine fraude et sine dolo et non de aliis nec alibi laborantibus.

And as touchinge the third Article of the said resolucion which doth concerne the extente of the Stannaries, but what place or places shalbe taken to be within the Stannaries and what places to be without, now controverted before us, wee are of the opinion that every village, Tythinge or Hamlett and all lands, Tenements, Commons, Moores, wastes and groundes within any of the said Villages, Tithinges and Hamletts wherein any such tynne worke now is, or at any time hereafter shalbe settled, found and wrought shalbe taken and accompted to be within the Stannaries duringe the continuance of any such Worke only and noe longer, and noe other place, which alsoe is confirmed by the said last mencioned resolucion in 8 R. 2., in these words. Et quod ad dictum articulum quod custos praedictus vel ejus locum tenens teneat omnia placita inter Stannatores emergentia extendet se clare Jurisdictio juxta vim verborum Chartae supradictorum, videlicet, in locis ubi iidem operarii operantur et non alibi nec alio loco prout in Cartis, confirmationibus et literis praedictis plenius continetur.

And haveinge perused all the rest of the articles of the said resolucion are of opinion that they are to be pursued and followed in all things and that if these explanacions shall not be pursued hereafter, that then all the proceedings in their Stannary Courts contrary to these explanacions shalbe voyd and Coram non Judice in such sorte as other things in the said Articles are appoynted to be when they are not pursued, and that the parties greived therein may take theire remedie at the Common Lawe and that prohibicions and Habeas Corpus may be granted in these Cases respectively when any such cause shall happen. Irrotuletur. Tho: Coventrye C. S., Ni: Hyde, Tho: Richardson, Io: Walter, Iohn Doddridge, Io: Denham, Richard Hutton, Will: Iones, Iames Whitelocke, Fr: Harvey, Geo: Croke, H. Yeluerton, Tho: Trevor.

*Close Roll*, 4 Charles I, part 2, No. 31 (also in Add. MS. 6713, fol. 347).

# I

## *Resolutions of the Privy Council*, 1632.

(Order of January 21, 1632.)

Whereas an humble peticion was heretofore presented to his Majestie by the Earle of Pembroke and Montgomery, Lord Warden of the Stanaryes concerning the Iurisdiccion and priviledges of the said Staneries and by his command sent to the Lords Cheife Iustices of both Benches with the rest of his Majesties Iustices there, to be by them pervsed, and considered, to the end some course might be setled for the distinguishing, regulateing and ordering of the limitts and priviledges of the seuerall Iurisdiccions of the said Courts, that his Majesties Subjects might the better know whether they were to resort for the Administracion of Iustice, and the heareing of their causes, and righting of their wrongs. Vpon a long heareing and debate of this business (his Majestie then sitting in Councell) and the said Iudges being present, as also his Majesties Atturney generall. It was resolued, and ordered that the said Iudges should search out, and peruse such Statutes, and other Records as might concerne that business And also that Mr. Atturney should doe the like, and conferr with the said Iudges for the cleareing of the Iurisdiccion of the said Staneries, that so if they could not reconsile and accommodate the differences aforesaid among themselues, then before, or at the longest on the 18th. of February next, they should attend his Majestie and make Report of the state of the cause, to the end that his Majestie may therevpon settle such a finall conclusion therein, as in his princely wisdome shall be fitt.

(February 18, 1632.)

This day (his Majestie being present in Councell) certaine Articles and Proposicions produced by his Majesties Attorney generall concerning the Iurisdiction of the Stannaries, were read and approued of by the Board; only some fewe particulars thought fitt to be added were by his Majestie recomended to his said Attorny generall; who is likewise required to cause a faire transcript thereof to be signed by the Iudges, before they goe theire Circuite and to retourne the same to this Board, to the end it may be kept in the Councell chest.

The Rules following to be observed in his Majesties Courts at Westminster and his Court of the Stanneries, were agreed of before the Board, his Majestie being present in Councell, and afterwards subsigned by the Lord Warden of the Stanneries and all the Iudges of his Majesties said Courts at Westminster and his Atturney Generall. And the Transcript thereof ordered to be entered into the Register of Councell Causes and the originall to remayne in the Councell chest.

The Workers about the Tynne, whether in Myne or Streame, the Carrier, Washer, and Blower of Tynne, and the necessarie Attendants aboute the workes haue priviledge that they ought not to be sued out of the Stannery (except it be in causes concerning Life, Member, or Freehould) for any cause aryseing within the Stannerie. And if they be sued elsewhere the warden may demand Conusans or the partie may plead his priviledge.

Besides theise there are other Tynners that doe noe handworke as are the owners of the Soyle, owners of the Bounds, owners of the Bloweing houses, and theire partners, buyers and sellers of Black Tynne, or Whyte Tynne before the deliuerance, theise may sue one an other, or working Tynners, or any other man, for any matter concerning Tynne, or Tynne works, in the Stannerie Courte.

Both theise Tynners and the workers may sue one an other in the Stannarye for all causes personall not concerning Freehold, Life or Member, ariseing within the Stannary or elsewhere aryseing.

One Tynner may sue a Forrayner in all lyke causes personall, aryseing within the Stannarye, but a Tynner may not sue a Forrayner, in the Stannarye for matters personall aryseing out of the Stannarye.

Of those later sorte of Tynners, such onely are intended as within some convenient tyme, make profitt or endeavour to make profitt to the Coynage.

For the manner of tryeing whether one be a Tynner or not, the use in Cornewall is by Plea, and if issue be ioyned, and found for the Plaintiffe it is not peremptory but a respondes.

In Devon it is by the oath of the partye.

For the Extent of the Stannaries.

We cannot yet discerne but that the Stannaries doe extend over the whole County of Cornwall.

In Devon there hath bin long Question concerning the extent of the Stannarie, as apeareth in sunderie Peticions in Parliament.

This is question of Fact and not of Lawe.

But for repose and quiettnes hereafter, whether it be convenient to award a Commission to some able persons who may enquire by oath of lawfull and indifferent men of the Bounds of each Stannarye for informacion onely, or whether it be more fitt to leaue it without further enquirie and as it hath byn heretofore wee humbly leaue it to your Majesties wisedome, with this; that vntill the matter of fact be further knowne, this Question concerning the Bounds of the Stannarye in the County of Devon may remayne without preiudice, by occacion of any former opinion deliuered concerning this question of facte. But

The exempcion of Tynners from Toll is over the whole county.

The power to digg and search for Tynne is over the whole county saueing under houses, orchards, gardens, etc.

The Tynne wrought in any parte of the county must be brought to the Coynage.

The priviledge of Empcion or preempcion is of Tynne gotten over the whole county.

Iudgements had in the Stannarye Courte are leaviable in all parts of the county.

Fynes and Amerciaments sett in the Stannary Courte may be leavied over the whole county by Proces of the Stannarie.

For trespasses in Tynne works, Proces may be executed in the whole county.

Water Courses for the Tynne works on Tynne Mills may be made in any place of the countye.

Signed:

PEMBROK & MONTGOMERIE.

THO. RICHARDSON
RO: HEATH
HUMFRAY DAVENPORT
JOHN DENHAM
RICH: HUTTON
WILLIAM IONES
GEORGE CROKE
THO: TREUOR
GEORGE VERNON
IAMES WESTON
RO. BARKLEY
FR: CRAWLEY
WM. NOYE

*Register of the Privy Council,* Charles I, vol. 8, pp. 412, 457, 485, 486. (Printed with omissions in Harrison's Report on the Jurisdiction of the Stannaries, pp. 158–160.)

# J

## *Production of Tin*

NOTE. * = Approximate. M. = Stannary thousand-weight of 1200 lbs. C. = Stannary hundred-weight of 120 lbs.

| YEAR ENDING MICHAELMAS | CORNWALL | | | DEVON | | | TOTAL | | |
|---|---|---|---|---|---|---|---|---|---|
| | M. | C. | lbs. | M. | C. | lbs. | M. | C. | lbs. |
| 1156–1160* | | | | 133 | | | 133 | | |
| 1161–1162* | | | | 160 | | | 160 | | |
| 1163–1168* | | | | 183 | 3 | 40 | 183 | 3 | 40 |
| 1169–1170* | | | | 533 | 3 | 40 | 533 | 3 | 40 |
| 1171–1189* | | | | 640 | | | 640 | | |
| 1195 | 256 | | | | | | | | |
| 1199 | | | | | | | 901 | 4 | 5 |
| 1200 | | | | | | | 807 | 1 | 6 |
| 1201 | | | | | | | 807 | 7 | 60 |
| 1206 | | | | | | | 607 | 0 | 2 |
| 1209 | | | | | | | 611 | 7 | 2 |
| 1211 | | | | | | | 813 | 3 | 90 |
| 1212 | | | | | | | 1,002 | 9 | 60 |
| 1214 | | | | | | | 1,198 | 5 | 110 |
| 1243 | | | | 74 | 4 | 72 | | | |
| 1288 | | | | 81 | 7 | 12 | | | |
| 1289 | | | | 82 | 6 | 3 | | | |
| 1290 | | | | 82 | 7 | 84 | | | |
| 1291 | | | | 87 | 7 | 85 | | | |
| 1292 | | | | 86 | 7 | 70 | | | |
| 1293 | | | | 86 | 0 | 79 | | | |
| 1294 | | | | 56 | 8 | 0 | | | |
| 1295 | | | | 46 | 2 | 65 | | | |
| 1296 | | | | 38 | 1 | 30 | | | |
| 1297 | | | | 43 | 0 | 56 | | | |
| 1298 | | | | 46 | 5 | 35 | | | |
| 1299 | | | | 52 | 9 | 20 | | | |
| 1300 | | | | 53 | 0 | 0 | | | |
| 1301 | 560 | 3 | 35 | 53 | 0 | 0 | 613 | 3 | 35 |
| 1302 | 636 | 4 | 34 | | | | | | |
| 1303 | 766 | 1 | 1 | 90 | 1 | 38 | 856 | 2 | 39 |
| 1304 | 774 | 6 | 10 | | | | | | |
| 1305 | 850 | 3 | 86 | | | | | | |
| 1306 | 863 | 2 | 34 | | | | | | |
| 1337 | 1,380 | 2 | 38 | | | | | | |
| 1338 | 1,228 | 4 | 78 | | | | | | |

| YEAR ENDING MICHAELMAS | CORNWALL | | | DEVON | | | TOTAL | | |
|---|---|---|---|---|---|---|---|---|---|
| | M. | C. | lbs. | M. | C. | lbs. | M. | C. | lbs. |
| 1355 | 496 | 6 | 76 | 0 | 0 | 0 | 496 | 6 | 76 |
| 1357 | 539 | 6 | 22 | | | | | | |
| 1368 | 853 | 7 | 14 | | | | | | |
| 1379 | 832 | 0 | 48 | 72 | 3 | 84 | 904 | 4 | 12 |
| 1382 | | | | 100 | 2 | 6 | | | |
| 1385 * | | | | 137 | 1 | 72 | | | |
| 1392 * | 1,154 | 7 | 0 | | | | | | |
| 1393 * | 1,090 | 0 | 0 | | | | | | |
| 1394 * | | | | 54 | 9 | 78 | | | |
| 1396 | | | | 97 | 9 | 11 | | | |
| 1397 * | | | | 100 | 1 | 12 | | | |
| 1398 | | | | 87 | 7 | 0 | | | |
| 1399 * | | | | 48 | 4 | 6 | | | |
| 1400 | 1,465 | 2 | 88 | 128 | 6 | 26 | 1,593 | 8 | 114 |
| 1412 | 1,305 | 2 | 66 | 107 | 8 | 52 | 1,413 | 0 | 118 |
| 1418 | 983 | 4 | 32 | | | | | | |
| 1435 | 801 | 2 | 47 | 73 | 3 | 37 | 874 | 5 | 84 |
| 1443 | 752 | 8 | 35 | 91 | 9 | 90 | 844 | 8 | 4 |
| 1447 | | | | 202 | 8 | 8 | | | |
| 1448 * | 1,021 | 6 | 4 | 35 | 2 | 2 | 1,056 | 8 | 6 |
| 1449 * | | | | 34 | 6 | 101 | | | |
| 1450 | 785 | 7 | 30 | 124 | 1 | 50 | 909 | 8 | 80 |
| 1455 | 687 | 2 | 117 | 112 | 8 | 10 | 800 | 1 | 7 |
| 1456 * | | | | 111 | 4 | 48 | | | |
| 1463 | | | | 130 | 6 | 82 | | | |
| 1469 | 817 | 9 | 9 | 189 | 6 | 30 | 1,007 | 5 | 39 |
| 1472 | 851 | 0 | 116 | 242 | 6 | 114 | 1,093 | 7 | 110 |
| 1477 | 682 | 7 | 102 | 205 | 5 | 70 | 888 | 3 | 52 |
| 1478 | 808 | 9 | 50 | 211 | 8 | 45 | 1,020 | 7 | 95 |
| 1489 | 810 | 2 | 113 | 243 | 7 | 71 | 1,054 | 0 | 64 |
| 1495 | 1,056 | 7 | 34 | 252 | 9 | 55 | 1,309 | 6 | 89 |
| 1496 | 1,017 | 2 | 60 | 218 | 4 | 34 | 1,235 | 6 | 94 |
| 1504 | 997 | 2 | 116 | 272 | 9 | 61 | 1,270 | 2 | 57 |
| 1517 | 1,207 | 9 | 99 | 468 | 1 | 23 | 1,676 | 1 | 2 |
| 1518 | 1,252 | 4 | 16 | 432 | 6 | 9 | 1,685 | 0 | 25 |
| 1524 | 1,200 | 2 | 114 | 470 | 2 | 48 | 1,670 | 5 | 42 |
| 1530 | 1,255 | 0 | 50 | 435 | 4 | 0 | 1,690 | 4 | 50 |
| 1540 | 1,377 | 2 | 82 | 312 | 1 | 60 | 1,689 | 4 | 22 |
| 1544 | 1,152 | 4 | 74 | 334 | 2 | 18 | 1,486 | 6 | 92 |
| 1545 | 1,279 | 0 | 35 | 337 | 7 | 29 | 1,616 | 7 | 64 |
| 1547 | 1,413 | 2 | 64 | 318 | 3 | 64 | 1,731 | 6 | 8 |
| 1549 | 988 | 4 | 96 | 201 | 2 | 96 | 1,189 | 7 | 72 |
| 1553 | 1,323 | 2 | 76 | 293 | 3 | 17 | 1,616 | 5 | 93 |

| YEAR ENDING MICHAELMAS | CORNWALL | | | DEVON | | | TOTAL | | |
|---|---|---|---|---|---|---|---|---|---|
| | M. | C. | lbs. | M. | C. | lbs. | M. | C. | lbs. |
| 1554 | 1,251 | 2 | 24 | 242 | 6 | 78 | 1,493 | 8 | 102 |
| 1555 | 1,133 | 4 | 24 | 272 | 7 | 62 | 1,406 | 1 | 86 |
| 1556 | 1,158 | 9 | 78 | 240 | 3 | 5 | 1,399 | 2 | 83 |
| 1561 | 1,162 | 3 | 26 | 171 | 7 | 100 | 1,334 | 1 | 6 |
| 1563 | 947 | 4 | 22 | 164 | 9 | 101 | 1,112 | 4 | 3 |
| 1564 | 1,106 | 2 | 56 | 161 | 3 | 5 | 1,267 | 5 | 61 |
| 1565 | 1,041 | 2 | 16 | 167 | 6 | 39 | 1,208 | 8 | 55 |
| 1566 | 1,082 | 7 | 68 | 170 | 5 | 43 | 1,253 | 2 | 111 |
| 1567 | 1,184 | 5 | 17 | 212 | 8 | 46 | 1,397 | 3 | 63 |
| 1568 | 793 | 5 | 64 | 180 | 5 | 93 | 974 | 1 | 37 |
| 1569 | 1,181 | 5 | 72 | 212 | 0 | 0 | 1,393 | 5 | 72 |
| 1570 | 818 | 3 | 90 | 248 | 1 | 99 | 1,066 | 5 | 69 |
| 1571 | 778 | 3 | 8 | 182 | 6 | 114 | 961 | 0 | 2 |
| 1572 | 772 | 1 | 32 | 178 | 7 | 0 | 950 | 8 | 32 |
| 1573 | 838 | 8 | 42 | 187 | 2 | 112 | 1,025 | 1 | 34 |
| 1574 | 801 | 3 | 114 | 187 | 9 | 96 | 989 | 3 | 90 |
| 1575 | 900 | 4 | 64 | 188 | 7 | 102 | 1,089 | 2 | 46 |
| 1576 | 1,001 | 4 | 52 | 196 | 9 | 92 | 1,198 | 4 | 24 |
| 1577 | 1,050 | 6 | 116 | 180 | 0 | 9 | 1,230 | 7 | 5 |
| 1578 | 895 | 3 | 102 | 173 | 7 | 54 | 1,069 | 1 | 36 |
| 1579 | 982 | 4 | 32 | 201 | 9 | 74 | 1,184 | 3 | 106 |
| 1580 | 1,182 | 8 | 19 | 188 | 1 | 59 | 1,370 | 9 | 78 |
| 1581 | 1,086 | 6 | 61 | 187 | 9 | 59 | 1,274 | 6 | 0 |
| 1582 | 1,088 | 1 | 60 | 167 | 9 | 100 | 1,256 | 1 | 40 |
| 1583 | | | | | | | | | |
| 1584 | 1,051 | 8 | 116 | 162 | 8 | 7 | 1,214 | 7 | 3 |
| 1585 | 1,078 | 8 | 26 | 164 | 9 | 114 | 1,243 | 8 | 20 |
| 1586 | 1,056 | 2 | 109 | 167 | 9 | 32 | 1,224 | 2 | 21 |
| 1587 | 1,015 | 9 | 57 | 175 | 9 | 71 | 1,191 | 9 | 8 |
| 1588 | 1,077 | 1 | 28 | 164 | 0 | 113 | 1,241 | 2 | 21 |
| 1589 | 987 | 7 | 92 | 160 | 1 | 61 | 1,147 | 9 | 33 |
| 1590 | 1,041 | 2 | 78 | 137 | 5 | 90 | 1,178 | 8 | 48 |
| 1591 | 1,152 | 9 | 60 | 128 | 0 | 26 | 1,280 | 9 | 86 |
| 1592 | 1,209 | 3 | 14 | 112 | 3 | 29 | 1,321 | 6 | 43 |
| 1593 | 1,071 | 2 | 118 | 112 | 3 | 114 | 1,183 | 6 | 112 |
| 1594 | 1,158 | 1 | 85 | 122 | 8 | 68 | 1,281 | 0 | 33 |
| 1595 | 1,200 | 1 | 116 | 157 | 5 | 28 | 1,357 | 7 | 24 |
| 1596 | 1,098 | 7 | 4 | 128 | 8 | 91 | 1,227 | 5 | 95 |
| 1597 | 932 | 4 | 87 | 113 | 6 | 117 | 1,056 | 1 | 84 |
| 1598 | 790 | 9 | 50 | 97 | 4 | 3 | 888 | 3 | 53 |
| 1599 | 877 | 9 | 69 | 131 | 0 | 42 | 1,008 | 9 | 111 |
| 1600 | 946 | 1 | 106 | 119 | 8 | 53 | 1,066 | 0 | 39 |
| 1601 | 1,067 | 0 | 112 | 135 | 8 | 116 | 1,202 | 9 | 108 |

| YEAR ENDING MICHAELMAS | CORNWALL | | | DEVON | | | TOTAL | | |
|---|---|---|---|---|---|---|---|---|---|
| | M. | C. | lbs. | M. | C. | lbs. | M. | C. | lbs. |
| 1602 | 1,250 | 6 | 12 | 126 | 5 | 13 | 1,377 | 1 | 25 |
| 1603 | 1,023 | 5 | 40 | 76 | 4 | 50 | 1,099 | 9 | 90 |
| 1604 | 1,051 | 3 | 76 | 99 | 6 | 88 | 1,151 | 0 | 44 |
| 1605 | 1,029 | 4 | 90 | 78 | 2 | 108 | 1,107 | 7 | 78 |
| 1606 | 986 | 6 | 60 | 60 | 4 | 48 | 1,047 | 0 | 108 |
| 1607 | 980 | 1 | 58 | 47 | 0 | 13 | 1,027 | 1 | 71 |
| 1608 | 1,011 | 2 | 70 | 70 | 9 | 62 | 1,082 | 2 | 12 |
| 1609 | 932 | 3 | 82 | 107 | 8 | 19 | 1,040 | 1 | 101 |
| 1610 | 970 | 2 | 100 | 50 | 7 | 65 | 1,021 | 0 | 45 |
| 1611 | 1,010 | 7 | 114 | 84 | 1 | 118 | 1,094 | 9 | 112 |
| 1612 | 984 | 8 | 50 | 100 | 1 | 57 | 1,085 | 4 | 107 |
| 1613 | 1,039 | 4 | 34 | 99 | 7 | 14 | 1,139 | 1 | 48 |
| 1614 | 1,079 | 0 | 100 | 82 | 6 | 74 | 1,161 | 7 | 54 |
| 1625 | 1,334 | 0 | 98 | 68 | 2 | 32 | 1,402 | 3 | 10 |
| 1638 | 972 | 1 | 84 | 28 | 3 | 61 | 1,000 | 5 | 25 |
| 1639 | 982 | 6 | 30 | 23 | 1 | 21 | 1,005 | 7 | 51 |
| 1640 | 983 | 6 | 7 | 18 | 0 | 44 | 1,001 | 6 | 51 |
| 1641 | 942 | 7 | 10 | 15 | 2 | 63 | 957 | 9 | 73 |
| 1642 | 1,001 | 5 | 18 | 12 | 6 | 73 | 1,014 | 1 | 91 |
| 1643 | 677 | 5 | 40 | 0 | 0 | 0 | 677 | 5 | 40 |
| 1644 | 227 | 5 | 6 | 0 | 0 | 0 | 227 | 5 | 6 |
| 1645 | 590 | 0 | 90 | 0 | 0 | 0 | 590 | 0 | 90 |
| 1646 | 293 | 5 | 4 | 0 | 0 | 0 | 293 | 5 | 4 |
| | lbs. | | | lbs. | | | lbs. | | |
| 1647 | | | | | | | 432,268 | | |
| 1648 | | | | | | | 9,458 | | |
| 1667 | | | | | | | 2,040,750 | | |
| 1669 | | | | | | | 1,581,552 | | |
| 1670 | | | | | | | 1,743,252 | | |
| 1671 | | | | | | | 1,992,894 | | |
| 1672 | | | | | | | 1,776,454 | | |
| 1673 | 2,141,064 | | | 640 | | | 2,141,704 | | |
| 1674 | 1,418,936 | | | 3,858 | | | 1,422,794 | | |
| 1675 | 2,489,854 | | | 6,708 | | | 2,496,562 | | |
| 1676 | 2,496,126 | | | 16,522 | | | 2,512,648 | | |
| 1677 | 2,996,082 | | | 13,700 | | | 3,009,782 | | |
| 1678 | 2,916,140 | | | 14,030 | | | 2,930,170 | | |
| 1679 | 2,362,610 | | | 15,588 | | | 2,376,198 | | |
| 1680 | 2,586,212 | | | 14,930 | | | 2,601,142 | | |
| 1681 | 2,633,966 | | | 11,896 | | | 2,645,862 | | |
| 1682 | 3,038,558 | | | 16,722 | | | 3,055,280 | | |
| 1683 | 3,133,920 | | | 18,266 | | | 3,152,186 | | |

| YEAR ENDING MICHAELMAS | CORNWALL | DEVON | TOTAL |
|---|---|---|---|
| | lbs. | lbs. | lbs. |
| 1684 | 2,693,282 | 20,892 | 2,714,174 |
| 1685 | 3,039,460 | 29,644 | 3,069,104 |
| 1686 | 3,427,730 | 27,902 | 3,455,632 |
| 1687 | 3,245,738 | 25,552 | 3,271,290 |
| 1688 | 3,109,476 | 26,124 | 3,135,600 |
| 1689 | 3,316,618 | 27,304 | 3,343,922 |
| 1690 | 2,813,728 | 26,398 | 2,840,126 |
| 1691 | 2,889,748 | 41,842 | 2,931,590 |
| 1692 | 2,725,090 | 36,134 | 2,761,224 |
| 1693 | 2,778,874 | 61,148 | 2,840,022 |
| 1694 | 2,629,974 | 52,138 | 2,682,112 |
| 1695 | 2,782,298 | 36,906 | 2,819,204 |
| 1696 | 2,643,700 | 32,696 | 2,676,396 |
| 1697 | 2,342,568 | 49,466 | 2,382,034 |
| 1698 | 2,772,978 | 45,666 | 2,818,644 |
| 1699 | 3,178,821 | 31,498 | 3,210,319 |
| 1700 | 3,151,504 | 47,384 | 3,198,888 |
| 1701 | 3,048,484 | 34,128 | 3,082,612 |
| 1702 | 2,460,266 | 34,294 | 2,494,560 |
| 1703 | 3,578,452 | 28,186 | 3,606,638 |
| 1704 | 3,312,416 | 26,230 | 3,338,646 |
| 1705 | 3,087,586 | 64,890 | 3,152,476 |
| 1706 | 3,200,016 | 123,636 | 3,323,652 |
| 1707 | 3,200,022 | 79,028 | 3,279,050 |
| 1708 | 3,200,000 | 57,914 | 3,257,914 |
| 1709 | 3,140,044 | 57,914 | 3,197,958 |
| 1710 | 4,800,060 | 73,862 | 4,873,922 |
| 1711 | 3,169,938 | 48,080 | 3,218,018 |
| 1712 | 3,199,860 | 24,048 | 3,223,908 |
| 1713 | 3,012,338 | 24,700 | 3,037,038 |
| 1714 | 2,464,756 | 27,064 | 2,491,820 |
| 1715 | 2,651,780 | 11,686 | 2,663,466 |
| 1716 | 2,417,864 | 15,838 | 2,433,702 |
| 1717 | 3,697,014 | 11,062 | 3,708,076 |
| 1718 | 3,639,388 | 15,136 | 3,654,524 |
| 1719 | | | |
| 1720 | 3,302,440 | 5,464 | 3,307,904 |
| 1721 | 2,563,622 | 1,382 | 2,565,004 |
| 1722 | 3,124,024 | 2,320 | 3,126,344 |
| 1723 | 3,085,430 | 2,860 | 3,088,290 |
| 1724 | 3,588,378 | 2,506 | 3,590,884 |
| 1725 | 3,719,896 | 6,144 | 3,726,040 |
| 1726 | 3,384,996 | 9,368 | 3,394,364 |

| YEAR ENDING MICHAELMAS | CORNWALL | DEVON | TOTAL |
|---|---|---|---|
| | lbs. | lbs. | lbs. |
| 1727 | 3,564,042 | 3,392 | 3,567,434 |
| 1728 | 3,262,730 | 12,784 | 3,275,514 |
| 1729 | 3,545,790 | 5,068 | 3,550,862 |
| 1730 | 3,462,692 | 930 | 3,463,622 |
| 1731 [1] | 1,555,986 | 0 | 1,555,986 |
| 1732 | 4,167,706 | 320 | 4,168,026 |
| 1733 | 3,647,454 | 104 | 3,647,558 |
| 1734 | 4,114,588 | 0 | 4,114,588 |
| 1735 | 3,942,512 | 0 | 3,942,512 |
| 1736 | 3,464,946 | 0 | 3,464,946 |
| 1737 | 3,777,130 | 0 | 3,777,130 |
| 1738 | 3,025,368 | 0 | 3,025,368 |
| 1739 | 3,996,748 | 0 | 3,996,748 |
| 1740 | 3,795,578 | 0 | 3,795,578 |
| 1741 | 3,462,078 | 1,508 | 3,463,586 |
| 1742 | 3,989,144 | 6,640 | 3,995,784 |
| 1743 | 4,227,998 | 6,424 | 4,234,422 |
| 1744 | 4,190,776 | 3,164 | 4,193,940 |
| 1745 | 3,886,258 | 0 | 3,886,258 |
| 1746 | 4,294,990 | 0 | 4,294,990 |
| 1747 | 4,127,910 | 506 | 4,128,416 |
| 1748 | 4,488,518 | 0 | 4,488,518 |
| 1749 | 2,584,836 | 0 | 2,584,836 |

| YEAR | AMOUNT | YEAR | AMOUNT |
|---|---|---|---|
| | tons | | tons |
| 1750 | 2,876 | 1763 | 2,736 |
| 1751 | 2,273 | 1764 | 2,618 |
| 1752 | 2,550 | 1765 | 2,757 |
| 1753 | 2,516 | 1766 | 3,055 |
| 1754 | 2,724 | 1767 | 2,850 |
| 1755 | 2,757 | 1768 | 2,667 |
| 1756 | 2,774 | 1769 | 2,898 |
| 1757 | 2,752 | 1770 | 2,997 |
| 1758 | 2,720 | 1771 | 2,823 |
| 1759 | 2,637 | 1772 | 3,150 |
| 1760 | 2,717 | 1773 | 2,852 |
| 1761 | 2,395 | 1774 | 2,458 |
| 1762 | 2,584 | 1775 | 2,619 |

[1] For a half-year only.

| YEAR | AMOUNT | YEAR | AMOUNT |
|---|---|---|---|
| | tons | | tons |
| 1776 | 2,652 | 1807 | 2,426 |
| 1777 | 2,770 | 1808 | 2,330 |
| 1778 | 2,515 | 1809 | 2,508 |
| 1779 | 2,678 | 1810 | 2,006 |
| 1780 | 2,926 | 1811 | 2,384 |
| 1781 | 2,610 | 1812 | 2,373 |
| 1782 | 2,546 | 1813 | 2,324 |
| 1783 | 2,570 | 1814 | 2,611 |
| 1784 | 2,685 | 1815 | 2,941 |
| 1785 | 2,885 | 1816 | 3,348 |
| 1786 | 3,399 | 1817 | 4,120 |
| 1787 | 3,204 | 1818 | 4,066 |
| 1788 | 3,352 | 1819 | 3,315 |
| 1789 | 3,405 | 1820 | 2,290 |
| 1790 | 3,193 | 1821 | 3,373 |
| 1791 | 3,470 | 1822 | 3,278 |
| 1792 | 3,809 | 1823 | 4,213 |
| 1793 | 3,202 | 1824 | 5,005 |
| 1794 | 3,351 | 1825 | 4,358 |
| 1795 | 3,440 | 1826 | 4,603 |
| 1796 | 3,061 | 1827 | 5,555 |
| 1797 | 3,240 | 1828 | 4,931 |
| 1798 | 2,820 | 1829 | 4,434 |
| 1799 | 2,862 | 1830 | 4,444 |
| 1800 | 2,522 | 1831 | 4,300 |
| 1801 | 2,328 | 1832 | 4,323 |
| 1802 | 2,627 | 1833 | 4,065 |
| 1803 | 2,914 | 1834 | 3,989 |
| 1804 | 2,993 | 1835 | 4,228 |
| 1805 | 2,742 | 1836 | 4,054 |
| 1806 | 2,855 | 1837 | 4,790 |

*References.*

1156–1214. Estimated from the taxation accounts in the Pipe Rolls of those years.
1243. Pipe R., 27 Hen. III, Devon.
1288–1290. *Ibid.*, 20 Edw. I, Devon.
1291–1295. *Ibid.*, 23 Edw. I, Devon.
1296–1301. Pipe R., 29 Edw. I, Devon.
1302, 1306. Duchy Accts., Excheq. Aug., port. 5.
1303. Accts. Excheq. K. R., bdle. 260, no. 24.
1304–1305. Pipe R., 33 Edw. I, Cornwall.

1337–1338. Duchy Accts., Excheq. Aug., port. 1.
1355–1357. Receiver's Rolls.
1368. Accts. Excheq. K. R., bdle. 263, no. 15.
1379. Receiver's Roll.
1382. Accts. Excheq. K. R., bdle. 263, no. 18.
1385. *Ibid.*, bdle. 265, no. 12.
1392. *Ibid.*, bdle. 263, nos. 21, 22.
1393. *Ibid.*, bdle. 263, no. 24.
1394. *Ibid.*, bdle. 263, no. 18.
1396. *Ibid.*, bdle. 263, no. 27.
1397. *Ibid.*, bdle. 263, no. 29.
1398. *Ibid.*, bdle. 264, no. 3.
1399. *Ibid.*, bdle. 264, no. 6.
1400. Receiver's Roll.
1412. Accts. Excheq. K. R., bdle. 265, no. 3.
1418–1443. Receiver's Rolls.
1447. Accts. Excheq. K. R., bdle. 265, no. 12.
1448. *Ibid.*, bdle. 265, no. 14.
1449. *Ibid.*, bdle. 265, no. 25; bdle. 266, nos. 1, 2.
1450–1455. Receiver's Rolls.
1456. Accts. Excheq., K. R., bdle. 266, nos. 4, 15.
1472–1625. Receiver's Rolls.
1638–1639. Receiver's Views.
1640–1641. Duchy Accts., Excheq. Aug., port. 3.
1642–1645. Receiver's Views.
1646–1648. Audit Accounts, Duchy of Cornwall.
1667–1749. Receiver's Views.
1750–1837. Hunt, 887.

## K

## *Coinage Duty.*

NOTE. * = approximate.

| YEAR ENDING MICHAELMAS | CORNWALL | | | DEVON | | | TOTAL | | |
|---|---|---|---|---|---|---|---|---|---|
| | £ | s. | d. | £ | s. | d. | £ | s. | d. |
| 1288 | | | | 63 | 6 | 6 | | | |
| 1289 | | | | 64 | 0 | 4 | | | |
| 1290 | | | | 64 | 3 | 0 | | | |
| 1291 | | | | 58 | 10 | 1 | | | |
| 1302 | 1,472 | 18 | 5 | | | | | | |
| 1303 | 1,532 | 4 | 1 | 69 | 17 | 1 | 1,602 | 1 | 2 |
| 1304 | 1,549 | 4 | 5 | | | | | | |
| 1305 | 1,700 | 15 | 5 | | | | | | |
| 1306 | 1,726 | 9 | 4 | | | | | | |

| YEAR ENDING MICHAELMAS | CORNWALL | | | DEVON | | | TOTAL | | |
|---|---|---|---|---|---|---|---|---|---|
| | £ | s. | d. | £ | s. | d. | £ | s. | d. |
| 1337 | 2,761 | 0 | 9 | 273 | 19 | 5 | 3,035 | 0 | 3 |
| 1338 | 2,427 | 8 | 3 | | | | | | |
| 1355 | 993 | 7 | 1 | | | | | | |
| 1357 | 1,079 | 9 | 10 | | | | | | |
| 1368 * | 1,707 | 8 | 7 | | | | | | |
| 1371 | | | | 52 | 5 | 9 | | | |
| 1374 | | | | | | | 1,016 | 1 | 4 |
| 1379 | 1,664 | 9 | 8 | 56 | 10 | 9 | 1,721 | 9 | 17 |
| 1382 * | | | | 78 | 4 | 9 | | | |
| 1385 * | | | | 107 | 2 | 11 | | | |
| 1392 * | 2,309 | 8 | 0 | | | | | | |
| 1393 * | 2,180 | 0 | 0 | | | | | | |
| 1394 | | | | 42 | 18 | 4 | | | |
| 1396 * | | | | 73 | 19 | 4 | | | |
| 1397 * | | | | 78 | 4 | 2 | | | |
| 1398 * | | | | 70 | 3 | 7 | | | |
| 1399 * | | | | 37 | 4 | 7 | | | |
| 1400 | 2,931 | 9 | 11 | 88 | 3 | 2 | 3,019 | 13 | 1 |
| 1412 | 2,613 | 18 | 7 | 85 | 5 | 11 | 2,699 | 4 | 6 |
| 1418 | 1,968 | 2 | 10 | 85 | 16 | 11 | 2,053 | 19 | 9 |
| 1435 | 1,604 | 3 | 7 | 103 | 9 | 10 | 1,707 | 13 | 5 |
| 1443 | 1,507 | 10 | 0 | 72 | 12 | 4 | 1,580 | 2 | 4 |
| 1447 * | | | | 158 | 8 | 9 | | | |
| 1448 * | 2,043 | 4 | 2 | 27 | 10 | 0 | 2,070 | 14 | 2 |
| 1449 * | | | | 27 | 1 | 9 | | | |
| 1450 | 1,573 | 8 | 7 | 98 | 2 | 4 | 1,671 | 10 | 11 |
| 1455 | 1,376 | 3 | 2 | 88 | 5 | 7 | 1,464 | 8 | 9 |
| 1471 | 1,618 | 11 | 1 | 150 | 2 | 9 | 1,768 | 13 | 10 |
| 1472 | 1,705 | 0 | 5 | 190 | 17 | 11 | 1,896 | 3 | 4 |
| 1477 | 1,574 | 8 | 7 | 161 | 15 | 8 | 1,736 | 4 | 3 |
| 1478 | 1,620 | 17 | 11 | 167 | 9 | 5 | 1,788 | 7 | 4 |
| 1479 | 1,620 | 17 | 11 | 166 | 6 | 5 | 1,787 | 4 | 4 |
| 1489 | 1,623 | 9 | 7 | 191 | 14 | 2 | 1,815 | 3 | 9 |
| 1495 | 2,114 | 18 | 3 | 199 | 14 | 7 | 2,314 | 12 | 10 |
| 1496 | 2,106 | 7 | 10 | 171 | 12 | 9 | 2,278 | 0 | 7 |
| 1504 | 1,997 | 19 | 8 | 221 | 15 | 2 | 2,219 | 14 | 10 |
| 1516 | 2,386 | 11 | 2 | 366 | 18 | 1 | 2,753 | 9 | 3 |
| 1517 | 2,386 | 11 | 2 | 366 | 18 | 1 | 2,753 | 9 | 3 |
| 1518 | 2,485 | 11 | 10 | 339 | 0 | 8 | 2,824 | 12 | 6 |
| 1523 | | | | | | | 2,771 | 3 | 9 |
| 1524 | 2,403 | 14 | 8 | 368 | 9 | 1 | 2,772 | 3 | 9 |
| 1530 | 2,510 | 2 | 1 | 340 | 3 | 2 | 2,850 | 5 | 3 |
| 1540 | 2,756 | 7 | 10 | 252 | 7 | 9 | 3,008 | 15 | 7 |

| YEAR ENDING MICHAELMAS | CORNWALL | | | DEVON | | | TOTAL | | |
|---|---|---|---|---|---|---|---|---|---|
| | £ | s. | d. | £ | s. | d. | £ | s | d. |
| 1544 | 2,306 | 17 | 6 | 262 | 3 | 1 | 2,569 | 0 | 7 |
| 1545 | 2,560 | 9 | 2 | 265 | 0 | 5 | 2,825 | 9 | 7 |
| 1547 | 2,828 | 2 | 10 | 248 | 15 | 2 | 3,076 | 18 | 0 |
| 1549 | 1,976 | 19 | 11 | 198 | 18 | 5 | 2,175 | 18 | 4 |
| 1553 | 2,649 | 8 | 6 | 229 | 3 | 3 | 2,878 | 11 | 9 |
| 1554 | 2,505 | 10 | 8 | 205 | 3 | 5 | 2,710 | 14 | 1 |
| 1555 | 2,269 | 8 | 4 | 213 | 0 | 6 | 2,482 | 8 | 10 |
| 1556 | 2,320 | 16 | 10 | 187 | 15 | 1 | 2,508 | 11 | 11 |
| 1557 | | | | | | | 2,350 | 10 | 10 |
| 1558 | | | | | | | 2,176 | 10 | 0 |
| 1559 | | | | | | | 2,108 | 4 | 0 |
| 1560 | | | | | | | 2,201 | 11 | 0 |
| 1561 | 2,326 | 19 | 2 | 134 | 3 | 10 | 2,461 | 3 | 0 |
| 1562 | | | | | | | 2,234 | 18 | 0 |
| 1563 | 1,896 | 17 | 9 | 128 | 18 | 5 | 2,025 | 16 | 2 |
| 1564 | 2,214 | 19 | 7 | 126 | 0 | 9 | 2,341 | 0 | 4 |
| 1565 | 2,084 | 13 | 9 | 130 | 19 | 5 | 2,236 | 7 | 0 |
| 1566 | 2,167 | 15 | 8 | 133 | 5 | 0 | 2,301 | 0 | 8 |
| 1567 | 2,371 | 10 | 10 | 166 | 5 | 8 | 2,538 | 6 | 6 |
| 1568 | 1,588 | 15 | 1 | 141 | 11 | 0 | 1,730 | 6 | 1 |
| 1569 | 2,367 | 15 | 2 | 164 | 17 | 6 | 2,532 | 12 | 8 |
| 1570 | 1,638 | 9 | 4 | 186 | 0 | 10 | 1,824 | 10 | 2 |
| 1571 | 1,558 | 4 | 9 | 142 | 15 | 0 | 1,700 | 19 | 9 |
| 1572 | 1,544 | 5 | 5 | 139 | 12 | 3 | 1,686 | 17 | 8 |
| 1573 | 1,679 | 15 | 8 | 146 | 4 | 0 | 1,825 | 19 | 8 |
| 1574 | 1,604 | 19 | 6 | 146 | 17 | 3 | 1,751 | 16 | 9 |
| 1575 | 1,802 | 6 | 5 | 163 | 2 | 6 | 1,965 | 8 | 11 |
| 1576 | 2,005 | 8 | 2 | 153 | 16 | 4 | 2,159 | 4 | 6 |
| 1577 | 2,103 | 12 | 8 | 140 | 12 | 10 | 2,244 | 5 | 6 |
| 1578 | 1,792 | 13 | 0 | 135 | 14 | 10 | 1,928 | 7 | 10 |
| 1579 | 1,967 | 7 | 7 | 157 | 15 | 10 | 2,125 | 3 | 5 |
| 1580 | 2,367 | 5 | 0 | 148 | 2 | 11 | 2,515 | 7 | 11 |
| 1581 | 2,175 | 13 | 9 | 146 | 16 | 9 | 2,322 | 10 | 6 |
| 1582 | 2,177 | 13 | 10 | 131 | 5 | 0 | 2,308 | 18 | 10 |
| 1583 | 2,212 | 15 | 5 | 124 | 5 | 11 | 2,337 | 1 | 4 |
| 1584 | 2,106 | 4 | 0 | 128 | 4 | 7 | 2,234 | 8 | 7 |
| 1585 | 2,161 | 9 | 0 | 129 | 5 | 5 | 2,290 | 14 | 5 |
| 1586 | 2,115 | 6 | 0 | 131 | 3 | 10 | 2,246 | 9 | 10 |
| 1587 | 2,033 | 10 | 8 | 137 | 13 | 1 | 2,171 | 3 | 9 |
| 1588 | 2,155 | 18 | 1 | 128 | 6 | 8 | 2,284 | 4 | 9 |
| 1589 | 2,076 | 18 | 2 | 125 | 1 | 4 | 2,201 | 19 | 6 |
| 1590 | 2,081 | 16 | 10 | 107 | 9 | 9 | 2,189 | 6 | 7 |
| 1591 | 2,307 | 4 | 10 | 96 | 2 | 10 | 2,403 | 7 | 8 |

| YEAR ENDING MICHAELMAS | CORNWALL | | | DEVON | | | TOTAL | | |
|---|---|---|---|---|---|---|---|---|---|
| | £ | s. | d. | £ | s. | d. | £ | s. | d. |
| 1592 | 2,419 | 17 | 3 | 87 | 15 | 4 | 2,507 | 12 | 7 |
| 1593 | 2,143 | 12 | 6 | 87 | 17 | 11 | 2,231 | 10 | 5 |
| 1594 | 2,317 | 11 | 3 | 92 | 1 | 6 | 2,409 | 12 | 9 |
| 1595 | 2,401 | 15 | 1 | 123 | 1 | 8 | 2,524 | 16 | 9 |
| 1596 | 2,198 | 17 | 4 | 100 | 13 | 9 | 2,299 | 11 | 1 |
| 1597 | 1,868 | 3 | 2 | 88 | 16 | 11 | 1,957 | 0 | 1 |
| 1598 | 1,582 | 19 | 2 | 76 | 2 | 11 | 1,659 | 2 | 1 |
| 1599 | 1,757 | 11 | 9 | 102 | 11 | 2 | 1,160 | 2 | 11 |
| 1600 | 1,893 | 11 | 1 | 93 | 12 | 9 | 1,987 | 3 | 10 |
| 1601 | 2,135 | 9 | 3 | 106 | 3 | 8 | 2,241 | 12 | 11 |
| 1602 | 2,502 | 10 | 9 | 98 | 17 | 11 | 2,601 | 8 | 8 |
| 1603 | 2,048 | 5 | 4 | 59 | 14 | 6 | 2,107 | 19 | 10 |
| 1604 | 2,103 | 19 | 9 | 77 | 16 | 4 | 2,181 | 16 | 1 |
| 1605 | 2,060 | 5 | 7 | 61 | 7 | 4 | 2,121 | 12 | 11 |
| 1606 | 1,974 | 10 | 11 | 47 | 4 | 1 | 2,027 | 15 | 0 |
| 1607 | 1,961 | 16 | 7 | 36 | 15 | 5 | 1,998 | 12 | 0 |
| 1608 | 2,023 | 10 | 6 | 55 | 8 | 8 | 2,079 | 9 | 2 |
| 1609 | 1,866 | 0 | 5 | 84 | 3 | 11 | 1,950 | 4 | 4 |
| 1610 | 1,941 | 16 | 10 | 75 | 6 | 7 | 2,017 | 3 | 5 |
| 1611 | 2,024 | 17 | 3 | 65 | 15 | 8 | 2,090 | 12 | 11 |
| 1612 | 2,171 | 0 | 11 | 78 | 13 | 0 | 2,249 | 13 | 11 |
| 1613 | 2,388 | 4 | 11 | 77 | 18 | 5 | 2,466 | 3 | 4 |
| 1614 | 2,159 | 8 | 7 | 64 | 11 | 8 | 2,224 | 0 | 3 |
| 1618 | | | | | | | 2,958 | 14 | 8 |
| 1625 | 2,670 | 3 | 1 | 53 | 6 | 2 | 2,723 | 9 | 3 |
| 1638 | 1,945 | 18 | 11 | 22 | 2 | 9 | 1,968 | 1 | 8 |
| 1639 | 1,966 | 16 | 8 | 18 | 1 | 4 | 1,657 | 18 | 0 |
| 1640 | 1,958 | 11 | 0 | 14 | 2 | 4 | 1,982 | 13 | 4 |
| 1641 | 1,897 | 4 | 8 | 11 | 18 | 3 | 1,909 | 2 | 11 |
| 1642 | 2,023 | 2 | 9 | 9 | 17 | 10 | 2,033 | 0 | 7 |
| 1643 | 1,356 | 6 | 8 | | | | 1,356 | 6 | 8 |
| 1644 | 457 | 6 | 7 | | | | 457 | 6 | 7 |
| 1645 | | | | | | | | | |
| 1667 | | | | | | | 4,144 | 0 | 5 |
| 1669 | | | | | | | 3,163 | 6 | 2 |
| 1670 | | | | | | | 3,486 | 10 | 2 |
| 1671 | | | | | | | 3,985 | 15 | 11 |
| 1672 | | | | | | | 3,552 | 18 | 3 |
| 1673 | | | | | | | 4,282 | 12 | 9 |
| 1674 | | | | | | | 2,840 | 18 | 0 |
| 1675 | | | | | | | 4,984 | 19 | 0 |
| 1676 | | | | | | | 5,005 | 3 | 7 |
| 1677 | 5,992 | 3 | 5 | 10 | 14 | 0 | 6,002 | 17 | 5 |

| YEAR ENDING MICHAELMAS | CORNWALL | | | DEVON | | | TOTAL | | |
|---|---|---|---|---|---|---|---|---|---|
| | £ | s. | d. | £ | s. | d. | £ | s. | d. |
| 1678 | 5,832 | 6 | 3 | 10 | 19 | 1 | 5,843 | 5 | 4 |
| 1679 | 4,725 | 4 | 8 | 10 | 12 | 0 | 4,735 | 16 | 8 |
| 1680 | 5,172 | 8 | 8 | 11 | 13 | 4 | 5,184 | 2 | 0 |
| 1681 | 5,267 | 18 | 10 | 9 | 5 | 10 | 5,277 | 4 | 8 |
| 1682 | 6,077 | 2 | 7 | 13 | 1 | 3 | 6,090 | 3 | 10 |
| 1683 | 6,267 | 16 | 11 | 14 | 5 | 4 | 6,282 | 2 | 3 |
| 1684 | 5,386 | 11 | 6 | 16 | 4 | 9 | 5,402 | 16 | 3 |
| 1685 | 6,082 | 2 | 8 | 23 | 3 | 2 | 6,105 | 15 | 10 |
| 1686 | 6,855 | 9 | 4 | 21 | 13 | 1 | 6,877 | 2 | 5 |
| 1687 | 6,491 | 9 | 7 | 19 | 19 | 3 | 6,511 | 8 | 10 |
| 1688 | 6,218 | 19 | 1 | 20 | 8 | 2 | 6,239 | 7 | 3 |
| 1689 | 6,633 | 4 | 9 | 21 | 6 | 9 | 6,654 | 19 | 6 |
| 1690 | 5,627 | 9 | 4 | 20 | 12 | 6 | 5,648 | 1 | 10 |
| 1691 | 5,779 | 10 | 0 | 32 | 13 | 9 | 5,812 | 3 | 9 |
| 1692 | | | | | | | 5,478 | 7 | 8 |
| 1693 | | | | | | | 5,605 | 10 | 11 |
| 1694 | 5,259 | 19 | 0 | 40 | 14 | 6 | 5,300 | 13 | 6 |
| 1695 | 5,564 | 12 | 0 | 28 | 16 | 6 | 5,593 | 8 | 6 |
| 1696 | 5,287 | 8 | 0 | 25 | 10 | 11 | 5,312 | 18 | 11 |
| 1697 | 4,685 | 2 | 8 | 38 | 12 | 10 | 4,723 | 15 | 7 |
| 1698 | 5,545 | 19 | 2 | 35 | 13 | 6 | 5,581 | 12 | 8 |
| 1699 | 6,357 | 13 | 2 | 24 | 12 | 2 | 6,382 | 5 | 4 |
| 1700 | 6,343 | 0 | 2 | 37 | 0 | 4 | 6,380 | 0 | 6 |
| 1701 | 6,095 | 19 | 5 | 26 | 13 | 3 | 6,123 | 12 | 8 |
| 1702 | 6,840 | 10 | 8 | 26 | 15 | 10 | 6,867 | 6 | 6 |
| 1703 | 7,156 | 18 | 1 | 22 | 0 | 5 | 7,178 | 18 | 6 |
| 1704 | 6,624 | 16 | 8 | 20 | 9 | 10 | 6,645 | 6 | 6 |
| 1705 | 6,175 | 3 | 6 | 50 | 13 | 11 | 6,225 | 17 | 5 |
| 1706 | 6,400 | 0 | 8 | 96 | 11 | 9 | 6,496 | 12 | 5 |
| 1707 | 6,400 | 0 | 10 | 63 | 14 | 9 | 6,463 | 15 | 7 |
| 1708 | 6,400 | 0 | 0 | 61 | 14 | 9 | 6,461 | 14 | 9 |
| 1709 | 6,400 | 1 | 10 | 45 | 4 | 11 | 6,445 | 6 | 9 |
| 1710 | 9,600 | 2 | 5 | 57 | 14 | 1 | 9,657 | 16 | 6 |
| 1711 | 6,339 | 17 | 7 | 37 | 11 | 3 | 6,377 | 8 | 10 |
| 1712 | 6,339 | 14 | 5 | 18 | 15 | 8 | 6,417 | 10 | 1 |
| 1713 | 6,024 | 13 | 3 | 19 | 5 | 11 | 6,043 | 19 | 2 |
| 1714 | 6,024 | 13 | 3 | 19 | 5 | 11 | 6,043 | 19 | 2 |
| 1715 | 5,803 | 11 | 3 | 9 | 2 | 6 | 5,312 | 13 | 9 |
| 1716 | 4,835 | 14 | 6 | 12 | 7 | 5 | 4,848 | 2 | 0 |
| 1717 | 7,394 | 0 | 8 | 8 | 12 | 10 | 7,402 | 13 | 6 |
| 1718 | 7,278 | 15 | 7 | 11 | 18 | 4 | 7,290 | 13 | 11 |
| 1720 | 6,604 | 17 | 8 | 4 | 5 | 2 | 6,609 | 2 | 10 |
| 1721 | 5,127 | 4 | 11 | 1 | 1 | 6 | 5,128 | 6 | 5 |

| YEAR ENDING MICHAELMAS | CORNWALL | | | DEVON | | | TOTAL | | |
|---|---|---|---|---|---|---|---|---|---|
| | £ | s. | d. | £ | s. | d. | £ | s. | d. |
| 1722 | 6,248 | 1 | 0 | 1 | 16 | 3 | 6,249 | 17 | 3 |
| 1723 | 6,170 | 17 | 3 | 2 | 4 | 8 | 6,173 | 1 | 11 |
| 1724 | 7,176 | 15 | 2 | 1 | 19 | 2 | 7,178 | 14 | 4 |
| 1725 | 7,439 | 15 | 11 | 4 | 15 | 11 | 7,444 | 11 | 10 |
| 1726 | 6,769 | 19 | 11 | 7 | 6 | 1 | 6,777 | 6 | 0 |
| 1727 | 7,128 | 1 | 9 | 2 | 12 | 11 | 7,130 | 14 | 8 |
| 1728 | 6,525 | 9 | 4 | 9 | 19 | 9 | 6,535 | 9 | 1 |
| 1729 | 7,091 | 11 | 11 | 3 | 19 | 1 | 7,095 | 11 | 0 |
| 1730 | 6,927 | 5 | 6 | 0 | 14 | 5 | 6,927 | 19 | 11 |
| 1731 | 3,111 | 19 | 6 | | | | 3,111 | 19 | 6 |
| 1732 | 8,335 | 8 | 4 | 0 | 5 | 0 | 8,335 | 13 | 4 |
| 1733 | 7,294 | 18 | 11 | 0 | 1 | 8 | 7,295 | 0 | 0 |
| 1734 | 8,229 | 2 | 5 | | | | 8,229 | 2 | 5 |
| 1735 | 7,885 | 0 | 7 | | | | 7,885 | 0 | 7 |
| 1736 | 6,929 | 17 | 11 | | | | 6,929 | 17 | 11 |
| 1737 | 7,554 | 5 | 4 | | | | 7,554 | 5 | 4 |
| 1738 | 6,050 | 14 | 10 | | | | 6,050 | 14 | 10 |
| 1739 | 7,993 | 10 | 1 | | | | 7,993 | 10 | 1 |
| 1740 | 7,591 | 3 | 3 | | | | 7,591 | 3 | 3 |
| 1741 | 6,924 | 3 | 3 | 1 | 3 | 6 | 6,925 | 6 | 9 |
| 1742 | 7,978 | 5 | 11 | 5 | 3 | 9 | 7,983 | 9 | 8 |
| 1743 | 8,456 | 0 | 2 | 5 | 0 | 5 | 8,461 | 1 | 3 |
| 1744 | 8,381 | 11 | 3 | 2 | 9 | 5 | 8,384 | 0 | 8 |
| 1745 | 7,772 | 10 | 5 | | | | 7,772 | 10 | 5 |
| 1746 | 8,589 | 19 | 9 | | | | 8,859 | 19 | 9 |
| 1747 | 8,255 | 16 | 7 | 0 | 8 | 1 | 8,256 | 4 | 8 |
| 1748 | 8,977 | 0 | 10 | | | | 8,977 | 0 | 10 |
| 1749 | 5,601 | 16 | 6 | | | | 5,601 | 16 | 6 |
| 1750 | 10,806 | 4 | 6 | | | | 10,806 | 4 | 6 |

*References.*

1288–1290. Pipe R., 20 Edw. I, Devon.
1291. *Ibid.*, 23 Edw. I, Devon.
1302, 1306. Duchy Accts., Excheq. Aug., port. 5.
1303. Accts. Excheq. K. R., bdle. 264, no. 24.
1304–1305. Pipe R., 33 Edw. I, Cornwall.
1337–1338. Duchy Accts., Excheq. Aug., port. 5.
1355–1357. Receiver's Rolls.
1368. Accts. Excheq. K. R., bdle. 263, no. 15.
1371. *Ibid.*, bdle. 263, no. 19.
1374 Delabeche, 587.

| | |
|---|---|
| 1379. | Receiver's Roll. |
| 1382. | Accts. Excheq. K. R., bdle. 263, no. 18. |
| 1385. | *Ibid.*, bdle. 265, no. 12. |
| 1392. | *Ibid.*, bdle. 263, nos. 21, 22. |
| 1393. | *Ibid.*, bdle. 263, no. 24. |
| 1394. | *Ibid.*, bdle. 263, no. 18. |
| 1396. | *Ibid.*, bdle. 263, no. 27. |
| 1397. | *Ibid.*, bdle. 263, no. 29. |
| 1398. | *Ibid.*, bdle. 264, no. 3. |
| 1399. | *Ibid.*, bdle. 264, no. 6. |
| 1400. | Receiver's Roll. |
| 1412. | Accts. Excheq. K. R., bdle. 265, no. 3. |
| 1418–1443. | Receiver's Rolls. |
| 1447. | Accts. Excheq. K. R., bdle. 265, no. 12. |
| 1448. | *Ibid.*, bdle. 265, no. 14. |
| 1449. | *Ibid.*, bdle. 265, no. 25; bdle. 266, nos. 1, 2. |
| 1450–1455. | Receiver's Rolls. |
| 1471. | Trans. Devon Assoc., iii, 316. |
| 1472–1504. | Receiver's Rolls. |
| 1516. | Lans. 19, fol. 99. |
| 1517–1518. | Receiver's Rolls. |
| 1523. | Doddridge, 98. |
| 1524. | Receiver's Roll. |
| 1530. | *Ibid.;* Lans. 19, fol. 99. |
| 1540–1556. | Receiver's Rolls. |
| 1557–1560. | *Ibid.;* Lans. 1215, fol. 226–230. |
| 1561. | Receiver's Roll. |
| 1562. | *Ibid.;* Lans. 1215, fol. 226–230. |
| 1563–1625. | Receiver's Rolls. |
| 1638–1639. | Receiver's Views. |
| 1640–1641. | Duchy Accts., Excheq. Aug., port. 3 |
| 1642–1645. | Receiver's Views. |
| 1667–1750. | Receiver's Views. |

## L

*Farm of the Tin.*

| YEAR ENDING MICHAELMAS | CORNWALL | | | DEVON | | | TOTAL | | |
|---|---|---|---|---|---|---|---|---|---|
| | £ | s. | d. | £ | s. | d. | £ | s. | d. |
| 1156–1160 | | | | 16 | 13 | 4 | 16 | 13 | 4 |
| 1161–1162 | | | | 20 | 0 | 0 | 20 | 0 | 4 |
| 1163–1168 | | | | 23 | 6 | 8 | 23 | 6 | 8 |
| 1169–1170 | | | | 66 | 13 | 4 | 66 | 13 | 4 |
| 1171–1189 | | | | 80 | 0 | 0 | | | |
| 1195 | 53 | 6 | 8 | | | | | | |
| 1197 | | | | 100 | 0 | 0 | | | |
| 1200–1214 | 66 | 16 | 0 | 100 | 0 | 0 | 166 | 16 | 0 |

*References:* The Pipe Rolls.

## M

*The King's Marks.*

| YEAR ENDING MICHAELMAS | | | | YEAR ENDING MICHAELMAS | | | |
|---|---|---|---|---|---|---|---|
| | £ | s. | d. | | £ | s. | d. |
| 1199 | 600 | 19 | 5 | 1209 | 407 | 16 | 2 |
| 1200 | 538 | 2 | 1 | 1211 | 542 | 5 | 0 |
| 1201 | 538 | 10 | 1 | 1212 | 668 | 12 | 9 |
| 1206 | 404 | 13 | 7 | 1214 | 799 | 1 | 3 |

*References:* The Pipe Rolls.

## N

*Black Rent.*

| YEAR ENDING MICHAELMAS | £ s. d. | YEAR ENDING MICHAELMAS | £ s. d. |
|---|---|---|---|
| 1243 | 1 4 10 | 1295 | 2 3 0 |
| 1288 | 2 10 0 | 1296 | 1 13 2 |
| 1289 | 2 11 4 | 1297 | 1 16 4 |
| 1290 | 2 13 10 | 1298 | 2 17 8 |
| 1291 | 3 15 0 | 1299 | 3 9 2 |
| 1292 | 3 16 2 | 1300 | 3 12 8 |
| 1293 | 3 15 6 | 1301 | 3 13 4 |
| 1294 | 2 15 8 | | |

*References.*

1243. Pipe R., 27 Hen. III, Devon.
1288–1290. *Ibid.*, 20 Edw. I, Devon.
1291–1295. *Ibid.*, 23 Edw. I, Devon.
1296–1301. *Ibid.*, 29 Edw. I, Devon.

## O

*Tribulage.*

| YEAR ENDING MICHAELMAS | PENWITH AND KERRIER | BLACKMORE | YEAR ENDING MICHAELMAS | PENWITH AND KERRIER | BLACKMORE |
|---|---|---|---|---|---|
| | £ s. d. | £ s. d. | | £ s. d. | £ s. d. |
| 1297 | 0 6 8 | 0 6 0 | 1386 | | 0 1 2 |
| 1301 | 0 13 8 | | 1394 | | 0 2 0 |
| 1302 | 0 12 8 | | 1398 | | 0 1 2 |
| 1303 | 0 16 0 | | 1417 | 13 6 8 | 0 4 4 |
| 1304 | 0 16 0 | | 1467 | 7 5 0 | 0 6 8 |
| 1305 | 1 1 4 | | 1472 | farmed | 0 6 8 |
| 1342 | 1 0 0 | 0 10 3 | 1475 | farmed | 0 6 8 |
| 1345 | 1 0 0 | 0 10 3 | 1484 | farmed | 0 6 8 |
| 1347 | 1 0 0 | 0 10 3 | 1489 | farmed | 0 6 8 |
| 1349 | 1 0 0 | 0 10 3 | 1491 | farmed | 0 6 8 |
| 1350 | 0 6 8 | 0 9 3 | 1498 | 10 0 0 | 0 7 8 |
| 1352 | 1 0 0 | 0 3 1 | 1502 | 10 0 0 | 0 7 8 |
| 1355 | 1 2 4 | 0 3 1 | 1503 | 10 0 0 | 0 7 8 |
| 1356 | 1 2 4 | 0 4 8 | 1507 | 10 0 0 | 0 7 8 |
| 1357 | 0 5 4 | 0 4 0 | 1524 | 12 0 0 | |
| 1362 | | 0 1 6 | 1545 | 11 0 0 | |
| 1363 | 1 0 0 | 0 1 6 | 1546–1563 | farmed | |
| 1369 | 1 0 0 | 0 0 10 | 1645 | 6 1 8 | |

*References.*

1297. Ministers' Accounts, Bailiff's Accounts of Edmund of Cornwall.
1301. Pipe R., 29 Edw. I, Cornw.
1302. *Ibid.*, 30 Edw. I, Cornw.
1303. Duchy Accts., Execheq. Aug., port. 5.
1304–1305. Pipe R., 33 Edw. I, Cornw.
1342–1507. Ministers' Accts., Duchy of Cornw.
1524–1563. Receiver's Rolls.
1645. Audit Accts., Duchy of Cornw.

## P

*Profits of the Large and Small Stamps.*

| YEAR ENDING MICHAELMAS | LARGE STAMP | | | SMALL STAMP | | | YEAR ENDING MICHAELMAS | LARGE STAMP | | | SMALL STAMP | | |
|---|---|---|---|---|---|---|---|---|---|---|---|---|---|
| | £ | s. | d. | £ | s. | d. | | £ | s. | d. | £ | s. | d. |
| 1243 | 60 | 14 | 11 | 8 | 0 | 16 | 1296 | 25 | 8 | 4 | 4 | 2 | 7 |
| 1291 | 57 | 10 | 10 | 9 | 9 | 9 | 1297 | 28 | 14 | 0 | 4 | 13 | 11 |
| 1292 | 57 | 16 | 9 | 9 | 8 | 0 | 1298 | 31 | 0 | 5 | 5 | 0 | 10 |
| 1293 | 57 | 7 | 7 | 9 | 6 | 6 | 1299 | 34 | 12 | 3 | 5 | 12 | 6 |
| 1294 | 37 | 18 | 8 | 6 | 3 | 3 | 1300 | 53 | 6 | 8 | 5 | 14 | 10 |
| 1295 | 33 | 16 | 9 | 5 | 0 | 2 | 1301 | 35 | 9 | 9 | 5 | 15 | 3 |

*References.*

1243. Pipe R., 27 Hen. III, Devon.
1291–1295. *Ibid.*, 23 Edw. I, Devon.
1296–1301. *Ibid.*, 29 Edw. I, Devon.

## Q

*White Rent.*

| YEAR ENDING MICHAELMAS | | | | YEAR ENDING MICHAELMAS | | | |
|---|---|---|---|---|---|---|---|
| | £ | s. | d. | | £ | s. | d. |
| 1288 | 2 | 6 | 2 | 1296 | 1 | 18 | 8 |
| 1289 | 2 | 7 | 10 | 1297 | 0 | 16 | 10 |
| 1290 | 2 | 8 | 6 | 1298 | 0 | 17 | 2 |
| 1291 | 2 | 14 | 4 | 1299 | 0 | 17 | 6 |
| 1292 | 2 | 12 | 0 | 1300 | 0 | 17 | 10 |
| 1293 | 2 | 12 | 4 | 1301 | 0 | 18 | 1 |
| 1294 | 2 | 5 | 2 | 1306 | 1 | 3 | 4 |

| YEAR ENDING MICHAELMAS | £ | s. | d. | YEAR ENDING MICHAELMAS | £ | s. | d. |
|---|---|---|---|---|---|---|---|
| 1400 | 4 | 6 | 8 | 1580 | 6 | 17 | 4 |
| 1418 | 7 | 9 | 1 | 1581 | 6 | 2 | 0 |
| 1435 | 6 | 10 | 0 | 1582 | 5 | 16 | 8 |
| 1443 | 7 | 18 | 8 | 1583 | 5 | 8 | 0 |
| 1450 | 8 | 11 | 4 | 1584 | 6 | 13 | 8 |
| 1455 | 7 | 11 | 4 | 1585 | 6 | 6 | 8 |
| 1469 | 9 | 2 | 0 | 1586 | 7 | 13 | 4 |
| 1472 | 12 | 10 | 8 | 1587 | 7 | 0 | 0 |
| 1477 | 11 | 6 | 0 | 1588 | 6 | 1 | 4 |
| 1478 | 10 | 12 | 0 | 1589 | 6 | 8 | 0 |
| 1489 | 12 | 0 | 0 | 1590 | 4 | 2 | 8 |
| 1495 | 12 | 16 | 8 | 1591 | 4 | 16 | 8 |
| 1496 | 12 | 4 | 8 | 1592 | 3 | 2 | 0 |
| 1504 | 14 | 14 | 8 | 1593 | 3 | 12 | 0 |
| 1517 | 16 | 17 | 4 | 1594 | 3 | 19 | 4 |
| 1518 | 16 | 18 | 0 | 1595 | 3 | 5 | 4 |
| 1523 | 14 | 2 | 8 | 1596 | 2 | 11 | 4 |
| 1524 | 14 | 2 | 8 | 1597 | 2 | 14 | 8 |
| 1540 | 10 | 5 | 4 | 1598 | 2 | 10 | 8 |
| 1544 | 8 | 14 | 8 | 1599 | 2 | 16 | 8 |
| 1545 | 8 | 10 | 8 | 1600 | 2 | 12 | 8 |
| 1547 | 8 | 2 | 8 | 1602 | 3 | 4 | 8 |
| 1549 | 7 | 0 | 16 | 1604 | 3 | 1 | 4 |
| 1553 | 7 | 16 | 0 | 1605 | 2 | 7 | 4 |
| 1554 | 7 | 6 | 0 | 1606 | 1 | 15 | 4 |
| 1555 | 6 | 9 | 4 | 1607 | 1 | 4 | 8 |
| 1556 | 5 | 14 | 8 | 1608 | 1 | 18 | 0 |
| 1561 | 4 | 8 | 8 | 1609 | 1 | 19 | 8 |
| 1563 | 4 | 11 | 4 | 1610 | 2 | 1 | 4 |
| 1564 | 4 | 5 | 4 | 1611 | 1 | 16 | 0 |
| 1565 | 3 | 15 | 4 | 1612 | 1 | 10 | 8 |
| 1566 | 3 | 12 | 0 | 1613 | 1 | 11 | 4 |
| 1567 | 3 | 17 | 0 | 1614 | 1 | 10 | 8 |
| 1568 | 3 | 14 | 8 | 1615 | 1 | 6 | 0 |
| 1569 | 4 | 1 | 4 | 1616 | 1 | 9 | 4 |
| 1570 | 4 | 6 | 8 | 1617 | 1 | 2 | 8 |
| 1571 | 3 | 12 | 0 | 1619 | 1 | 2 | 8 |
| 1573 | 4 | 14 | 8 | 1620 | 1 | 1 | 4 |
| 1574 | 5 | 2 | 8 | 1621 | 1 | 2 | 0 |
| 1575 | 5 | 16 | 8 | 1622 | 1 | 3 | 4 |
| 1576 | 5 | 11 | 4 | 1625 | 1 | 3 | 4 |
| 1577 | 4 | 4 | 8 | 1626 | 1 | 4 | 0 |
| 1579 | 5 | 10 | 8 | 1627 | 1 | 3 | 4 |

| YEAR ENDING MICHAELMAS | £ | s. | d. | YEAR ENDING MICHAELMAS | £ | s. | d. |
|---|---|---|---|---|---|---|---|
| 1628 | 1 | 2 | 0 | 1635 | 0 | 14 | 8 |
| 1629 | 0 | 19 | 4 | 1638 | 0 | 13 | 4 |
| 1630 | 0 | 19 | 4 | 1639 | 0 | 12 | 0 |
| 1631 | 0 | 14 | 0 | 1640 | 0 | 9 | 4 |
| 1633 | 0 | 13 | 8 | 1641 | 0 | 11 | 8 |
| 1634 | 0 | 13 | 8 | 1642 | 0 | 11 | 8 |

*References.*

1288–1290. Pipe R., 20 Edw. I, Devon.
1291–1294. *Ibid.*, 23 Edw. I, Devon.
1296–1301. *Ibid.*, 29 Edw. I, Devon.
1306. Duchy Accts., Excheq. Aug., port. 5.
1400–1625. Receiver's Rolls.
1626–1642. Audit Accts., Duchy of Cornwall.

## R

### *Fine of Tin.*

| YEAR ENDING MICHAELMAS | £ | s. | d. | YEAR ENDING MICHAELMAS | £ | s. | d. |
|---|---|---|---|---|---|---|---|
| 1297 | 5 | 8 | 0 | 1576 | 13 | 9 | 8 |
| 1302 | 5 | 8 | 0 | 1577 | 17 | 0 | 0 |
| 1303 | 5 | 8 | 0 | 1578 | 23 | 14 | 3 |
| 1304 | 5 | 2 | 0 | 1579 | 33 | 15 | 2 |
| 1305 | 5 | 8 | 0 | 1580 | 41 | 13 | 4 |
| 1306 | 5 | 8 | 0 | 1581 | 49 | 18 | 4 |
| 1517 | 8 | 11 | 0 | 1582 | 47 | 0 | 3 |
| 1518 | 4 | 3 | 4 | 1583 | 39 | 11 | 7 |
| 1547 | 0 | 14 | 6 | 1584 | 31 | 13 | 2 |
| 1549 | 1 | 12 | 5 | 1585 | 42 | 8 | 0 |
| 1555 | 1 | 0 | 9 | 1586 | 42 | 12 | 8 |
| 1561 | 15 | 8 | 0 | 1587 | 60 | 13 | 10 |
| 1563 | 0 | 8 | 1 | 1588 | 61 | 3 | 9 |
| 1564 | 13 | 11 | 0 | 1589 | 32 | 4 | 2 |
| 1565 | 20 | 14 | 4 | 1590 | 51 | 1 | 3 |
| 1566 | 26 | 12 | 4 | 1591 | 51 | 5 | 11 |
| 1573 | 18 | 6 | 8 | 1592 | 67 | 2 | 11 |
| 1574 | 15 | 3 | 4 | 1593 | 57 | 19 | 9 |
| 1575 | 21 | 10 | 0 | 1594 | 55 | 15 | 10 |

| YEAR ENDING MICHAELMAS | £ | s. | d. | YEAR ENDING MICHAELMAS | £ | s. | d. |
|---|---|---|---|---|---|---|---|
| 1595 | 52 | 5 | 1 | 1619 | 80 | 19 | 11 |
| 1596 | 65 | 13 | 3 | 1620 | 83 | 8 | 8 |
| 1597 | 42 | 1 | 4 | 1621 | 85 | 2 | 0 |
| 1598 | 45 | 2 | 6 | 1622 | 91 | 16 | 11 |
| 1599 | 48 | 13 | 2 | 1633 | 78 | 5 | 1 |
| 1600 | 53 | 17 | 4 | 1634 | 79 | 13 | 8 |
| 1602 | 84 | 14 | 5 | 1638 | 77 | 1 | 10 |
| 1603 | 68 | 4 | 9 | 1639 | 78 | 13 | 10 |
| 1604 | 63 | 8 | 2 | 1642 | 77 | 3 | 8 |
| 1605 | 69 | 18 | 8 | 1643 | 50 | 12 | 7 |
| 1606 | 62 | 2 | 10 | 1644 | 18 | 8 | 10 |
| 1607 | 63 | 0 | 2 | 1645 | 38 | 8 | 6 |
| 1608 | 69 | 19 | 0 | 1646 | 27 | 2 | 3 |
| 1609 | 59 | 11 | 8 | 1674 | 21 | 3 | 8 |
| 1610 | 60 | 18 | 4 | 1675 | 124 | 14 | 10 |
| 1611 | 54 | 12 | 11 | 1699 | 20 | 0 | 0 |
| 1612 | 75 | 16 | 2 | 1700 | 10 | 0 | 0 |
| 1613 | 76 | 11 | 1 | 1710 | 10 | 0 | 0 |
| 1614 | 72 | 12 | 11 | 1720 | 8 | 10 | 0 |
| 1615 | 60 | 1 | 3 | 1730 | 8 | 15 | 0 |
| 1616 | 57 | 2 | 6 | 1740 | 10 | 0 | 0 |
| 1617 | 64 | 7 | 9 | 1750 | 10 | 0 | 0 |

*References.*

1297. Ministers' Accts., Acct. of Bailiff of Edmund of Cornwall.
1302. Pipe R., 30 Edw. I, Cornw.
1303. Duchy Accts., Excheq. Aug., port. 3.
1304–1306. Pipe R., 33 Edw. I, Cornw.
1517–1622. Receiver's Rolls.
1633–1646. Audit Accts., Duchy of Cornw.
1674–1750. Receiver's Views.

## S

### *Profits of the Cornish Courts.*

| YEAR ENDING MICHAELMAS | £ | s. | d. | YEAR ENDING MICHAELMAS | £ | s. | d. |
|---|---|---|---|---|---|---|---|
| 1297 | 21 | 8 | 8 | 1553 | 17 | 0 | 3 |
| 1302 | 9 | 14 | 4 | 1554 | 19 | 1 | 2 |
| 1303 | 17 | 3 | 2 | 1555 | 17 | 4 | 4 |
| 1304 | 13 | 18 | 2 | 1556 | 17 | 19 | 8 |
| 1305 | 14 | 17 | 11 | 1561 | 18 | 3 | 4 |
| 1306 | 21 | 12 | 7 | 1563 | 18 | 5 | 4 |
| 1307 | 22 | 3 | 11 | 1564 | 23 | 2 | 6 |
| 1337 | 17 | 9 | 0 | 1565 | 19 | 2 | 8 |
| 1342 | 24 | 2 | 11 | 1566 | 9 | 18 | 5 |
| 1345 | 39 | 14 | 11 | 1567 | 19 | 13 | 10 |
| 1347 | 35 | 3 | 3 | 1568 | 13 | 14 | 0 |
| 1349 | 25 | 7 | 5 | 1569 | 13 | 12 | 3 |
| 1350 | 14 | 5 | 9 | 1570 | 12 | 10 | 11 |
| 1355 | 40 | 9 | 6 | 1571 | 6 | 9 | 0 |
| 1356 | 53 | 9 | 2 | 1572 | 8 | 17 | 5 |
| 1357 | 46 | 17 | 7 | 1574 | 12 | 7 | 5 |
| 1362 | 70 | 12 | 1 | 1575 | 13 | 15 | 2 |
| 1363 | 29 | 1 | 8 | 1576 | 11 | 7 | 10 |
| 1369 | 40 | 0 | 11 | 1577 | 11 | 11 | 11 |
| 1418 | 32 | 7 | 3 | 1578 | 22 | 7 | 2 |
| 1435 | 29 | 12 | 1 | 1579 | 18 | 2 | 10 |
| 1443 | 14 | 16 | 7 | 1580 | 31 | 5 | 8 |
| 1450 | 3 | 0 | 0 | 1581 | 17 | 14 | 9 |
| 1455 | 44 | 7 | 8 | 1582 | 20 | 12 | 0 |
| 1469 | 17 | 0 | 6 | 1583 | 18 | 2 | 11 |
| 1472 | 26 | 18 | 1 | 1584 | 27 | 6 | 2 |
| 1477 | 33 | 6 | 5 | 1585 | 17 | 16 | 9 |
| 1478 | 19 | 14 | 5 | 1586 | 21 | 4 | 10 |
| 1489 | 16 | 17 | 6 | 1587 | 23 | 0 | 15 |
| 1495 | 64 | 16 | 6 | 1588 | 23 | 7 | 8 |
| 1496 | 25 | 13 | 11 | 1589 | 23 | 2 | 0 |
| 1504 | 21 | 4 | 1 | 1590 | 21 | 1 | 10 |
| 1517 | 35 | 13 | 10 | 1591 | 20 | 2 | 10 |
| 1518 | 35 | 5 | 1 | 1592 | 19 | 11 | 2 |
| 1524 | 26 | 18 | 0 | 1593 | 19 | 0 | 9 |
| 1540 | 28 | 8 | 6 | 1594 | 20 | 13 | 2 |
| 1544 | 31 | 13 | 8 | 1595 | 16 | 17 | 8 |
| 1545 | 26 | 13 | 8 | 1596 | 19 | 6 | 2 |
| 1547 | 23 | 4 | 8 | 1597 | 19 | 2 | 7 |
| 1549 | 19 | 7 | 4 | 1598 | 17 | 2 | 7 |

| YEAR ENDING MICHAELMAS | £ | s. | d. | YEAR ENDING MICHAELMAS | £ | s. | d. |
|---|---|---|---|---|---|---|---|
| 1599 | 17 | 1 | 8 | 1616 | 12 | 0 | 1 |
| 1600 | 17 | 5 | 1 | 1617 | 18 | 12 | 5 |
| 1602 | 15 | 17 | 10 | 1619 | 29 | 10 | 3 |
| 1603 | 11 | 1 | 5 | 1620 | 16 | 4 | 4 |
| 1604 | 12 | 7 | 10 | 1621 | 17 | 3 | 7 |
| 1605 | 13 | 6 | 8 | 1622 | 19 | 1 | 1 |
| 1606 | 13 | 11 | 5 | 1623 | 19 | 6 | 0 |
| 1607 | 10 | 18 | 5 | 1624 | 17 | 4 | 10 |
| 1609 | 22 | 16 | 8 | 1625 | 17 | 9 | 8 |
| 1610 | 18 | 10 | 0 | 1626 | 15 | 11 | 7 |
| 1611 | 16 | 17 | 2 | 1633 | 21 | 1 | 4 |
| 1612 | 19 | 4 | 10 | 1634 | 22 | 4 | 11 |
| 1613 | 19 | 16 | 8 | 1661 | 10 | 2 | 4 |
| 1614 | 24 | 1 | 6 | 1700–1750 | 9 | 14 | 0 |
| 1615 | 21 | 9 | 8 | | | | |

*References.*

1297. Ministers' Accts., Bailiff's Acct. of Edmund of Cornwall.
1302. Pipe R., 30 Edw. I, Cornwall.
1304–1305. *Ibid.*, 33 Edw. I, Cornwall.
1303, 1306. Duchy Accts., Excheq. Aug., port. 5.
1337. *Ibid.*, port. 1.
1342–1369. Ministers' Accts., Duchy of Cornwall.
1418–1626. Receiver's Rolls.
1633–1750. Receiver's Views.

## T

## *Profits of the Devon Courts.*

| YEAR ENDING MICHAELMAS | £ | s. | d | YEAR ENDING MICHAELMAS | £ | s. | d. |
|---|---|---|---|---|---|---|---|
| 1243 | 0 | 4 | 0 | 1293 | 2 | 18 | 4 |
| 1288 | 2 | 0 | 6 | 1294 | 3 | 2 | 0 |
| 1289 | 2 | 1 | 4 | 1295 | 3 | 12 | 2 |
| 1290 | 2 | 0 | 6 | 1296 | 5 | 14 | 0 |
| 1291 | 2 | 10 | 9 | 1297 | 1 | 13 | 5 |
| 1292 | 2 | 10 | 3 | 1298 | 2 | 2 | 7 |

| YEAR ENDING MICHAELMAS | £ | s. | d. | YEAR ENDING MICHAELMAS | £ | s. | d. |
|---|---|---|---|---|---|---|---|
| 1299 | 0 | 3 | 3 | 1576 | 3 | 17 | 10 |
| 1300 | 2 | 9 | 8 | 1577 | 6 | 6 | 6 |
| 1301 | 3 | 5 | 2 | 1578 | 6 | 15 | 0 |
| 1369 | 47 | 3 | 10 | 1579 | 4 | 13 | 0 |
| 1400 | 2 | 14 | 5 | 1580 | 5 | 7 | 0 |
| 1418 | 5 | 15 | 6 | 1581 | 2 | 4 | 8 |
| 1435 | 9 | 8 | 5 | 1582 | 3 | 2 | 10 |
| 1443 | 4 | 5 | 7 | 1583 | 5 | 8 | 8 |
| 1450 | 5 | 15 | 6 | 1584 | 6 | 10 | 6 |
| 1455 | 6 | 17 | 11 | 1585 | 4 | 5 | 8 |
| 1469 | 9 | 9 | 5 | 1586 | 5 | 11 | 2 |
| 1472 | 16 | 3 | 1 | 1587 | 4 | 3 | 0 |
| 1477 | 17 | 19 | 8 | 1588 | 5 | 0 | 8 |
| 1478 | 20 | 13 | 8 | 1589 | 2 | 2 | 6 |
| 1489 | 10 | 8 | 2 | 1590 | 2 | 6 | 8 |
| 1495 | 12 | 2 | 10 | 1591 | 2 | 3 | 6 |
| 1496 | 39 | 17 | 0 | 1592 | 1 | 10 | 2 |
| 1504 | 127 | 6 | 1 | 1593 | 1 | 8 | 2 |
| 1517 | 48 | 6 | 3 | 1594 | 1 | 12 | 10 |
| 1518 | 34 | 15 | 0 | 1595 | 1 | 4 | 10 |
| 1524 | 27 | 11 | 1 | 1596 | 3 | 7 | 0 |
| 1540 | 41 | 4 | 11 | 1597 | 2 | 13 | 2 |
| 1544 | 15 | 7 | 4 | 1598 | 2 | 2 | 0 |
| 1545 | 20 | 13 | 4 | 1599 | 1 | 16 | 9 |
| 1547 | 7 | 3 | 10 | 1600 | 1 | 19 | 4 |
| 1549 | 3 | 6 | 6 | 1602 | 1 | 15 | 2 |
| 1553 | 3 | 11 | 10 | 1603 | 1 | 7 | 6 |
| 1554 | 8 | 5 | 2 | 1604 | 1 | 2 | 6 |
| 1555 | 8 | 1 | 17 | 1605 | 1 | 4 | 6 |
| 1556 | 18 | 16 | 0 | 1606 | 1 | 4 | 8 |
| 1561 | 7 | 16 | 0 | 1607 | 1 | 10 | 0 |
| 1563 | 3 | 7 | 0 | 1608 | 0 | 11 | 4 |
| 1564 | 5 | 6 | 10 | 1609 | 2 | 5 | 8 |
| 1565 | 5 | 6 | 8 | 1610 | 1 | 0 | 0 |
| 1566 | 6 | 0 | 4 | 1611 | 0 | 12 | 8 |
| 1567 | 34 | 8 | 8 | 1612 | 3 | 1 | 2 |
| 1568 | 6 | 11 | 8 | 1613 | 0 | 19 | 6 |
| 1569 | 7 | 11 | 0 | 1614 | 0 | 15 | 2 |
| 1570 | 5 | 16 | 2 | 1615 | 1 | 4 | 0 |
| 1571 | 5 | 8 | 11 | 1616 | 0 | 17 | 7 |
| 1573 | 7 | 16 | 6 | 1617 | 1 | 8 | 6 |
| 1574 | 6 | 9 | 2 | 1619 | 1 | 1 | 6 |
| 1575 | 6 | 0 | 12 | 1620 | 6 | 18 | 3 |

| YEAR ENDING MICHAELMAS | £ | s. | d. | YEAR ENDING MICHAELMAS | | £ | s. | d. |
|---|---|---|---|---|---|---|---|---|
| 1621 | 1 | 11 | 2 | 1633 | farmed | 1 | 13 | 6 |
| 1622 | 0 | 15 | 10 | 1634 | farmed | 1 | 13 | 6 |
| 1623 | 6 | 16 | 2 | 1660 | farmed | 0 | 16 | 9 |
| 1624 | 4 | 15 | 2 | 1661 | farmed | 1 | 13 | 6 |
| 1626 | 4 | 16 | 6 | 1750 | farmed | 1 | 13 | 6 |

*References.*

1243. Pipe R., 27 Hen. III, Devon.
1288–1290. *Ibid.*, 20 Edw. I, Devon.
1291–1295. *Ibid.*, 23 Edw. I, Devon.
1296–1301. *Ibid.*, 29 Edw. I, Devon.
1369. Accts. Excheq. K. R., bdle. 263, no. 16.
1400–1626. Receiver's Rolls.
1633–1750. Receiver's Views.

## U

## *Prices of Tin.*

NOTE. * = taken from Roger's "History of Agriculture and Prices." M = 1200 lbs. The price is to be understood as per pound, unless otherwise indicated.

| | |
|---|---|
| 1199. | £3 M. |
| 1266.* | 1⅔d. (Woodstock). |
| 1270.* | 1½d. (Berkhampstead). |
| 1280.* | 2d. (Corfe Castle). |
| 1282.* | 2½d. (Dover). |
| * | 3d. (Oxford). |
| 1283.* | 2½d. (Dover). |
| * | 2d. (Dover). |
| 1285.* | 2d. (Corfe Castle). |
| 1287.* | 2d. (Dover). |
| 1288.* | 1½d. (Oxford). |
| 1290.* | 2d. (Pevensey). |
| * | 1d. (Bosham). |
| 1291.* | 1½d. (Strugull). |
| | 9s. cwt. (white rent). |
| 1292. | 10s. cwt. (white rent). |
| * | 3½d. (Pevensey). |
| 1293. | 9s. cwt. (white rent). |
| 1294. | 8s. cwt. (white rent). |
| 1295. | 8s. cwt. (white rent). |
| * | 2d. (Oxford). |
| 1296. | 8s. 6d. cwt. (white rent). |
| 1296.* | 2d. (Langley). |
| * | 2½d. (Ledes). |
| 1297. | 10s. cwt. (white rent). |
| | £4 13s. 4d. M. |
| 1298. | 10s. cwt. (white rent). |
| 1299. | 10s. cwt. (white rent). |
| 14th cent. | 13s. 4d. cwt. (Cornwall). |
| 1300. | 10s. cwt. (white rent). |
| 1301. | 10s. cwt. (white rent). |
| * | 2½d. (Cuxham). |
| 1303.* | 1½d. (Oxford). |
| * | 22s. 4¼d. cwt. (Maldon). |
| 1308.* | 2d. (Oxford). |
| 1310.* | 3¼d. (Oxford). |
| 1313? | £4 M. (Cornwall). |
| 1318.* | 3d. (Caerleon). |
| * | 2½d. (Southampton). |
| 1319.* | 3½d. (Beaumaris). |
| * | 3d. (Beaumaris). |
| 1320.* | 3d. (Usk). |
| * | 2⅓d. (Southampton). |
| 1323.* | 2½d. (Southampton). |

| | |
|---|---|
| 1324.* | 2¼d. (Castle Barnard). |
| 1326.* | 2d. (Clarette). |
| * | 1⅞d. (Clarette). |
| 1334.* | 2½d. (Oxford). |
| 1337.* | 2½d. (Boxley). |
| 1340.* | 5d. (London). |
| 1341.* | 7⅜d. (Bichyndon). |
| * | 1½d. (Oxford). |
| 1343.* | 2d. (Bichyndon). |
| 1351.* | 2½d. (Clare). |
| 1357.* | 4d. (Acornbury). |
| | 3d. (?) |
| 1365.* | 3d. (St. Briavels). |
| * | 3d. (Sheppey). |
| * | 4d. (Woodstock). |
| 1367–68. | £13 6s. 8d. M. |
| 1368.* | 3½d. (Eton). |
| * | 4d. (Islip). |
| 1371.* | 4d. (Bicester). |
| 1372.* | 3d. (Hadley). |
| * | 3½d. (Oxford). |
| * | 4d. (Woodstock). |
| 1373. | 2d. (Exeter). |
| 1374.* | 3d. (St. Briavels). |
| 1381.* | 3½d. (Pershore). |
| 1391. | 13s. 4d. cwt. (Cornwall). |
| 1392. | 20s. cwt. (Southampton customs). |
| 1402. | 2⅞d. (London) (tin vessel). |
| | 25s. cwt. (London) (tin vessel). |
| * | 3d. (York). |
| 1404.* | 3d. (York). |
| 1045.* | 3d. (Windsor). |
| 1407.* | 3½d. (Denbigh Castle). |
| * | 3d. (Denbigh Castle). |
| 1415.* | 3d. (York). |
| 1418.* | 3d. (York). |
| 1419.* | 3d. (York). |
| 1421. | 26s. 10d. cwt. (London) (tin vessel). |
| | 4d. (London) (tin vessel). |
| | 5d. (London) (tin vessel). |
| | 28s. 8d. cwt. (London) (tin vessel). |
| * | 3d. (York). |
| 1422.* | 3d. (York). |
| 1432.* | 3d. (York). |
| 1433.* | 3d. (York). |
| 1441.* | 2¾d. (York). |
| 1442.* | 2¾d. (York). |
| 1444.* | 2¾d. (York). |
| 1445.* | 2¾d. (York). |
| 1446.* | 3½d. (York). |
| 1450. | 33s. 4d. cwt. (Southampton) (tin vessel). |
| 1456.* | 3d. (York). |
| 1457.* | 3d. (York). |
| 1459.* | 2d. (York). |
| 1461.* | 3d. (Oxford). |
| 1470.* | 8d. (York). |
| 1471.* | 3¾d. (York). |
| 1472.* | 4d. (York). |
| 1478.* | 4d. (York). |
| 1481.* | 4d. (York). |
| 1483. | 25s. 8d. cwt. (London). |
| | 26s. 4d. cwt. (London). |
| | 27s. cwt. (London). |
| | 23s. cwt. (London). |
| 1485.* | 4d. (York). |
| 1492. | 29s. cwt. (London). |
| 1497.* | 4d. (York). |
| 1498.* | 4d. (York). |
| 1503.* | 3d. (Oxford). |
| 1504.* | 4½d. (York). |
| 1512.* | 6d. (Oxford). |
| 1515.* | 4d. (Oxford). |
| 1520. | 4d. |
| 1521.* | 4d. (Sion). |
| 1529.* | 33s. 4d. cwt. (Sion). |
| * | 4d. (Sion). |
| * | 5d. (York). |
| 1530.* | 5d. (Oxford). |
| 1532. | 20s. cwt. (Exeter customs). |
| 1533.* | 4½d. (Stonor). |
| 1535.* | 5d. (York). |
| * | 4d. (Oxford). |
| 1542. | 4d. (St. Michael's). |
| 1544.* | 4d. (London). |
| 1559.* | 7d. |
| 1561. | 33s. 4d. cwt. (Southampton customs). |
| 1565. | 46s. 8d. cwt. (Southampton customs) (pewter). |
| 1568. | 7½d. |
| 1569.* | 6d. (Cambridge). |
| 1571.* | 55s. cwt. (London). |
| * | 50s. cwt. (London). |

| | |
|---|---|
| 1571.* | 58s. cwt. (London). |
| | 54s. 7d. cwt. (London). |
| 1573. | 46s. 8d. cwt. (Southampton customs) (pewter) |
| 1579. | 52s. cwt. |
| 1580. | 5d. (Devon?). |
| | 3d. (Devon?). |
| | 6d. (Devon?). |
| | 1¾d. (Devon?). |
| 1588 | 40s. cwt. (Midsummer). |
| | 46s. cwt. (Michaelmas). |
| 1589. | 44s. cwt. |
| 1590. | 46s. cwt. |
| 1591. | 44s. cwt. (Midsummer). |
| | 43s. cwt. (Michaelmas). |
| 1595. | £3 10s.–£3 12s. cwt. |
| | £40 M. |
| | 63s. cwt. (Midsummer). |
| 1599. | £29 M. |
| 1600. | 40s. cwt. |
| 1604. | 72s. cwt. (for pewterers). |
| 1605.* | 72s. cwt. (Cambridge). |
| 1606. | 65s. cwt. |
| 1607. | 5½d. |
| 1608. | 70s. cwt. (for pewterers). |
| 1611.* | 11d. (Cambridge). |
| 1615. | 87s. cwt. |
| 1626?. | 100s. cwt. |
| 1630. | 100s. cwt. |
| 1635. | 112s. cwt. |
| | 104s. 10d. cwt. |
| 1650. | 65s. cwt. |
| 1652–60. | 125s. cwt. (selling price for tinners). |
| 1666. | 82s. per cwt. (selling price for tinners). |
| 1689. | 80s. cwt. (Truro). |
| | 67s. cwt. (London). |
| 1690. | 50s. cwt. (selling price for tinners). |
| 1692. | 64s. cwt. |
| 1693. | 59s. cwt. |

| | |
|---|---|
| 1694–97.* | Houghton's prices.[1] |
| 1697. | 63s. 9d. cwt. |
| 1698. | 63s. 9d. cwt. |
| 1699. | 63s. 9d. cwt. |
| 1700. | 63s. 9d. cwt. |
| 1701. | 63s. 9d. cwt. |
| 1702. | 63s. 9d. cwt. |
| | 45s.–60s. cwt. |
| 1709.* | 80s. cwt. (London). |
| * | 81s. 6d. cwt. (London) (bars). |
| 1710. | 69s. cwt. |
| 1713. | 73s. cwt. (Cornwall). |
| 1714. | 71s. cwt. (London). |
| | 75s. cwt. (Cornwall). |
| 1715. | 59s. 6d. cwt. (Cornwall). |
| | 68s. cwt. (London). |
| 1716. | 76s. cwt. |
| 1717. | 76s. cwt. (London). |
| | 60s.–61s. cwt. (Cornwall) (November). |
| | 72s.–74s. cwt. (London). |
| | 59s. cwt. (Cornwall) (Christmas). |
| 1718. | 66s. 6d. cwt. |
| | 56s.–57s. cwt. (Cornwall) (Christmas). |
| | 59s. cwt. (Cornwall) (Michaelmas). |
| | 66s.–68s. cwt. (London) (July to Christmas). |
| | 60s.–60s. 6d. cwt. (Cornwall) (Midsummer). |
| | 59s.–59s. 6d. cwt. (Cornwall) (Lady Day). |
| 1719. | 56s.–57s. cwt. (Cornwall). |
| 1728,* Oct. | 79s. cwt. (London). |
| | 81s. cwt. (London). (bars). |
| 1729,* Jan. | 79s. cwt. (London). |

[1] Rogers quotes Houghton's prices for tin at London (per cwt.): —

1694, 1st qr., 72s., 62s.; 2d qr., 62s.; 3d qr., 62s.; 4th qr., 62s.

1695, 1st. qr., 62s.; 2d qr., 62s. and 59s.; 3d qr., 59s., 63s., 66s.; 4th qr., 66s., 69s., 72s., 74s., 76s., 74s.

1696, 1st qr., 74s., 86s.; 2d qr., 80s., 78s.; 3d qr., 78s., 75s., 72s.; 4th qr., 68s.

1697, 1st qr., 66s., 60s.; 2d qr., 60s., 61s.; 3d qr., 60s.

1729.* Jan. 81s. cwt. (London) (bars).
* July. 78s. cwt. (London) (bars).
* 80s. cwt. (London) (bars).
* Sept. 77s. cwt. (London).
* 79s. cwt. (London) (bars).
1730,* Oct. 78s. cwt. (London).
* 79s. cwt. (London) (bars).
1731.* Jan. 80s. cwt. (London).
82s. cwt. (London) (bars).
1733.* Nov. 78s. cwt. (London).
* 80s. cwt. (London) (bars).
1735,* April. 74s. cwt. (London).
* 76s. cwt. (London) (bars).
* June. 73s. 6d. cwt. (London).
* 75s. 6d. cwt. (London) (bars).
1750. 55s. cwt. (selling price for tinners).
1762,* Jan. 65s. cwt. (London).
1763.* 65s. cwt. (London).

# BIBLIOGRAPHY

# BIBLIOGRAPHY

This list does not aim at completeness. It merely brings together, with a fuller statement of title and with a few additions from other sources used, the references to manuscript and printed materials scattered through the footnotes. Unless otherwise stated, London is to be understood as the place of publication for the English books here listed.

## MANUSCRIPT SOURCES


British Museum.

Additional Manuscripts, nos. 6681, 6682, 6713, 24744, 24746, 28079, 36767. Nos. 6682 and 6713 are especially valuable compendiums of stannary law and custom. They contain the proceedings of most of the stannary parliaments.

Harleian MSS., nos. 1507, 6380, 6696. No. 6380 is an important sixteenth century account of the stannaries, written by a tinner.

Information as to the stannaries will also be found in the Cotton (Titus B. v.), Hargrave (no. 321), and Lansdowne (nos. 18, 19, 22, 24, 26, 76, 81, 86, 1215) manuscript collections.

Library of Lincoln's Inn.

Hales MSS., no. 83.

Bodleian Library, Oxford.

Additional MSS.

Cambridge University Library.

Manuscript Collection. Ff. vi.

Privy Council Office.

Council Registers.

Public Record Office.

Patent Rolls, Close Rolls, Fine Rolls, and Signet Office Docquet Books. These furnish a large part of the direct information concerning the stannaries during the Middle Ages.

Pipe Rolls. Useful as indicating for the earlier years the receipts from stannary taxes as well as some details of stannary administration.

Exchequer, Augmentation Department, Duchy of Cornwall Accounts and Stannary Rolls. The latter give the coinage accounts, showing from the fourteenth to the seventeenth centuries the amounts of tin coined and the names of the owners.

Exchequer, Treasury of the Receipt. Council Book of the Duchy of Cornwall, 25–39 Edw. III, known as the "White Book of Tenure in Cornwall." It contains entries of writs and warrants directed to officials of the Duchy under Edward the Black Prince and of petitions and other proceedings before the Council of the Duchy.

Letters and Papers, Foreign and Domestic (for Henry VIII), and State Papers, Domestic Series. For the sixteenth and seventeenth centuries, these furnish a large part of our material for the history of the mines and miners of Cornwall, but with especial emphasis upon the relations between the tinners and the Crown. They are continued by the Home Office Papers, the Treasury Papers, and the News Letter Entry Book.

Use has also been made of the following classes of documents in the Record Office.

Ancient Petitions.
Black Book of the Exchequer.
Chancery Proceedings.
Chancery Enrolments.
Charter Rolls.
Court Rolls (Stannary Courts).
Customs Accounts.
Durham, Palatinate of, Cursitor's Records.
Escheators' Accounts, Enrolments of.
Exchequer Accounts, Miscellaneous.
Lancaster, Duchy of, Miscellanea.
Lord Treasurer's Remembrancer. Memoranda Rolls.
Ministers' Accounts.
Originalia Rolls.
Subsidy Rolls, Lay
Warrant Books.

DUCHY OF CORNWALL OFFICE.

Manuscript Volume. A large folio volume, filled with stannary documents mainly of the sixteenth century.

Receivers' Rolls. This valuable series, running from early in the fourteenth century to the seventeenth century, gives data as to the production of tin and as to stannary revenue. They are continued to the latter part of the eighteenth century by the Receivers' Views.

Audit Accounts.
Ministers' Accounts.

## PRINTED RECORDS. PARLIAMENTARY DOCUMENTS. LEGAL REPORTS

*Acts of the Privy Council.* New Series, ed. J. R. Dasent. 32 vols., 1890–1907.
*Acts of the Parliament of Scotland.* 12 vols., 1814–1875.
Bainbridge, W. *Laws of Mines and Minerals.* 5th ed., 1900.
*Chancery Proceedings* (*Calendars of the*) *in the reign of Queen Elizabeth.* 3 vols., 1827–1832.
Coke, E. *The fourth part of the Institutes of the Laws of England.* 1644.
Coke, E. *Reports.* 7 vols., ed. 1777.

*Compleat Mineral Laws of Derbyshire.* 1734.

Concanen, G. *A report of the trial at bar, Rowe v. Brenton . . . as to the right to minerals in the assessional lands of the Duchy of Cornwall.* 1830.

(Dean Forest.)

*Reports of the Dean Forest Commissioners.* Parliamentary Documents, 1835, vol. 36. Of the five reports, the fourth is especially valuable.

*Report from the Select Committee* appointed to inquire into the laws and rights affecting Dean Forest. Parliamentary Documents, 1874, vol. 7.

See below, Sopwith, *Award of Dean Forest Commissioners.*

Delebecque, A. J. *Pasinomie, ou collection complète des lois . . . qui peuvent être invoqués en Belgique.* 1833, etc.

D'Ewes, Sir Symonds, *A Compleat Journal of the Votes, Speeches, and Debate both of the House of Lords and House of Commons throughout the whole reign of Queen Elizabeth.* 1693.

*Domesday Book.* 4 vols., 1783–1816.

Dugdale, W. *Monasticon Anglicanum.* 6 vols., ed. 1846.

Hansard. *Parliamentary Debates.*

Harrison, Sir George. *Substance of a Report on the Laws and Jurisdiction of the Stannaries in Cornwall.* 1835.

Historical Manuscripts Commission. *Reports.*

Houghton, Thomas. *The Compleat Miner;* or a Collection of the Laws, Liberties, ancient Customs . . . of the several Mines and Miners in the Counties of Derby, Gloucester, and Somerset. 1688.

Kemble, J. M. *Codex diplomaticus aevi Saxonici.* 6 vols., 1839–1848.

Loisel, A. *Institutes coustumières.* Paris, 1646.

Louvrex, M. G. de. *Recueil contenant les édits et réglements faits pour le pays de Liège et comté de Looz.* Liege, [1749–]1752.

Northcote, Sir John. *Notebook containing memoranda of proceedings in the House of Commons during the first Session of the Long Parliament,* 1640. Ed. A. H. A. Hamilton, 1877.

Patrick, R. W. C. *Early Records relating to Mining in Scotland.* Edinburgh, 1878.

Pearce, T. *The Laws and Customs of the Stannaries.* 1725.

*Pipe Rolls (The) for the Counties of Cumberland, Westmoreland and Durham.* Newcastle, 1847.

*Placita de quo Warranto.* Edw. I–Edw. III. 1818.

Plowden, E. *Commentaries.* 1761.

*Register of the Privy Council of Scotland.* Edinburgh, 1877, etc.

*Rotuli hundredorum.* 2 vols., 1812–1818.

*Rotuli parliamentorum.* 6 vols. with index, 1767–1832.

*Rotuli Scotiae.* 2 vols., 1814–1819.

Rymer, T. *Foedera.* (Record Commission edition.) 4 vols., 1816–1869.

Smirke, E. *The Case of Vice against Thomas.* 1843. Valuable for its appendix, in which are printed many stannary documents.

(Sopwith, T.). *The Award of the Dean Forest Mining Commissioners as to the coal and iron mines . . . and the rules and regulations for working the same,* with preliminary observations by T. Sopwith. 1841.

(Stannary Law.)

*The old lawes and statutes of the Stannarie of Devon* [1575?].

*Statement of the Proceedings of the Convocation for the Stannary of Cornwall.* 1750.
*Laws of the Stannaries of Cornwall made at the convocation or parliament of Tinners at Truro, Sept.* 13. *Anno* 27 *Geo. II.* [*ca.* 1760?].
*The Laws of the Stannaries of Cornwall.* Truro, 1808.
*Procedure of the Court of the Vice Warden of the Stannaries.* 1856.
*Reports from the Select Committee on Stannaries Act* (1869) *Amendment Bill.* Parliamentary Documents, 1887, vol. 12.
See above, Concanen, Harrison, Pearce.
*State Papers, Calendars* (Domestic, Colonial and Venetian series).
*Statutes of the Realm.* 11 vols., 1810–1828.

## OTHER BOOKS

Abbott, G. *An Essay on the Mines of England.* 1833.
Achenbach, Hermann. *Das französische Bergrecht.* Bonn, 1869.
Achenbach, Hermann. *Das gemeine deutsche Bergrecht.* Bonn, 1871, etc.
*Aggravii Venetiani, or the Venetian and other Grievances, together with a Proposal for raising the Price of Tin,* etc. 1697.
Agricola, G. *De re metallica.* Ed. 1556.
Anglicus, Bartholomæus. See *Glanvilla.*
*Annales Monasterii de Burton.* In *Annales Monastici,* ed. H. R. Luard. Rolls Series. 1864.
Arndt, Adolf. *Bergbau und Bergbaupolitik.* Leipzig, 1894.
Atkinson, Stephen. *The Discoverie and Historie of the Gold Mynes in Scotland* (1619). Edited by G. L. Meason for the Bannatyne Club. Edinburgh, 1825.
Balducci Pegolotti, F. *La Practica della Mercatura.* (Vol. iii in G. F. Pagnini dal Ventura. *Della Decima.* 1765–1766.)
Bapst, Germain. *Études sur l'étain dans l'antiquité et au moyen âge.* Paris, 1884.
Bartlett, T. *A Treatise on British Mining: with a Digest of the Cost-Book System.* 1850.
Becher, J. J. *Alphabetum Minerale,* etc. Frankfurt-a.-M., 1689.
Beer, Adolf. *Allgemeine Geschichte des Welthandels.* 3 vols., 1860–1864.
Bernhard, Ludwig. *Die Entstehung und Entwicklung der Gedingeordnungen im deutschen Bergrecht.* Leipzig, 1899.
Boase, G. C., and Courtney, W. P. *Bibliotheca Cornubiensis.* 3 vols., 1874–1882.
Borlase, William. *Observations on the Antiquities of Cornwall.* 2d ed., 1769.
Borlase, William. *Observations on the Ancient and Present State of the Islands of Scilly.* Oxford, 1756.
Borlase, W. C. *Tin Mining in Spain, Past and Present.* 1897.
Bushell, Thomas. *The Case of Thomas Bushell truly stated.* 1649.
Calvert, J. *The Gold Rocks of Great Britain and Ireland.* 1853.
Camden, William. *Britannia.* Translated by Richard Gough, 2d ed., 4 vols., 1806.
Carew, Richard. *Survey of Cornwall.* Ed. 1811.
Charleton, A. G. *Tin: Describing the Chief Methods of Mining, Dressing, and Smelting it Abroad.* 1884.
Childrey, Joshua. *Britannia Baconica.* 1660.

Collins, John. *A Plea for the Bringing in of Irish Cattel and Keeping out of Fish caught by Foreigners*, etc. 1680.
*Cornish Mining: Tin*. By the Special Correspondent of the *Mining Journal*. 1902.
Cunningham, W. *The Growth of English Industry and Commerce in Modern Times*. 1903.
Daenell, E. *Die Blütezeit der Deutschen Hanse*. Vol. i, 1905.
*Declaration of Sundry Grievances concerning Tin and Pewter*. 1646. (By I. S.)
Delabeche, Sir H. T. *Report on the Geology of Cornwall, Devon, and West Somerset*. 1839.
Del Mar, A. *History of the Precious Metals*. 2d ed., 1901.
Diest, P. H. van. *Banca and its Tin Stream-Works*. Translated from the Dutch by C. LeN. Foster. Truro, 1867.
Doddridge, Sir John. *The History of the Ancient and Moderne Estate of the Principality of Wales, Dutchy of Cornwall, and Earldome of Chester*. 1630.
Dudley, Dud. *Metallum Martis*. 1665.
Eberstadt, R. *Das französische Gewerberecht*. Leipzig, 1899.
Edwards, E. *The Life of Sir Walter Ralegh*. 2 vols., 1868.
*Egerton Papers, The*. Ed. J. P. Collier. Camden Society Publications, 1840.
Ellis, Sir Henry. *A General Introduction to Domesday Book*. 1833.
Ercker, Lazarus. *Fleta Minor. The laws of art and nature in knowing, judging, assaying . . . Metals*. 1683.
Fairbairn, Sir William. *Iron: Its History, Properties, and Processes of Manufacture*. 3d ed., Edinburgh, 1869.
Farey, J. *General View of the Agriculture and Minerals of Derbyshire*. 3 vols., 1811–1817.
Flower, P. W. *A History of the Trade in Tin; a Short Description of Tin Mining and Metallurgy*, etc. 1800.
(Forest of Dean.) *A True Narrative concerning the Woods and Ironworks of the Forest of Dean; and how they have been disposed of since the Year 1635* [*ca*. 1670].
Francis, G. G. *The Smelting of Copper in the Swansea District from the Time of Elizabeth to the Present Day*. 2d ed., 1881.
Fuller, Thomas. *The History of the Worthies of England*. 1662.
Galloway, R. L. *A History of Coal Mining in Great Britain*. 1882. 2d ed., 1898.
Gierke, Otto. *Das deutsche Genossenschaftsrecht*. 2 vols., Berlin, 1868–1873.
Glanvilla, Bartholomæus de. *Batman uppon Bartholome his Booke de Proprietatibus Rerum . . . with additions*. 1582. (J. Trevisa's translation.)
Gmelin, J. F. *Beyträge zur Geschichte des teutschen Bergbaus*. Halle, 1783.
Gross, Charles. *The Gild Merchant*. 2 vols., Oxford, 1890.
Guicciardini, Ludovico. *The Description of the Low Countreys*. 1593.
Heton, Thomas. *Some Account of Mines and the Advantages of them to the Kingdom*. 1707.
Heyd, W. *Histoire du commerce du Levant au moyen-âge*. 2 vols., Leipzig, 1885–1886.
Holinshed, Raphael. *Chronicles of England, Scotland, and Ireland*. 6 vols., ed. 1807–1808.
Houghton, John. *A Collection for the Improvement of Husbandry and Trade*. 1692, etc.

Hudson, W. H. *Leet Jurisdiction in the City of Norwich.* (Selden Society Publications, No. 5.) 1892.
Hüllman, C. D. *Geschichte des Ursprungs der Regalien in Deutschland.* Frankfurt-an-der-Oder, 1806.
Hunt, Robert. *British Mining.* 2d ed., 1887.
Inama-Sternegg, K. T. von. *Deutsche Wirthschaftsgeschichte.* 3 vols., Leipzig, 1879–1901.
James, G. C. *Pseudo Cost-Book Mines.* 1855.
Jars, Gabriel. *Voyages metallurgiques.* 3 vols., Lyon, 1774–1781.
Jorden, E. *A Discourse of Naturall Bathes.* 1631.
Kemble, J. M. *The Saxons in England.* 2 vols., 1849.
Lambert, J. M. *Two Thousand Years of Gild Life.* Hull, 1891.
Lapsley, G. T. *The County Palatine of Durham.* 1900.
Leontius, *Vita Sancti Joannis Eleemosynarii.* In Migne's *Patrologia* (Greek Series), vol. xciii.
Louis, Henry. *The Production of Tin.*
Lysons, Daniel. *Magna Britannia.* 6 vols., [1806–]1822.
Macpherson, D. *Annals of Commerce.* 4 vols., 1805.
Madox, Thomas. *The History and Antiquities of the Exchequer of England.* 2d ed., 2 vols., 1769.
Maitland, F. W. *Select Pleas in Manorial Courts.* (Selden Society Publications, No. 2.) 1888.
Malden, H. E. *History of Surrey.* 1900.
Malynes, Gerard de. *Consuetudo vel Lex Mercatoria.* 1622.
*Narrative concerning the Forest of Dean.* See above, *Forest of Dean.*
Nicholls, H. G. *The Forest of Dean.* 1858.
Nicholls, H. G. *Iron Making in the Olden Times; as instanced in the ancient mines, forges and furnaces of the Forest of Dean.* 1866.
Norden, John. *Speculi Britanniae Pars; a topographical and chorographical description of Cornwall.* 1728.
Pegolotti, F. B. See Balducci.
Percy, J. *Metallurgy: Art of extracting Iron and Steel.* 1864.
Percy, J. *The Metallurgy of Lead.* 1870.
Pettus, Sir John. *Fodinae Regales: or the History, Laws, and Places of the Chief Mines and Mineral Words in England, Wales,* etc. 1670.
Pettus, J. *Essays on Metallick Words.* (Part ii of Ercker's *Fleta Minor.*) 1683.
(Pewterers.) *The Case of the Pewterers of England as to the Exportation of Pewter* [1690?].
Phillips, J. A. and Darlington, J. *Records of Mining and Metallurgy.* 1857.
Pigeonneau, H. *Histoire du Commerce de la France.* 2 vols., Paris, 1885–1897.
Pike, J. R. *Britain's Metal Mines: a Complete Guide to their Laws,* etc. 1860.
Pilkington, James. *A View of the Present State of Derbyshire.* Derby, 1789.
Plot, Robert. *The Natural History of Staffordshire.* Oxford. 1686.
Pollock, Sir F. and Maitland, F. W. *The History of English Law.* 2d ed. 1898.
Polwhele, R. *Historical Views of Devonshire.* Exeter, 1793.
Polwhele, R. *The History of Cornwall.* 7 vols., ed. 1816.
Postlethwaite, J. *Mines and Mining in the Lake District.* Leeds, 1889.
Price, W. H. *The English Patents of Monopoly.* Boston, 1906.
Pryce, W. *Mineralogia Cornubiensis.* 1778.

Pulsifer, W. H. *Notes for a History of Lead.* New York, 1888.

Ray, John. *A Compleat Collection of English Proverbs.* Ed. 1737.

Ray, John. *A Collection of English Words not generally used . . . and an account of the preparing and refining such metals as are gotten in England.* 1674.

*Reasons for raising the Price of Tin at Present.* [1697?].

Reyer, E. *Zinn. Eine geologisch-montanistisch-historische Monografie.* Berlin, 1881. Translated by R. Symons, *History of Tin.* Truro, 1881.

Rogers, J. E. T. *A History of Agriculture and Prices in England.* 7 vols., 1866–1902.

Rogers, J. E. T. *A Complete Collection of the Protests of the Lords.* 3 vols., Oxford, 1875.

Rovenzon, John. *A Treatise of Metallica.* 1613.

Ruding, R. *Annals of the Coinage of Britain.* 3 vols., 1817–1819.

Schanz, G. *Englische Handelspolitik.* 2 vols., Leipzig. 1881.

Schaube, A. *Handelsgeschichte der Romanischen Völker des Mittelmeergebiets bis zum Ende der Kreuzzüge.* München and Berlin, 1906.

Scrivenor, H. *A Comprehensive History of the Iron Trade.* 1841.

Sinclair, G. *The Hydrostaticks . . . with a Short History of Coal.* Edinburgh, 1672.

Smiles, S. *Industrial Biography; Iron Workers and Tool Makers.* 1863.

Sombart, Werner. *Der moderne Kapitalismus.* 2 vols., Leipzig, 1902.

Sopwith, Thomas. *An Account of the Mining Districts of Alston Moor, Weardale and Teesdale.* Alnwick, 1833.

(Stannaries.)

*The Case of the Stannaries Stated.* [1650?].

*The Case of the Tinners Plainly Stated.* 1636.

Stevens, N. *A Guide to Cheddar and the Neighbourhood.* Cheddar, 1869.

Stow, John. *A Survey of the Cities of London and Westminster.* Enlarged by J. Strype, 2 vols., 6th ed., 1754–1755.

Stringer, Moses. *English and Welsh Mines and Minerals Discovered in some Proposals to the . . . House of Commons.* 1699.

Sturtevant, Simon. *Metallica.* 1612.

Swank, J. M. *History of the Manufacture of Iron in all Ages.* 2d ed., Philadelphia, 1892.

*Tin Duties, The.* 1833.

*Tinners' Grievance.* See Aggravii Venetiani.

Unwin, George. *Letters, Remarks, etc., with a View to open an Extensive Trade in the Article of Tin from the County of Cornwall to India, Persia, and China.* 1790.

Unwin, George. *Industrial Organization in the Sixteenth and Seventeenth Centuries.* Oxford, 1904.

Vinogradoff, P. *Villainage in England.* Oxford, 1892.

Waller, William. *The Answer of Mr. Waller to Mr. Hawkins's Report, Given in at the General Court of the Governour and Company of the Mine-Adventurers of England.* 1710.

Waller, William. *The Mine-Adventure Laid Open.* 1710.

Watson, J. Y. *A Compendium of British Mining, with Statistical Notices of the Principal Mines in Cornwall; to which is added . . . a Glossary of the Terms and Usages of Mining.* 1843.

Welch, C. *History of the London Pewterers.* 1902.
Westcote, T. *A View of Devonshire in* 1630. Exeter, 1845.
Worth, R. N. *Historical Notes Concerning the Progress of Mining Skill in Devon and Cornwall.* Falmouth, 1872.
Yarranton, Andrew. *England's Improvement by Sea and Land.* 1677-1681.
Zycha, Adolf. *Das böhmische Bergrecht des Mittelalters auf Grundlage des Bergrechts von Iglau.* 2 vols., Berlin, 1900.

## PERIODICALS

Archæological Journal.
Cornwall Polytechnic Society (Falmouth). Reports and Transactions.
Derbyshire Archæological and Natural History Society (Derby). Journal.
Devonshire Association for the Advancement of Science, Literature, and Art (Plymouth). Transactions.
Edinburgh Review (Edinburgh).
English Historical Review.
Ethnological Society of London. Transactions, continued as Journal.
Gentleman's Magazine.
Jahrbuch für Gesetzgebung, Verwaltung und Volkswirthschaft im deutschen Reich (Leipzig).
Jahrbücher für Nationalökonomie und Statistik (Jena).
Law Quarterly Review.
The Literary Panorama.
Mining Almanack for 1849 (1850). Compiled by Henry English.
Mining and Smelting Magazine. Edited by H. C. Salmon.
Mining Association and Institute of Cornwall (Redruth). Transactions.
Mining Institute of Cornwall (Truro). Proceedings.
Notes and Queries.
Penzance Natural History and Antiquarian Society. Report and Transactions.
Philosophical Magazine. Edited by A. Tilloch.
Plymouth Institution. Report and Transactions.
Quarterly Mining Review. Edited by Henry English.
Quarterly Journal of Science. Continued as the Journal of Science.
Records of the London and West Country Chamber of Mines.
Royal Geological Society of Cornwall (Penzance). Transactions.
Royal Institution of Cornwall (Truro). Journal and Reports.
Royal Society. Philosophical Transactions.
Royal Statistical Society. Journal.
Society for the Encouragement of Arts, Manufactures, and Commerce. Transactions.
Vierteljahrschrift für Social- und Wirtschaftsgeschichte (Stuttgart).
Zeitschrift der Savigny-Stiftung für Rechtsgeschichte (Weimar).
Zeitschrift für die gesammte Staatswissenschaft (Tübingen).

# INDEX

# INDEX